AF789935

Mechanics of Carbon Nanotubes

Mechanics of Carbon Nanotubes

Fundamentals, Modeling and Safety

Vasyl Harik

Nanodesigns Consulting, Wilmington, DE, United States

ELSEVIER

ACADEMIC PRESS

An imprint of Elsevier

Academic Press is an imprint of Elsevier
125 London Wall, London EC2Y 5AS, United Kingdom
525 B Street, Suite 1650, San Diego, CA 92101, United States
50 Hampshire Street, 5th Floor, Cambridge, MA 02139, United States
The Boulevard, Langford Lane, Kidlington, Oxford OX5 1GB, United Kingdom

Copyright © 2018 Elsevier Inc. All rights reserved.

No part of this publication may be reproduced or transmitted in any form or by any means, electronic or mechanical, including photocopying, recording, or any information storage and retrieval system, without permission in writing from the publisher. Details on how to seek permission, further information about the Publisher's permissions policies and our arrangements with organizations such as the Copyright Clearance Center and the Copyright Licensing Agency, can be found at our website: www.elsevier.com/permissions.

This book and the individual contributions contained in it are protected under copyright by the Publisher (other than as may be noted herein).

Notices
Knowledge and best practice in this field are constantly changing. As new research and experience broaden our understanding, changes in research methods, professional practices, or medical treatment may become necessary.

Practitioners and researchers must always rely on their own experience and knowledge in evaluating and using any information, methods, compounds, or experiments described herein. In using such information or methods they should be mindful of their own safety and the safety of others, including parties for whom they have a professional responsibility.

To the fullest extent of the law, neither the Publisher nor the authors, contributors, or editors, assume any liability for any injury and/or damage to persons or property as a matter of products liability, negligence or otherwise, or from any use or operation of any methods, products, instructions, or ideas contained in the material herein.

British Library Cataloguing-in-Publication Data
A catalogue record for this book is available from the British Library

Library of Congress Cataloging-in-Publication Data
A catalog record for this book is available from the Library of Congress

ISBN: 978-0-12-811071-3

For Information on all Academic Press publications
visit our website at https://www.elsevier.com/books-and-journals

ELSEVIER
Book Aid International
Working together to grow libraries in developing countries
www.elsevier.com • www.bookaid.org

Publisher: Matthew Deans
Acquisition Editor: Brian Guerin
Editorial Project Manager: Joshua Bayliss
Production Project Manager: Sruthi Satheesh
Cover Designer: Victoria Pearson

Typeset by MPS Limited, Chennai, India

To my family and friends

Contents

Foreword

This monograph on mechanics of carbon nanotubes presents fundamentals about material and structural properties of carbon nanotubes, classification of their atomic lattice structures into 20 classes of distinct nanostructures, nanoscale laws of motion and friction, atomic potentials, nanoscale expressions for stress, nanoscale homogenization criteria for the atomic-continuum representation and their effect on Cauchy–Born approximation, scaling analysis of atomic lattice structures and scaling inequalities, and ranges of applicability of the continuum beam and shell models and their limitations at nanoscale. This monograph also examines molecular dynamics (MD) simulations of carbon nanotubes, nanoscale interfacial structures and nanocomposites, and includes examples of such simulations and their results for the stress–strain dependence in nanocomposites, the pullout process of a carbon nanotube, and its interfacial sliding. Nanoscale averaging of MD simulations, criteria for nanoscale homogenization of MD models and their results have been discussed, along with applicability of various continuum models and macroscale laws at nanoscale.

University, graduate and postgraduate students and academic and industrial researchers will learn about a rigorous definition of nanomaterials and under what conditions their material properties attain the unique and well-defined values independent of a nanoscale phenomenon of the size-dependency of material properties of atomic structures. Postdoctoral students will be fascinated by the so-called "effective thickness paradox" for carbon nanotubes associated with the multiple estimates for the effective thickness of a single layer of atoms, and its resolution based on the balance between the in-plane elastic interactions and transverse van der Waals forces. A solution of the thickness paradox for carbon nanotubes has been presented with a review of different estimates for their thickness. Academic researchers and scholars will find out about the recent discovery of a nanoscale analog of Pauli's principle associated with the spatial exclusion of electrons between atomic interfaces and its role in the interfacial friction and sliding.

A new matrix analysis for the atomic registry between adjacent atomic lattices has been developed for the hexagonal lattices of carbon nanotubes and graphene. That is the atomic registry matrix analysis, which includes the matrix-based description for the commensurate and noncommensurate registry between adjacent carbon rings with different matrices for distinct registries and the interregistry transition matrices. The value of coefficients in each registry matrix depends on local interatomic interactions and atomic potentials, while those in the interregistry transition matrices reflect more complex interactions associated with the changes from each atomic registry. Atomic registry matrix analysis allows one to describe the interfacial interactions and sliding between adjacent atomic lattices by taking into account local interactions. Interfacial sliding between graphene sheets can be described by nanoscale analogs of Newton's friction laws.

An updated classification of carbon nanotubes includes 20 classes of single-wall carbon nanotube (SWCNT) and multi-wall carbon nanotube (MWCNT) with up to 10 classes of the long beam-like atomic structures, four classes of thin and thick lattice shells with cylindrical symmetry, and several classes of the carbon nanotube nanobeams and nanocrystal rods with polygonal symmetry. Different classes of carbon nanotubes are defined by the separation criteria based on scaling analysis of their atomic lattice structure. It is shown that MWCNTs may also behave as thin shells for a certain range of geometric parameters, which is represented in the materials map. Degeneration of the cylindrical symmetry, the circumferential buckling and corrugation of carbon rings, and increase in the bond length and internal van der Waals forces characterize the unique structural and material properties of the carbon-nanotube nanobeams and nanocrystal rods. A topological transition separates the hollow carbon nanotube nanobeams with the polygonal symmetry and the nanocrystal rods having a tighter atomic lattice structure.

Considering a wide variety of carbon nanotube geometries and their atomic structures, it is understandable that some carbon nanotube lattice shells require the thin shell theories for their analysis and the data reduction of MD simulations results, while other polygonal lattices are better represented by atomic-continuum beam models. Classification of SWCNTs and the appropriate models are represented in a parametric map, which is useful for the design of nanoscale probes (e.g., for atomic force microscopy) and predicting their deformation modes. This monograph presents chiral formulation of several formulae for the critical buckling strain of carbon nanotube lattice shells and nanobeams in the atomistic-continuum form. A new lattice-sensitive formula for the critical strain of chiral SWCNTs is evaluated, along with its predictions for nanoscale buckling of SWCNTs for different estimates for their effective thickness. Numerical results compare well with the available data of MD simulations. Equivalent-continuum shell models for carbon nanotubes have been reviewed, along with the ranges of applicability of the continuum shell theories to the mechanics of carbon nanotubes.

Unique material properties of MWCNTs have been analyzed to demonstrate how their structure changes between different classes and influences Poisson's affects in some cases. A nanoscale phenomenon of simultaneous beam- and shell-like deformation modes occurring during a complex buckling of short MWCNTs is described. Timoshenko beam model is presented for the deformation of MWCNTs. Interfacial sliding in MWCNTs and in the carbon nanotube-reinforced nanocomposites is also analyzed. Interfacial friction models based on nanoscale friction laws have been presented for MWCNTs and nanocomposites. A friction model is introduced for the characterization of the strength of interfacial adhesion during the pullout of a carbon nanotube from the polymer matrix. Interfacial structure in nanocomposites is examined by the scaling analysis, atomic registry analysis and reflected in the modified rules of mixture for the evaluation of elastic properties of the carbon nanotube-reinforced nanocomposites. Tsai–Halpin relation and its modified nanoscale form are also presented for the approximation of elastic properties of nanocomposites. Constitutive models of multifunctional nanocomposites have been presented to describe the electromechanical, thermomechanical, and elastomagnetic effects. Recent trends in publications on mechanics of carbon nanotubes and their safety have been reviewed.

The author gratefully acknowledges resources of Nanodesigns Consulting, color images of the shutterstock database, along with the help and support of Elsevier staff and Nanodesigns Press.

Vasyl Harik

Preface

Carbon nanotubes have superior mechanical and electrical properties which can be used for a variety of nanotechnology applications. Carbon nanotubes have been some of the most popular nanomaterials including graphene. Atomic lattice structures of carbon nanotubes and graphene have many similarities and differences associated with the effects of curvature. Mechanical behavior and material properties of carbon nanotubes, graphene and other nanomaterials, which are subjected to various physical forces, are studied by nanomechanics and nanoscale mechanics, molecular mechanics, and molecular dynamics, as well as in other disciplines. Nanoscale mechanics may include micromechanics with some nanoscale effects. Foundations of nanomechanics are based on different branches of nanoscale physics, nanoscale chemistry, and the areas of solid state physics and continuum mechanics models, which are capable of describing the nanoscale effects through new methods, new experimental techniques and devices, as well as appropriate extensions of existing methodologies and instrumentation. The mechanics of carbon nanotubes is a part of nanomechanics and nanoscale mechanics of nanomaterials, nanocomposites, and nanostructured materials. The first course on mechanics of carbon nanotubes was developed by the author in 2001 at NASA Langley Research Center in Hampton, Virginia, and presented later in the year at Annual Congress of American Society of Mechanical Engineers in New York.

Carbon nanotubes are being used in computer industry (e.g., 9 nm size transistor). The carbon nanotube-based technology has surpassed the silicon-based technology in the computer chip industry. Carbon nanotubes can be also used for different nanoscale devices, sensors, and actuators, nanoscale probes and microelectronics. The use of carbon nanotubes and their bundles in nanocomposites and the carbon fiber-reinforced composite materials is still being optimized in order to improve the materials properties of nanocomposites, the strength of interfaces, and safety. Long high aspect ratio multi-wall carbon nanotubes may have toxic effects, depending on the exposure and the dose. Reducing the health risks associated with the use of carbon nanotubes is very important for stable development of nanotechnology and the nanotechnology driven industrial revolution. This monograph addresses both the mechanical aspects of the carbon nanotube technology and reduction of health risks by controlling the size and structure of carbon nanotubes and their use in nanocomposites.

Vasyl Harik

CHAPTER 1

Nanotechnology of Carbon Nanotubes: Sensors, Transistors and Nanocomposites

Chapter Outline

1.1 Properties of Carbon Nanotubes

The discovery of carbon nanotubes in 1991 by S. Iijima made these nanostructures the most popular type of nanomaterial and nanoparticles used in novel applications; they have since been studied by academic and industrial researchers, and college students (Iijima, 1991). The name "carbon nanotubes" is derived from the material named carbon (C) and the hollow, tubular structure of the wall, formed by the one-atom-thick layer of carbon lattice (Dresselhaus et al., 1996). The single-wall carbon nanotubes (SWCNTs) have one layer of carbon atoms, while the multiwall carbon nanotubes (MWCNTs) have multiple layers of carbon lattice (Iijima and Ichihashi, 1993). Carbon nanotubes have extraordinary material properties, such as high stiffness and strength, and exceptional electrical and thermal conductivities when compared to other promising carbon materials (Gogotsi, 2006, 2017). Their elastic modulus is much higher than that of steel and carbon fiber-reinforced

Mechanics of Carbon Nanotubes.
DOI: https://doi.org/10.1016/B978-0-12-811071-3.00001-9

© 2018 Elsevier Inc. All rights reserved.

composites (CFRC). The superior strength of carbon nanotubes comes from their hyperelastic material properties and high elastic moduli.

Material properties of carbon nanotubes, including their stiffness and conductivity, have a unique dependence on their atomic structure, size, and geometry. Carbon nanotubes may have different physical properties, depending on the orientation of hexagonal carbon rings in the structure of their atomic lattice (Fig. 1.1). The armchair SWCNTs are metallic and conductive, while zig-zag SWCNTs are brittle and semiconductive (Dresselhaus et al., 1996; Harris, 1999). The stiffness and strength of SWCNTs is influenced by their chirality, i.e., the orientation of their atomic lattice structure. Elastic properties of metallic and semiconductive SWCNTs are also affected by their radius. Smaller carbon nanotubes have higher elastic modulus. Small fragments and short shells of carbon nanotubes have different properties from their longer counterparts, depending on their size or the number of carbon rings spanning their length (Harik, 2001a, 2011).

The geometry of carbon nanotubes has a significant effect on the elastic properties of their atomic lattice structures. Carbon nanotubes with small radii behave like beams, while SWCNTs with large diameters behave like lattice shells (Yakobson et al., 1996; Harik, 2001b). Curvature of carbon nanotubes, which is inversely proportional to their radius, increases the stiffness of carbon nanotubes and their elastic modulus. In fact, carbon nanotubes become a new type of nanomaterial, the so-called nanotube crystals, when their radius is smaller than the size of a carbon ring (Harik, 2001a). The uniqueness of these carbon nanotubes is associated with their unique material properties, which are different from other SWCNTs with larger radii. The length of the atomic structure of SWCNTs can also influence their mechanical properties and mechanical behavior (Harik, 2001b,c).

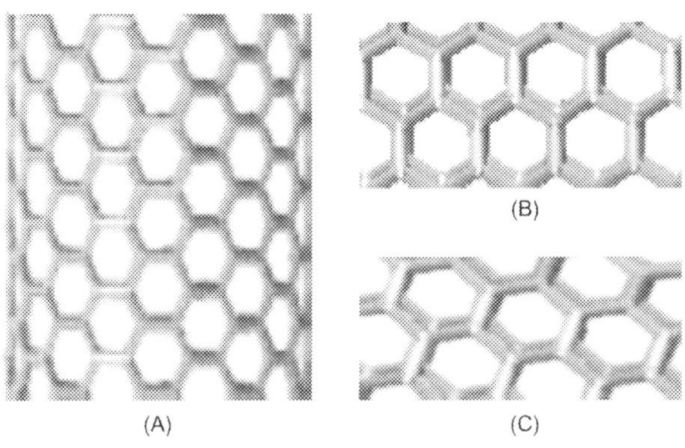

Figure 1.1
Atomic structure of the armchair (10,10) SWCNT (A), a zig-zag (n,0) SWCNT (B) with $n > 10$, and a chiral SWCNT (C) with different orientations of the hexagonal carbon rings. *SWCNT*, single-wall carbon nanotube. © *2011 by Nanodesigns Consulting, reproduced with permission.*

The most successful application of carbon nanotubes has been in the computer industry. IBM has developed its famous 9-nm transistor by using electrically conductive carbon nanotubes. The current use of 22, 45, and 90-nm size parts by Intel in computer chips ensures that the computer industry is capable of using a variety of carbon nanotubes in its applications. Other industries, especially the microelectronics industry, are also capable of utilizing carbon nanotubes in various phone applications, microelectronics, and sensors. Mechanical properties of carbon nanotubes are important for nanoscale designs. Understanding of the mechanical behavior of carbon nanotubes is critical for the optimization of nanodevices, sensors, and probes (Harik and Salas, 2003). Nanoscale phenomena associated with the mechanics of carbon nanotubes are also important for predicting behavior of other nanostructures and nanomaterials.

1.1.1 Discovery of Carbon Nanotubes in the 1950s and 1990s

Carbon nanotubes were discovered by Iijima in 1991 at NEC Laboratories in Japan. In 1996, he has received the Nobel Prize for the discovery of MWCNTs in 1991, and of SWCNTs in 1993. Later it turned out that L. V. Radushkevich and V. M. Lukjanovich (Ukrainians or Belarusians) had reported a discovery of MWCNTs in 1952 (see Fig. 1.2). However, the unique material properties of carbon nanotubes and other nanostructures were recognized much later, in the early 1990s. A brief historical timeline is provided below:

- **1952**—L. V. Radushkevich and V. M. Lukjanovich observed carbon nanotubes (Fig. 1.2).
- **1991**—S. Iijima (NEC Labs) rediscovered MWCNTs as "microtulules."
- **1992**—M. S. Dresselhaus's team described the chirality of carbon nanotubes.
- **1992**—D. Brenner's team examined the elasticity of carbon nanotubes.

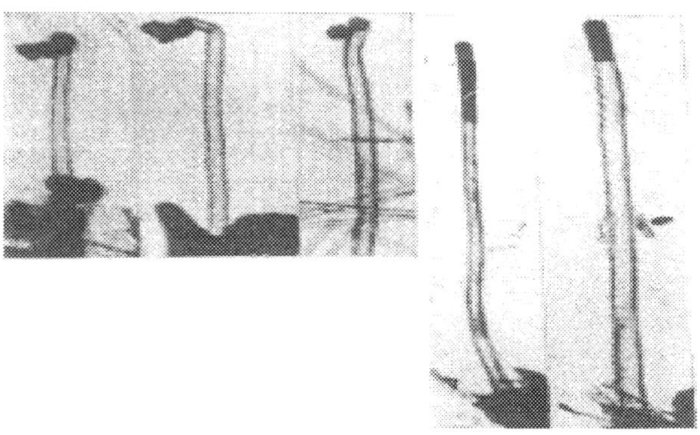

Figure 1.2
Images of the multiwall carbon nanotubes discovered by Radushkevich, L. V., Lukjanovich, V. M., 1952. Rus. J. Phys. Chem., 26 (1), 88–95.

- **1993**—S. Iijima and T. Ichihashi (NEC Labs) discovered SWCNTs and used the concept of "shell" to describe carbon nanotubes.
- **1993**—D. Tomanek's team in Michigan studied the low-frequency vibrational modes of long "carbon tubules," i.e., carbon nanotubes (Overney et al., 1993).
- **1993/94**—R. S. Ruoff and J. Tersoff team at IBM simulated deformation of carbon nanotubes in carbon nanotube crystals (Ruoff et al., 1993).
- **1996**—M. M. J. Treacy, T. W. Ebbesen, and J. M. Gibson carried out one of the first experimental testing of carbon nanotubes with the atomic force microscope (AFM).
- **1996**—B. I. Yakobson, C. J. Brabec, and J. Bernholc performed molecular dynamics (MD) simulations of axial buckling and twisting of carbon nanotubes.
- **1997**—M. R. Falvo and R. Superfine's team at the University of North Carolina conducted experiments on bending and buckling of carbon nanotubes under large strains.
- **1997**—C. M. Lieber and his team at Harvard carried out experimental testing of vibrating carbon nanotubes, similar to the earlier AFM experiments.
- **1998**—Many scientists (e.g., Ajayan, Avouris, Brenner, Dai, Dresselhouse, Lieber, Lordi, Lu, Ru, Ruoff, Rubio, Schadler, Sinnott, Smalley, Superfine, Treacy, Wagner, White, and others) tested, modeled, and analyzed SWCNTs.
- **2001**—V. H. Crespi's group at Pennsylvania State University and V. M. Harik [Institute for Computer Applications in Science and Engineering) (ICASE) (NASA)], independently predicted the uniqueness of carbon nanotubes with ultrasmall radii.[1]
- **2001**—V. M. Harik (ICASE Institute, NASA Langley Research Center) introduced classification of SWCNTs into four classes: thin and thick lattice shells, long high-aspect-ratio nanotubes, and the beam-like carbon nanotube crystals of small radii.

Carbon nanotubes turned out to be quite a diverse group of nanostructures, with markedly different physical and material properties. The mechanics of carbon nanotubes provides a systematic analysis of their mechanical properties, their mechanical behavior during deformation, and the mechanical phenomena occurring with carbon nanotubes at nanoscale. The first nationally distributed course, *Mechanics of Carbon Nanotubes*, was developed in 2001 by the author at ICASE Institute at NASA Langley Research Center in Hampton, Virginia, for the continuing education institute of American Society of Mechanical Engineers (ASME) in New York (Harik, 2001c). The original course was presented in 2001 at the Annual Congress of ASME in New York. The course was then presented in 2002, at

[1] In 2001 V. H. Crespi's group at Penn State University and V. M. Harik at NASA Langley Research Center have independently predicted the degeneration of SWCNT lattice shells into thin nanobeams around the critical value of the normalized SWCNT radius, $R_{NT}/a \approx 1$. Crespi had predicted the breaking of "the symmetry of sp^3 bonds in tubular geometries" in the smallest nanotubes. See Harik, V. M. 2001a., Solid State Communications., 120 (7–8), 331–335 (2001a) and D. Stojkovic D., P. Zhang P., and Crespi V. H. 2001., Physical Review Letters, 87 (12), 125502 (2001).

the Nanosystems Conference, and again in 2004 at the ASME Annual Congress. Later the ASME continuing education institute distributed this course nationwide on CDs and videos as basic introductory material on the mechanics of carbon nanotubes.

1.1.2 Atomic Structure of Carbon Nanotubes: Chirality and Its Effects

Carbon nanotubes consist of carbon atoms arranged into an atomic lattice of hexagonal carbon rings at different orientation angles (Fig. 1.1). Orientation of carbon rings in the structure of carbon nanotubes affects the physical properties of SWCNTs. Atomic lattice of armchair SWCNTs, shown in Fig. 1.1A, is associated with metallic properties of carbon nanotubes, e.g., electrical conductivity. Atomic zig-zag structure of the zig-zag SWCNTs (Fig. 1.1B) is associated with brittle material properties and semiconductivity. Elastic properties of SWCNTs, such as stiffness and strength, are also influenced by the orientation of their atomic lattice (Fig. 1.3).

Carbon nanotubes are called "chiral" if their atomic lattice has any orientation other than armchair or zig-zag (Fig. 1.3). Chirality is the property of the atomic structure of carbon nanotubes associated with a skewed orientation of hexagonal carbon rings. Chirality of

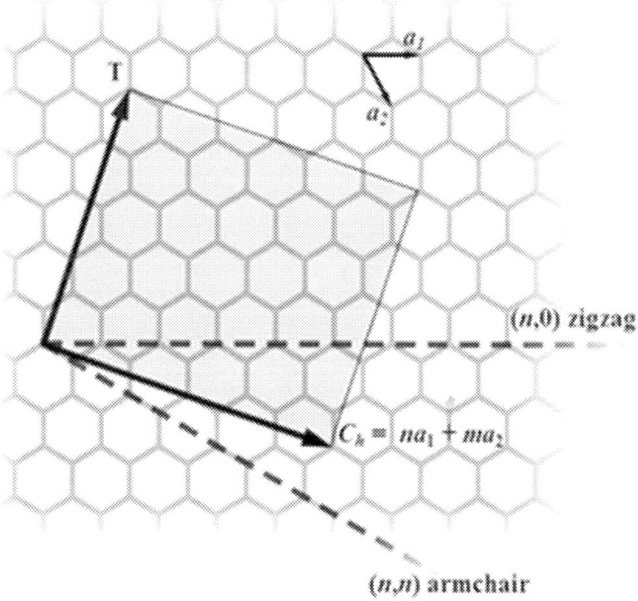

Figure 1.3

A schematic describing the chirality of carbon nanotubes by using the chirality vector, $\mathbf{C}_h$, with components *n* and *m* in a coordinate system with the unit vectors $\mathbf{a}_1$ and $\mathbf{a}_2$ (after Fujita et al., 1992). Edges of armchair (*n,n*) and zig-zag (*n,0*) carbon nanotubes with the axis vector, **T**, are shown by the dashed line identifying the chirality of metallic and semiconductive lattices.

carbon nanotubes can be characterized by an angle specifying a particular orientation of hexagonal carbon rings in the atomic structure of SWCNTs. Fujita et al. (1992) have introduced a description of chirality in the atomic structure of carbon nanotubes by using a coordinate system with the unit vectors $\mathbf{a}_1$ and $\mathbf{a}_2$ (Fig. 1.3), and integers m and n identifying the components of the so-called chirality vector. The chirality vector ($\mathbf{C}_h$) identifies orientation of the atomic structure of carbon nanotubes. As a result, the chirality of carbon nanotubes is typically described by a pair of integer numbers (n,m), which also define vector $\mathbf{C}_h$ as noted above (Fig. 1.3). This pair of indices (n,m), and vector $\mathbf{C}_h$ also describe the structure of an open end of a carbon nanotube, e.g., a zig-zag edge is denoted by the pair $(n,0)$ while an armchair edge is denoted as (n,n).

The armchair carbon nanotubes have the chiral angle of 30° for their metallic atomic lattice (Fig. 1.4), while the chiral angle for the semiconducting zig-zag nanotubes is 0° as indicated by the top horizontal row in the periodic table of carbon nanotubes. Moreover, the relation between the atomic structure of carbon nanotubes, their chirality, curvature and diameter, metallic and semiconducting properties are illustrated in the periodic table of SWCNTs (Fig. 1.4). Metallic SWCNTs have vanishing energy gap, E_g, while semimetallic SWCNTs have small energy gap, i.e., E_g <1 eV (see www.quantumwise.com). Semiconducting SWCNTs have a noticeable energy gap, such that E_g >1 eV.

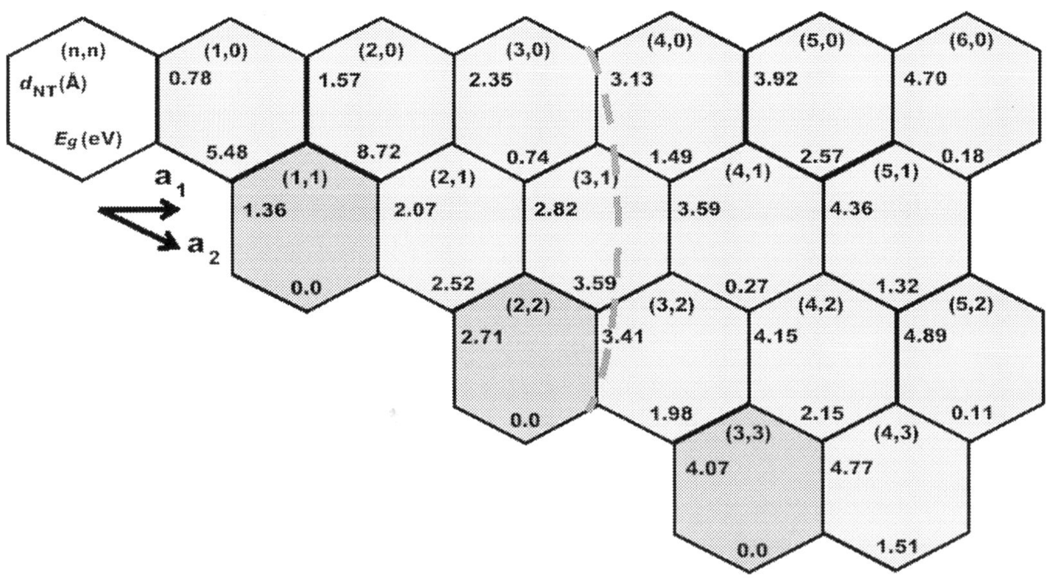

Figure 1.4

A periodic table of carbon nanotubes showing zig-zag SWCNTs (top row) with their diameters and the energy gap, E_g. Armchair SWCNTs (side row along direction $\mathbf{a}_2$) are indicated by dark metallic color, while semimetallic SWCNTs are shown in gray color. Semiconducting SWCNTs are shown in light sand color. SWCNT crystals have very small diameters (marked by the dashed line). *SWCNT*, single-wall carbon nanotube.

1.1.3 Geometry of Carbon Nanotubes and Their Properties

Carbon nanotubes have been discovered as tubular nanostructures or nanoscale shells (Iijima, 1991; Iijima and Ichihashi, 1993). It has been noted that changes in the diameter of carbon nanotubes result in noticeable changes in the material properties of these nanoscale fibers. Geometry of SWCNTs has a significant effect on the elastic properties of their atomic lattices. Smaller carbon nanotubes have a higher elastic modulus (Stojkovic et al., 2001). Curvature of SWCNTs is inversely proportional to the radius, so its higher values increase the stiffness of atomic structures and their elastic moduli. The effect of high curvature in atomic lattices is so important for carbon nanotubes that these nanostructures become a new type of nanomaterial, called SWCNT crystals. The radius of a SWCNT crystal is smaller than the size of a carbon ring (Harik, 2001a,b, 2002). In the periodic table of carbon nanotubes the group of SWCNT crystals has been identified by the dashed line (Fig. 1.4). The uniqueness of these carbon nanotubes is associated with their unique material properties, which are quite different from other SWCNTs of larger radii. SWCNT crystals do not have a very high strain-to-failure, shown by most carbon nanotubes (Fig. 1.5).

Mechanical behavior of SWCNT crystals is similar to that of other nanoscale crystals, e.g., they deform as atomic or nanoscale beams. Carbon nanotubes with small diameters, which are clearly larger than the size of a carbon ring, also behave like beams. However, in contrast to SWCNT crystals these beams are clearly hollow inside. One could ask: "Are carbon nanotubes the single-wall nanoscale shells, or nanoscale beams? SWCNTs with large diameters behave as lattice shells" (Yakobson et al., 1996; Harik, 2001a). The length of the atomic structure of SWCNTs can also influence their mechanical properties and mechanical behaviors (Harik, 2001a,b). In fact, most SWCNT shells behave as nanoscale

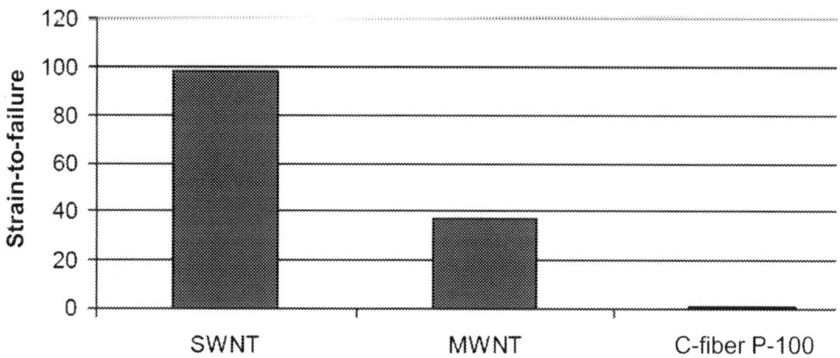

Figure 1.5
Comparison of the strain-to-failure of SWCNTs and MWCNTs with respect to the strain-to-failure of a carbon fiber P-100 (unit value). *SWCNT*, single-wall carbon nanotube; *MWCNTs*, multiwall carbon nanotubes.

beams, when their aspect ratio becomes large enough (Harik, 2002). In the case of large diameters, SWCNT shells may collapse as membranes into nanoscale ribbons, and then deform further under multiaxial loading. Small fragments and short shells of carbon nanotubes also have unique variations in their properties that are different from their longer counterparts. The magnitude of such differences depends on the size of short shells, or on the number of carbon rings spanning their length (Harik, 2011). The mechanics of carbon nanotubes answers these and other questions about the mechanical behavior of various SWCNTs, as well as the carbon nanotube-based probes, nanodevices, and nanocomposites.

1.1.4 Effective Thickness Paradox for Carbon Nanotubes

Thickness of carbon nanotubes is the basic geometric parameter describing the shape and size of all SWCNT lattice shells. This parameter is critical in any nanoscale design involving carbon nanotubes. The value of effective thickness of SWCNTs is also important for the evaluation of their elastic moduli, which are the key material parameters (Fig. 1.6). Therefore, any discrepancy in the estimates of SWCNT thickness affects the predicted value of their elastic moduli, and brings potential errors and problems with nanoscale designs. The thickness of discrete atomic lattices can be evaluated by using different methods. The initial efforts in the development of continuum models for carbon nanotubes have yielded both fascinating results and an unexplained paradox associated with the effective thickness, h_{NT}, of the atomic lattices of carbon nanotubes (Yakobson et al., 1996; Harik, 2001a,b; Huang et al., 2006).

The first theoretical modeling results of carbon nanotube crystals have demonstrated that the effective thickness of the carbon nanotube lattices seems to be the same as the effective

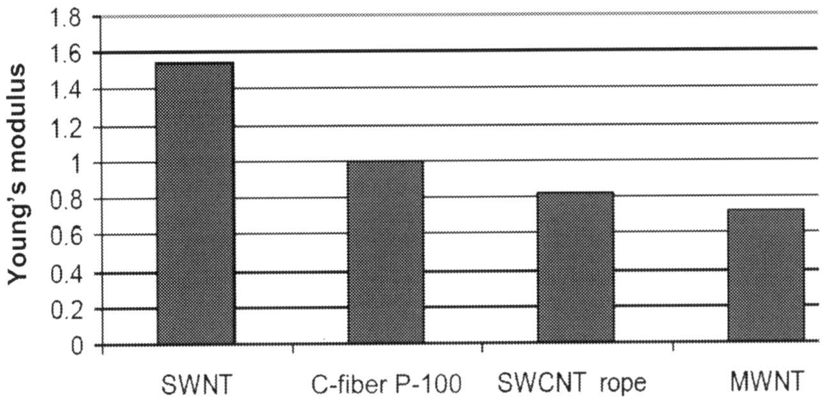

Figure 1.6
Comparison of Young's modulus of SWCNTs, SWCNT rope and MWCNTs with respect to Young's modulus of a carbon fiber P-100 (unit value). *SWCNT*, single-wall carbon nanotube; *MWCNTs*, multiwall carbon nanotubes.

thickness of graphene sheets in the graphite, that is, 3.4 Å (Ruoff et al., 1993). Moreover, these modeling results have shown that the effective thickness of carbon nanotubes, h_{NT}, is the same for all nanotube lattices, regardless of their diameter, d_{NT}, which seemed reasonable as a first-order approximation. The unexplained paradox with the effective thickness of carbon nanotube lattices came about when a continuum shell model was successfully used for the analysis of axial buckling of one carbon nanotube, the (10,10) armchair nanotube (Yakobson et al., 1996). MD modeling of Yakobson et al. (1996, 2003) and their continuum shell model have shown that the effective thickness of the (10,10) carbon nanotube is about 0.66 Å. This is the Yakobson, Brabec, Bernholc (YBB) estimate for the thickness of a carbon nanotube, which is about 7/15 l_{C-C}, where l_{C-C} is the bond length between two carbon atoms.

The estimate of 0.66 Å for the effective SWCNT thickness is five times smaller than the earlier predicted thickness of nanotube lattices and the thickness of graphene sheets. This discrepancy in the evaluation of the thickness of carbon nanotubes constitutes the "carbon nanotube thickness paradox" (Harik, 2011, 2014): the effective thickness of the hexagonal atomic lattice of carbon nanotubes may vary from 3.4 Å, which is the effective thickness of the graphene, to 0.66 Å, which is the YBB estimate of about 7/15 l_{C-C}. The effective thickness paradox is important for the understanding of the equivalent continuum models for a variety of atomic lattices, not just carbon nanotubes. This paradox is also critical for the fundamental understanding of the equivalent continuum, as the continuum representation of any single-, double-, or triple-layer atomic lattices in the still-emerging nanoscale mechanics of crystalline and nanocrystalline materials.

The effective thickness of carbon nanotubes was not an issue in the application of the continuum beam model for the vibrating nanotubes in the analysis of experimental data from the AFM observations (Treacy et al., 1996; Wong et al., 1997). In a simple analysis, a hollow cylinder model was used for the nanotube geometry, and the corresponding continuum beam model was utilized to predict Young's modulus of a vibrating carbon nanotube. These were some of the first experiments on the mechanical behavior of carbon nanotubes. The effective thickness of carbon nanotubes in these beam models was assumed to be 3.4 Å as in the majority of other studies. The first vibration experiments have demonstrated that carbon nanotubes vibrate similarly to a string or the high aspect ratio beams. These lateral vibration experiments have also posed a new question: does carbon nanotube vibrate as a continuum beam or as a thick lattice shell? The role of the effective thickness estimates was not addressed in the vibration experiments, along with the puzzling effective thickness paradox.

One part of the effective thickness paradox for carbon nanotubes has to do with the question, do carbon nanotubes behave similarly to continuum beams or continuum shells?

The material model that describes the mechanical behavior of any material during deformation processes is called the constitutive model. After the first theoretical and experimental investigations of carbon nanotube behavior it was clear that the carbon nanotubes seem to be elastic or hyperelastic. However, it was not clear when they behave as the shell-like structures or as the beam-like crystals (Harik, 2001a,b). The answer to one part of the question about the constitutive model came in the 2001 NASA report, "The ranges of applicability of the continuum beam model in the constitutive modeling of carbon nanotubes: nanotubes or nanobeams?" soon after the National Nanotechnology Initiative began (Harik, 2001b). The question, "nanotubes or nanobeams?" was answered in such a way that the new theoretical results were supporting both theoretical continuum shell models, as well as the continuum beam models used in the AFM experiments. This was quite intriguing, since beam-like SWCNTs had extraordinary strength (Fig. 1.7).

The novelty of the presented analysis was in the rigorous explanation of the facts that carbon nanotubes can behave both as shells and as beams, depending on the geometric parameters of their molecular structures. The new analysis of the constitutive behavior of carbon nanotubes was sensitive to the molecular structure of nanotube lattices and their geometric parameters, as it was based on the scaling analysis. The new scaling analysis has also rigorously linked the continuum beam and shell models for carbon nanotubes with their molecular structure and their effective thickness. However, the scaling analysis has considered all existing estimates for the effective thickness of carbon nanotubes, and presented a new estimate for the effective thickness, 0.72 Å adjusted later to 0.71 Å or 1/2 l_{C-C} (Harik, 2011, 2014), which further exposed the existence of the effective thickness paradox with the effective thickness of graphene sheets, 3.4 Å. The new estimate for the effective thickness of carbon nanotubes was based on an approximation for the thickness

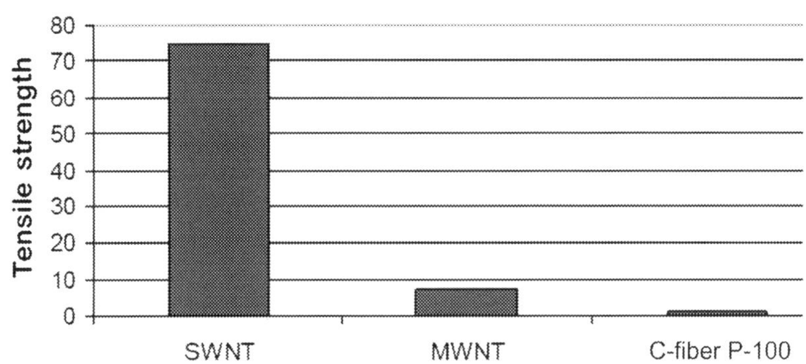

Figure 1.7
Comparison of the tensile strength of SWCNTs and MWCNTs with respect to the tensile strength of a carbon fiber P-100 (unit value). *SWCNT*, single-wall carbon nanotube; *MWCNTs*, multiwall carbon nanotubes.

of the load-transferring bonds, that is, the half-length of the covalent bond, 1/2 l_{C-C}, which was close to 0.66 Å as estimated by Yakobson's group (Yakobson and Dimitrica, 2003). The effective thickness paradox for carbon nanotubes became more explicit with the two independent estimates, pointing to much thinner effective thickness of the nanotube lattices.

The same NASA report (Harik, 2001b, 2002) has also presented new parametric maps for the continuum beam and the continuum shell models, which have identified the ranges of geometric parameters for when these models can be applied for the atomic lattices of carbon nanotubes. These ranges of applicability of the continuum beam and continuum shell models have been shown to depend on the estimates of the effective thickness of the carbon nanotube lattices. However, the existing estimates of the effective thickness of carbon nanotubes are varied between 0.66 and 3.4 Å, or five times in magnitude. A new edited volume on the trends in nanoscale mechanics (Harik, 2014) presented more details on the analysis of the effective thickness of carbon nanotubes, as well as the resolution of the effective thickness paradox for the atomic lattices with the hexagonal and nonhexagonal structures. The proposed resolution of the effective thickness paradox is based on a new methodology for the evaluation of contributions from the two limiting values of the effective thickness of atomic lattices of carbon nanotubes, which depends on the applied loads. Resolution of the effective thickness paradox is stated as follows (Harik, 2014):

> The effective thickness of the hexagonal atomic lattice of carbon nanotubes may vary from 3.4 Å, which is the effective thickness of the graphene under the transverse loading through van der Waals forces, to 0.66 Å, which is the so-called YBB estimate for the axial loading through the load transferring covalent bonds; these variations are load dependent and proportional to the number of carbon rings subjected to a particular load.

Estimates of the effective thickness of the single-, double-, and triple-layer atomic lattices are critical for the design and optimization of novel nanodevices. These estimates are applicable to the lattice shells, the atomic structure of which is not affected by the high curvature effects, and the internal van der Waals forces, such as in the case of thin nanotube crystals.

1.1.5 Material Properties of Carbon Nanotubes

A nanomaterial is usually defined as a material with at least one of its dimensions being less than 100 nanometers (Harris, 1999; Harik and Salas, 2003; Gogotsi, 2006). Carbon nanotubes satisfy this requirement. All SWCNTs have diameters smaller than 100 nm, while some MWCNTs have only internal shells with the diameters in this range. Therefore, some MWCNTs can be considered as nanostructured submicroscopic and microscopic fibers. Nanostructured materials consist of nanoscale structural elements comprised of nanoparticles, nanoscale shells, crystals and grains or nanostructures. Sometimes, a lower dimensional limit, i.e., 1 nm, is cited for the size limit of nanoparticles or nanomaterials. A special homogenization criterion can be used to clearly define the lower size limit for any nanomaterials, all nanoparticles and nanostructures having well-defined material properties, and uniquely defined mechanical properties (Harik, 2001a,b,c; Harik, 2011).

Carbon nanotubes have superior material properties with respect to their macroscopic counterparts: carbon fibers and the CFRCs. A number of universities, and NASA Langley[2] Research Center have been studying carbon nanotubes, namely SWCNTs and MWCNTs, for developing new composite materials and nanocomposites for aerospace and space applications. Several groups at NASA Langley Research Center and ICASE Institute have been working since 2000 on NASA Nanotechnology programs for Multifunctional Materials and Structures (Fig. 1.8). In order to support these programs, a nanotechnology database has been established, with reference materials on material properties of carbon nanotubes and related research.

Carbon nanotubes have extraordinary strength and strain-to-failure, high elastic moduli, and electrical conductivity. Such material properties seemed quite useful for the development of novel multifunctional nanocomposites, which could replace the CFRC. The tensile strength of SWCNTs and MWCNTs is considerably higher than the tensile strength of carbon fiber P-100 (Fig. 1.7). The strain-to-failure of hyperelastic SWCNTs and MWCNTs is also much higher than the strain-to-failure of carbon fiber P-100 (Fig. 1.5). Young's modulus of SWCNTs, SWCNT ropes and MWCNTs is also superior with respect to Young's modulus of carbon fiber P-100 (Fig. 1.6). However, in 2004–05, it was indicated that some carbon nanotubes may be toxic, and much more care may be required for the robust implementation of SWCNTs and MWCNTs in composite materials.

Carbon nanotubes have exceptional electrical and thermal conductivities that can be used in aerospace composites for lightening protection, as well as in nanoscale devices such as sensors and actuators for monitoring and controls. Thermal conductivity of SWCNTs and aluminum Al 2219-T87 are compared in Fig. 1.9 with the thermal conductivity of carbon fiber P-100 used in composites. These materials are used in aerospace applications, where thermal and electrical conductivities are also important for the structural health monitoring by the built-in failure sensors, strain sensors and special microelectromechanical systems (MEMS). Superior electrical and thermal properties of carbon nanotubes can be used in nanoscale devices without significant health risks caused by the presence of large number of long SWCNTs and MWCNTs (Harik, 2014). The use of carbon nanotubes in nanocomposites and CFRC can be justified only for applications where the crack or damage size is such that the cumulative dose

[2] In 2002, NASA has funded the Princeton University based multi-university institute for bio-inspired nanostructured materials, which continues to provide journal publications for the public. Nanodesigns Consulting is a 2004 spin off from NASA Langley Research Center, which was formed to support the research of this institute.

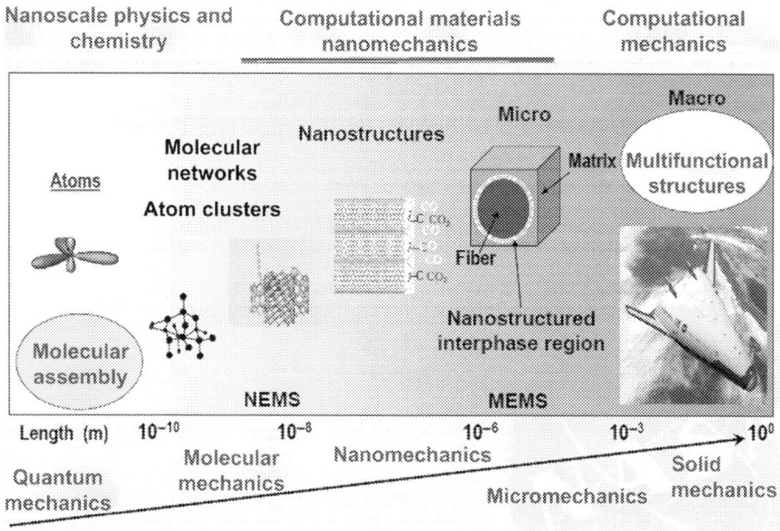

Figure 1.8

Schematic[3] of the length scales involved in the mechanics of carbon nanotubes and nanostructured materials (e.g., nanocomposites, nanotube-modified polymers and multifunctional membranes) and the relevant areas of science.

of released carbon nanotubes during a material failure is well below the critical dose, with the carbon nanotubes (SWCNTs or MWCNTs) smaller than the critical length (Harik, 2014).

1.2 Applications of Carbon Nanotubes in Nanotechnology

Nanotechnology deals with the manufacturing, processing, manipulation, deformation, and use of various nanomaterials, which are made of nanostructured material elements whose size is less than 100 nm. Carbon nanotubes are very promising nanomaterials, due to their extraordinary material properties. SWCNTs and MWCNTs are also critical nanoparticles used for reinforcement and modification of material properties in nanocomposites with polymer, metal and ceramic matrices. Carbon nanotubes can be also used as nanoscale probes for AFMs and sensors, because of their high stiffness and high aspect ratios. Other examples of carbon nanotube-based applications in nanotechnology include: nanoelectromechanical systems (NEMS) devices, transistors, quantum dots, nanoscale bio-probes, bio-sensors, the nanotube-enhanced polymer membranes, thin film sensors and multifunctional nanocomposites. This chapter presents a review of typical nanotechnology applications of carbon nanotubes.

[3] This figure was developed in 2000 during the establishment of new nanotechnology programs at NASA Langley research center, and ICASE Institute, which has been re-established in 2002 as National Institute of Aerospace near NASA Langley research center (NASA LaRC) in Hampton, Virginia.

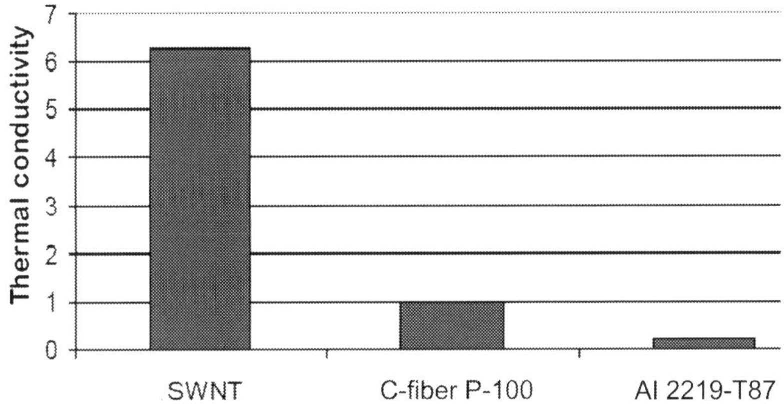

Figure 1.9

Comparison of thermal conductivity of SWCNTs and aluminum Al 2219-T87 with respect to the thermal conductivity of a carbon fiber P-100 (unit value). *SWCNT*, single-wall carbon nanotube.

Carbon nanotube-based sensors and actuators can be used in many industrial applications, instrumentation and medical devices. SWCNTs are used by IBM in computer chips in transistors. In fact, the carbon nanotube-based computer chips have now outperformed the silicon-based computer chips. Therefore, advances in computer technology are now ensured by the use of carbon nanotubes. In the future, graphene-based transistors will support further developments in the carbon-based computer technologies and electronics. Understanding of material properties and mechanical behavior of these materials is critical for the optimization of nanoscale designs and for biomedical applications.

1.2.1 Carbon Nanotube Probes for Atomic Force Microscopes

The high aspect ratios of carbon nanotubes and their stiffness allow the design of nanoscale tips (Wong et al., 1998) for AFMs and nanoscale manipulators (see Fig. 1.10). Nanotechnology is based on the ability to use and manipulate various nanoparticles, nanostructures and nanostructured materials by utilizing the advancement of nanoscale probes (Harik and Luo, 2004; Schaefer, 2010). Carbon nanotube-based probes allow for characterization of the atomic scale structure and material properties of nanostructured materials and nanoparticles. AFM tips, which are made of carbon nanotubes, can be functionalized for measuring molecular binding forces (Wong et al., 1998). The development of nanoscale instrumentation for manipulation and control of nanostructures requires fundamental understanding of various mechanical phenomena occurring with the nanoscale probes (Harik, 2011) and their functionalized tips (Wong et al., 1998).

AFMs with nanoscale probes (Fig. 1.10) are important instruments for the nanoscale characterization of material properties of various nanostructures, nanocrystals, and molecules (Wong et al., 1998). Carbon nanotubes, including both SWCNTs and MWCNTs

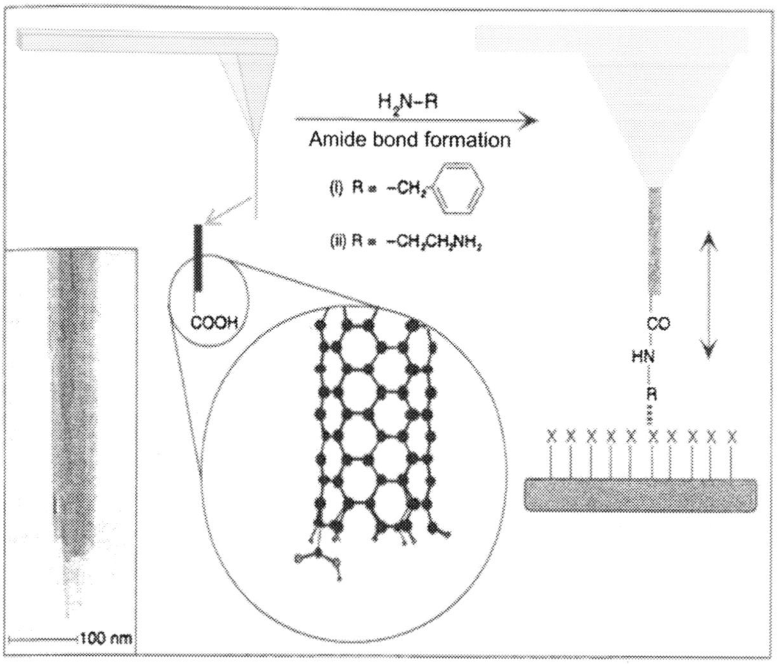

Figure 1.10
AFM probes with the high-aspect-ratio carbon nanotube tip, which can be functionalized for measuring molecular binding forces (after Wong et al., 1998). *AFM*, atomic force microscope.

(Fig. 1.11), have been used as the high aspect ratio tips in the pyramidal AFM probes. Because of the smaller diameters (e.g., 1–2 nm for SWCNTs and ~50–70 nm for MWCNTs), resolution of the nanotube-based AFM probes is much higher as opposed to the pyramidal tip with diameters between 1 and 50 μm. AFM probes allow the characterization of nanoscale surface profiles and the degree of roughness, corrugation of atomic lattices, surface adhesion and friction, surface sliding and stiction, adhesion of single molecules and groups of atoms to the atomic lattices, the pullout and pulloff processes at the nanoscale, interfacial friction and atomic scale sliding or rolling of nanostructures, vibrational and material properties of carbon nanotubes (including nanoscale buckling) as well as various manipulation processes. The carbon nanotube-based AFM probes may also utilize the electrical and semiconducting properties of the armchair and zig-zag carbon nanotubes, electromechanical and magnetic properties of various atomic lattices in order to implement sophisticated characterization methodologies for multiphysics[4] phenomena. The constitutive equations for the electromechanical and the magnetomechanical coupling in the nanotube-enhanced polymer materials will be presented later in one of the chapters.

[4] Multiphysics may imply the presence of electro-mechanical or magneto-mechanical interactions (e.g., MEMS or NEMS or fluidic devices).

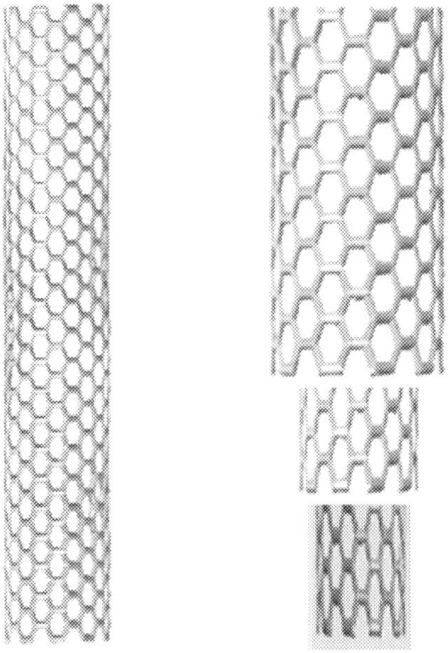

Figure 1.11
Atomic lattice structure of a single wall (10,10) carbon nanotube with the length $L_{NT} = 52.5$ Å and diameter, $d_{NT} = 13.6$ Å, and a schematic of a multiwall carbon nanotube. © *2011 by Nanodesigns Consulting, reproduced with permission.*

1.2.2 Carbon Nanotube-Based Sensors and Actuators

Carbon nanotube applications in nanotechnology also include the high aspect ratio resonators and sensors (Fig. 1.12). Carbon nanotube resonators make use of the elongated structure of SWCNTs and their high stiffness (Treacy et al., 1996; Wong et al., 1997). Carbon nanotube sensors, which monitor changes in the frequency of vibrations of SWCNTs (Overney et al., 1993) after the absorption of different molecules on nanotube surface, have similar requirements for the length of SWCNTs and their stiffness. Molecules may be also attached at the SWCNT tip near the dangling bonds. SWCNT tips can be functionalized to improve sensing capabilities of SWCNT sensors (Fig. 1.12). Typical dimensions of SWCNT diameters vary between 0.4 and $\sim$2 nm and the length varies between a few nanometers and 100s of micrometers, for device applications, and even a few centimeters and meters, for composite materials applications (Harik, 2011). The size of SWCNT diameters determines the size of molecules, which can get attached at the open SWCNT tip.

In nanoscale sensor applications, particular atomic lattice structures of carbon nanotubes are selected in order to ensure desired electrical conductivity. An example of the armchair (10,10) carbon nanotube with metallic conductive properties is shown in Fig. 1.11. Its atomic lattice is composed of carbon atoms in the periodic hexagonal arrangement in the armchair configuration, which ensures high electrical conductivity. Electrical conductivity

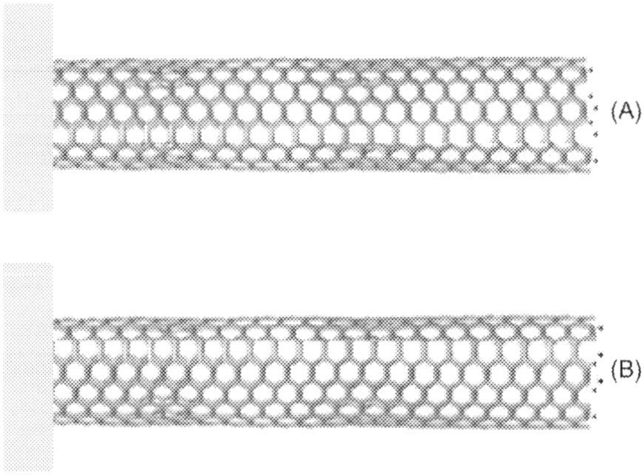

Figure 1.12
Schematic of carbon nanotube resonators used as sensors with the functionalized SWCNT tips (A) and (B). *SWCNT*, single-wall carbon nanotube. © *2011 by Nanodesigns Consulting, reproduced with permission.*

of carbon nanotubes also allows for the design of electromagnetic actuation by applying voltage to SWCNT cantilever beams (Fig. 1.12) suspended over a conductive plate. The pull-in voltage for SWCNT cantilever beams depends on the bending stiffness of SWCNTs, which can be optimized by choosing appropriate geometry and size of SWCNTs (Harik and Luo, 2004). Carbon nanotube actuators are important for NEMS and nanodevices.

MWCNTs have higher stiffness than that of SWCNTs, which makes them even better resonators. A schematic of a MWCNT is also presented in Fig. 1.11. The lattice structure of MWCNTs consists of several layers of the cylindrical lattice shells of carbon atoms separated by the van der Waals spacing. MWCNTs are robust nanoscale beams of high stiffness, which makes them useful as mechanical pins and nanoscale rods. As a result, MWCNTs can be also used as nanoscale probes. It should be noted, however, that long MWCNTs or the high aspect ratio MWCNTs can be toxic, and should be handled with care (Harik, 2014), especially when there is a large number of these thin pin-like nanoparticles.

1.2.3 Carbon Nanotube-Based Transistors for Computers

Transistors are considered one of the greatest inventions of the 20th century, allowing the creation of computers and leading to the information-based economy. According to Moore's law, the transistor's size will decrease exponentially, while its speed will increase exponentially. Miniaturization of material components in computer chips is a key for the development of faster and faster computers. The physical barriers of making computer parts smaller arise from our limited ability to efficiently manufacture and manipulate nanoscale material components. Technologies for manufacturing carbon nanotubes with certain properties and sizes help to remove some of these limitations.

Carbon nanotubes have been used by IBM to build 9 nm transistors for computer chips. This is the most successful application of carbon nanotubes in nanotechnology and computer industries. Carbon nanotube computer chips have now outperformed the silicon-based computer chips. New top gated devices are being also developed at universities (Fig. 1.13). The smaller size of transistors and other parts in computer chips (e.g., 22, 45, and 90 nm) results in faster processing in computers. Intel and IBM uses such small parts in

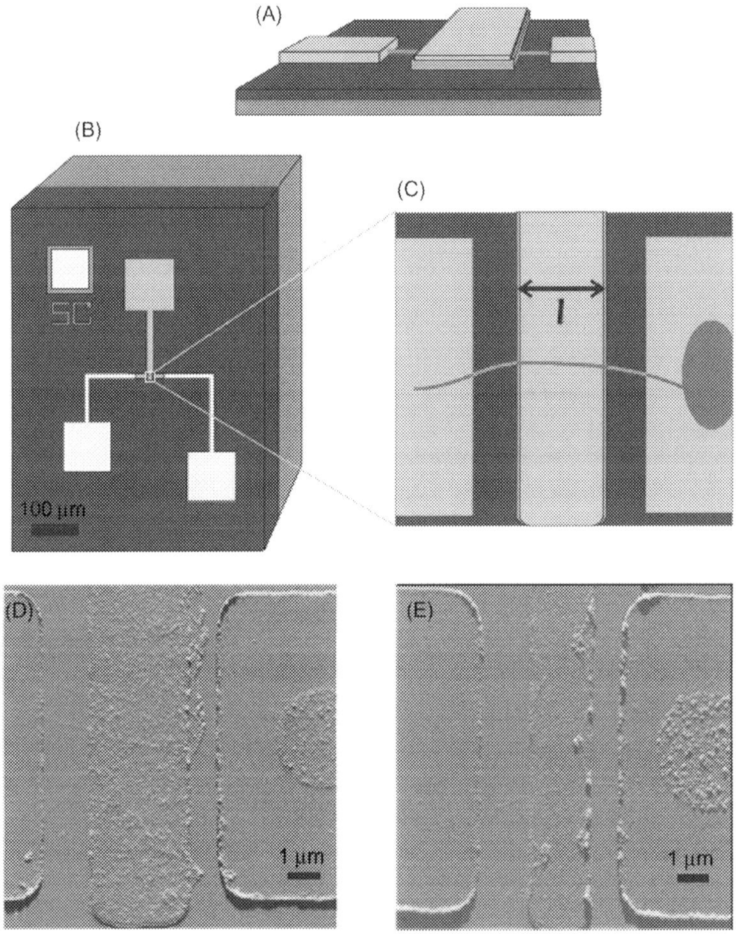

Figure 1.13
Schematic and atomic force microscope images of the top-gated devices with a source and drain connected by a carbon nanotube: (A) schematic side view of the top-gated device; (B) schematic side view of the top-gated device as it appears on wafer; (C) zoom in of the carbon nanotube area, showing catalyst in purple, palladium in orange, a carbon nanotube in red, and the top gate barrier oxide in light blue. Carbon nanotube is a thin line in all images, while colors indicate the shades. The distance *l* is the length of the top-gated section of a straight nanotube, and (D) AFM image of the top-gated device with a carbon nanotube, and (E) another AFM image of the top-gated device with a carbon nanotube. Scale bars are 1 μm. An illustration of the research done by Luke Donev (Cornell University). *AFM*, atomic force microscope.

current computer chips, which ensures that the computer industry is now capable of using a variety of carbon nanotubes in its applications. Other industries, especially the microelectronics industry, are also capable of utilizing carbon nanotubes of different sizes in various phone applications and consumer microelectronics.

1.2.4 Carbon Nanotubes in Nanocomposites

Carbon nanotubes have higher values of elastic modulus (Fig. 1.6), strength (Fig. 1.7) and failure strain (Fig. 1.5) than that of carbon fibers, which are used for reinforcing polymer matrix and carbon matrix composite materials (Schadler et al., 1998; Thostenson and Chou, 2002; Ounaies et al., 2008). Therefore, carbon nanotube polymer nanocomposites should have superior material properties with respect to CFRC. Uniform dispersion of carbon nanotubes in polymer matrix is critical for manufacturing nanocomposites with the well distributed reinforcement, i.e., SWCNTs or MWCNTs. Interfacial adhesion between polymer matrix and carbon nanotubes is important for the effective stress or load transfer, and for improving material properties of nanocomposites (Harik, 2011).

Interfacial shear strength of the polymer matrix–carbon nanotube interface can characterize the adhesion strength of the interface (Frankland and Harik, 2003). Interfacial adhesion can be based on covalent or noncovalent bonding. Noncovalent bonding is a result of van der Waals interactions between adjacent atomic surfaces (Fig. 1.14). The discrete surface of

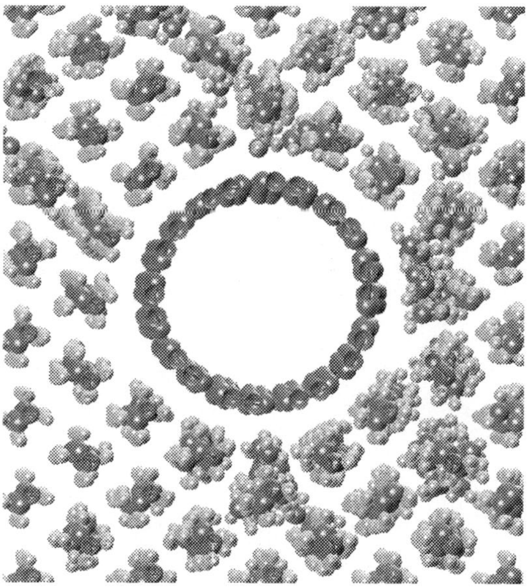

Figure 1.14
Molecular lattice structure of the crystalline part of the PE polymer with the $(CH_2)_n$ zig-zag chains surrounding the armchair (10,10) carbon nanotube (Frankland and Harik, 2003). *PE*, polyethylene.

carbon nanotubes is covered by the local distributions of π-electrons, which interact with the surrounding molecular structure of the matrix interface. Carbon nanotubes can be functionalized by different molecular groups in order to form the covalent bonding between polymer chains in the matrix and the surface atoms of SWCNTs or MWCNTs. The strength of interfacial adhesion associated with the covalent bonding depends on the density of covalent bonds along the surface of carbon nanotubes. The higher number of covalent bonds per unit area the higher strength of nanoscale interfaces.

Carbon nanotube-based nanocomposites require large numbers of SWCNTs or MWCNTs to be used for reinforcement. MWCNTs are known to be toxic, while some SWCNTs may also have some toxic effects, depending on the dose and exposure (Harik, 2014). Therefore, the use of carbon nanotubes in composite materials has to be carefully controlled and optimized to minimize the potential health risks. Nanocomposites and CFRC can be modified by carbon nanotubes only for applications where the crack or damage size will be such that the cumulative dose of released carbon nanotube fibers during a failure of a material component will be well below the critical dose. Moreover, these carbon nanotubes (SWCNTs or MWCNTs) should be smaller than the critical length (Harik, 2014). Later, one of the chapters provides detailed information about the critical length of carbon nanotubes, which increases the potential toxic effects associated with SWCNTs and MWCNTs.

1.2.5 Carbon Nanotubes in Biomedical Applications

Carbon nanotubes can be used for biomedical applications and regenerative medicine involving tissue engineering and development of implants (Paratala and Sitharaman, 2011). Carbon nanotubes have unique physical and chemical properties, which are promising for biocompatible nanocomposites and implant materials capable of carrying large loads (e.g., for joint implants). Tissue engineering can be considered a subfield of regenerative medicine, since it combines materials and engineering principles in order to improve the biological and bio-mechanical properties of the bio-engineered tissues such as bone tissues. Scaffolds are porous biomaterials, which are important in the tissue engineering. Porous scaffolds provide temporary load-carrying structural support, which guides cells to grow and assist the transport of essential nutrients. Scaffolds facilitate the formation of functional bio-engineered tissues (e.g., bone tissue). For example, the SWCNT–alginate nanocomposite-based scaffold showed greater cellular and endothelial adhesion in comparison to the non-SWCNT alginate scaffold (Paratala and Sitharaman, 2011). Small additions of SWCNTs have also increased the mechanical strength and integrity of these scaffolds.

Polypropylene fumarate (PPF) is biodegradable polyester, which is a promising polymeric biomaterial that can be used for tissue regeneration and polymeric scaffolds. Carbon nanotubes can be used for improving the load-carrying capability of polymeric scaffolds in tissue engineering. Addition of SWCNTs into PPF polymer matrix can substantially improve mechanical properties of the SWCNT-PPF nanocomposite (Paratala and Sitharaman, 2011). Very low concentrations of SWCNTs in PPF polymer (e.g., less than 0.5% by weight) increase compressive and flexural properties of the SWCNT-PPF nanocomposite up to two (or three) times with respect to pure PPF matrix (Paratala and Sitharaman, 2011).

Carbon nanotubes may induce inflammation and some toxic effects (Harik, 2014), due to their biopersistence and oxidative stresses as they interact with biological tissues and cells or cell membranes. It has been reported that inflammatory cells may appear after 4 weeks within the scaffolds made of the SWCNT-PPF nanocomposites (Paratala and Sitharaman, 2011). However, the level of inflammatory cells may decrease in these scaffolds after 12 weeks. Research results indicate that the SWCNT-PPF scaffolds may be potentially bioactive and could promote osteogenesis. Carbon nanotubes have been also claimed rather prematurely to show promise as contrast agents for noninvasive in vivo molecular imaging. Noninvasive in vivo imaging with SWCNTs requires penetration of these nanoscale fibers into biological fluids, where they can be transported into different organs and affect different biological tissues. The use of carbon nanotubes for biological and biomedical applications is a new area and requires special attention and continuous oversight.

1.2.6 Nanotechnology Literacy Problems

Carbon nanotubes have been around since their recent rediscovery in 1991, although, they had been encountered much earlier, as stipulated by different researchers. SWCNTs and MWCNTs have been studied in numerous colleges and universities and talked about far beyond the academic world. Therefore, nanotechnology literacy problems are not expected to occur among scientists and engineers dealing with carbon nanotubes. However, there are interesting cases, for instance, when scientists or engineers make surprising mistakes in their R&D proposals involving new conductive materials with carbon nanotubes (see Fig. 1.15). A design of new conductive materials with the proposed structure is scientifically impossible, because continuum copper wires are simply inconceivable at this atomic scale within the framework of existing laws of physics and chemistry. The size of a single carbon ring is about 0.245 nm; therefore, the proposed width of a copper wire is also about that size, which is impossible. One could ask, what is the approximate size of an atom in a copper metal? What is the size of the discrete atomic lattice in the crystalline structure of copper? One could also ask, how many lattice cells of the discrete atomic lattice of copper metal would it take to make the continuum copper material having continuous isotropic macroscopic properties? Scientists or engineers may not know about the nanoscale homogenization criterion discussed later in one of the chapters. However, the basic appreciation of possible adjacent scales in the atomic

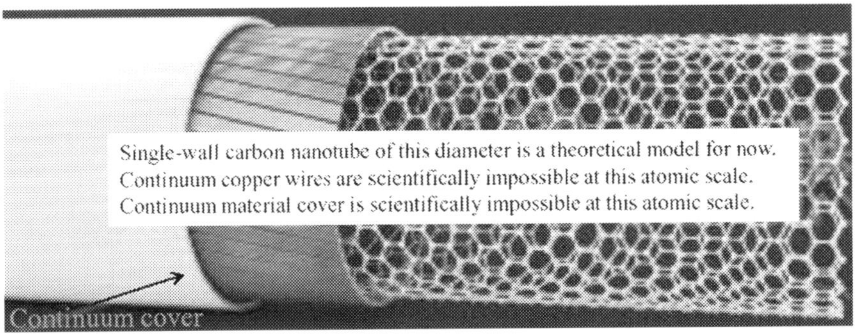

Figure 1.15
Schematics showing atomic structure of a zig-zag carbon nanotube of a large diameter with atomic crystals inside and a continuum material over the copper wires covering the hexagonal lattice of carbon rings (an illustration of nanotechnology literacy problems).

structure of materials constitutes a clear case of illiteracy regarding nanoscale sciences or nanotechnology literacy problems.

In order to address some nanotechnology literacy problems involving carbon nanotubes, this chapter presents several sample problems below. However, in order to avoid some misconceptions about carbon nanotubes or how to use them, readers are urged to read all of the chapters, since they present unique material that is unavailable in other books on this subject. In order to understand this statement, readers may note important concepts such as nanoscale homogenization criteria, the classification of carbon nanotubes into 20 different classes, viscosity of surface electrons and of nanoscale interfaces, effective thickness paradox for carbon nanotubes, various thickness estimates for carbon nanotubes, parametric maps for the beam and shell models for carbon nanotubes, nanoscale friction laws, potential toxicity of some carbon nanotubes, and criteria for carbon nanotubes regarding the potential toxic effects and many other concepts (Harik, 2011, 2014). Some of these concepts will become a requirement for nanotechnology literacy of university students entering the workplace after the nanotechnology-driven industrial revolution.

Sample Problems

1. List approximate dimensions of known nanostructures and microscopic objects.
2. List approximate dimensions of the atomic lattice and microscopic elements in the material structure of any material.
3. Draw a schematic representation of atomic scale for any material structure, along with a representation of this nanoscale material fragment of 10 nm in size or larger.
4. Draw a schematic representation of different known geometries and sizes of SWCNTs.

5. Make a Powerpoint viewgraph with a schematic representation of different known geometries and sizes of SWCNTs.
6. Draw a schematic representation of the atomic structure of any SWCNT next to the atomic structure of any other metal material.
7. Draw a schematic representation of the atomic structure of any SWCNT next to the atomic structure of any other polymer material.
8. Make a Powerpoint viewgraph with a schematic representation of the atomic structure of any SWCNT next to the atomic structure of any other polymer material.
9. Draw a schematic representation of the atomic structure of any zig-zag and armchair carbon nanotube.
10. Make a Powerpoint viewgraph with a schematic representation of the atomic structure of any zig-zag and armchair carbon nanotube.
11. Draw a schematic representation of the atomic structure of any armchair SWCNT along the description of its chirality with a coordinate system and unit vectors.
12. Draw a schematic representation of the atomic structure of any zig-zag SWCNT along the description of its chirality with a coordinate system and unit vectors.
13. Draw a schematic representation of the atomic structure of any chiral SWCNT along the description of its chirality with a coordinate system and unit vectors.
14. Draw a schematic representation of the atomic structure of graphene along with the system describing the chirality of carbon nanotubes with a coordinate system and unit vectors.
15. What are the approximate dimensions of different material elements in Fig. 1.14?
16. What are the approximate dimensions of different material elements in Fig. 1.15?
17. What are the approximate dimensions of different material elements in Figs. 1.11 and 1.12?

References

Dresselhaus, M.S., Dresselhaus, G., Eklund, P.C., 1996. Science of Fullerenes and Carbon Nanotubes. Academic Press, San Diego.

Frankland, S.J.V., Harik, V.M., 2003. Carbon nanotube pull-out from a polymer matrix. Surf. Sci. Lett. 525 (L103), 2003.

Fujita, M., Saito, R., Dresselhaus, G., Dresselhaus, M.S., 1992. Formation of general fullerenes by their projection on a honeycomb lattice. Phys. Rev. B 45, 13834–13836.

Gogotsi, Yu (Ed.), 2006. Nanomaterials Handbook. CRC Press, New York.

Gogotsi, Y. (Ed.), 2017. Nanomaterials Handbook. second ed CRC Press, New York.

Harik, V.M., 2001a. Ranges of applicability for the continuum beam model in the mechanics of carbon nanotubes and nanorods. Solid State Comm. 120 (7–8), 331–335.

Harik, V.M., 2001b. Ranges of applicability for the continuum beam model in the constitutive analysis of carbon nanotubes: nanotubes or nano-beams? NASA/CR-2001-211013. NASA Langley Research Center, Hampton, VA.

Harik, V.M., 2001c. Mechanics of carbon nanotubes, Lecture notes: a short course,. Continuing Education Institute, American Society of Mechanical Engineers. ASME, New York, NY.

Harik, V.M., 2002. Mechanics of carbon nanotubes: applicability of the continuum-beam models. Comput. Mater. Sci. 24 (3), 328–342.
Harik, V.M., Salas, M.D. (Eds.), 2003. Trends in Nanoscale Mechanics. Kluwer Academic Press, The Netherlands (Reprinted by Springer).
Harik, V.M., Luo, L.-S. (Eds.), 2004. Micromechanics and Nanoscale Effects. Kluwer Academic Press, The Netherlands (Reprinted by Springer).
Harik, V.M., 2011. Mechanics of Carbon Nanotubes. Nanodesigns Press, Newark, DE.
Harik, V.M. (Ed.), 2014. Trends in Nanoscale Mechanics. Springer, Dordrecht.
Harris, P.J.F., 1999. Carbon Nanotube and Related Structures: New Materials for the 21st Century. Cambridge University Press, Cambridge, United Kingdom.
Huang, Y., Wu, J., Hwang, K.C., 2006. Thickness of graphene and single-wall carbon nanotubes. Phys. Rev. B 74 (24), 24541.
Iijima, S., 1991. Helical microtubules of graphitic carbon. Nature 354, 56.
Iijima, S., Ichihashi, T., 1993. Single-shell carbon nanotubes of low diameters. Nature 363, 603.
Ounaies, Z., Park, C., Harrison, J., Lillehei, P., 2008. Evidence of piezoelectricity in SWNT-polyimide and SWNT-PZT-polyimide composites. J. Thermoplast. Compos. Mater 21 (5), 393–409.
Overney, G., Zhong, W., Tomanek, D., 1993. Structural rigidity and low-frequency vibrational-modes of long carbon tubules. Z. Phys. D: At., Mol. Clusters 27, 93–96.
Paratala, B.S., Sitharaman, B., 2011. In: Klingeler, R., Sim, R.B. (Eds.), Carbon Nanotubes in Regenerative Medicine, in Carbon Nanotubes for Biomedical Applications, Carbon Nanostructures. Springer-Verlag, Berlin.
Ruoff, R.S., Tersoff, J., Lorents, D.C., Subramoney, S., Chan, B., 1993. Radial deformation of carbon nanotubes by van der Waals' forces. Nature 364, 514–516.
Schadler, L.S., Giannaris, S.C., Ajayan, P.M., 1998. Load transfer in carbon nanotube epoxy composites. Appl. Phys. Lett. 73, 3842–3844.
Schaefer, H.-E., 2010. Nanoscience. Springer-Verlag, Berlin, Heidelberg.
Stojkovic, D., Zhang, P., Crespi, V.H., 2001. Smallest nanotubes: breaking the symmetry of sp^3 bonds in tubular geometries. Phys. Rev. Lett. 87 (12), 125502.
Thostenson, E.T., Chou, T.-W., 2002. Aligned multi-walled carbon nanotube-reinforced composites: processing and mechanical characterization. J. Phys. D: Appl. Phys., 35 (16), L77–L80.
Treacy, M.M.J., Ebbesen, T.W., Gibson, J.M., 1996. Exceptionally high Young's modulus observed for individual carbon nanotubes. Nature 381, 680.
Wong, E.W., Sheehan, P.E., Lieber, C.M., 1997. Nanobeam mechanics: elasticity, strength, and toughness of nanorods and nanotubes. Science 277, 1971.
Wong, S.S., Joselevich, E., Wooley, A.T., Cheung, C.L., Lieber, C.M., 1998. Covalently functionalized nanotubes as nanometer-sized probes in chemistry and biology. Nature 394, 52.
Yakobson, B.I., Dimitrica, T., 2003. Nanomechanics: engineering between physics and chemistry. In: Harik, V. M., Salas, M. (Eds.), Trends in Nanoscale Mechanics. Kluwer Academic Publishers, The Netherlands, pp. 3–33. , Chapter 1.
Yakobson, B.I., Brabec, C.J., Bernholc, J., 1996. Nanomechanics of carbon tubes: instabilities beyond linear response. Phys. Rev. Lett. 76, 2511.

CHAPTER 2

Nanoscale Laws of Motion and Governing Equations for Atomic and Continuum Scales

Chapter Outline

2.1 The Atomic Structure of Nanomaterials and Its Length Scales

Carbon nanotubes and graphene sheets consist of a single layer of carbon atoms arranged into a periodic hexagonal lattice of carbon rings. Graphene platelets and multiwall carbon nanotubes may consist of several layers of atomic lattice. Other nanomaterials, such as crystalline metals and ceramics, also have periodic structures with different sizes and forms of the smallest atomic structural elements. The size of periodic cells, a, in the atomic structure of nanomaterials is important for the determination of the smallest size of a material particle, which can be considered as a nanoparticle of a particular nanostructured material with the uniquely defined physical properties. In general, the smaller the size of

Mechanics of Carbon Nanotubes.
DOI: https://doi.org/10.1016/B978-0-12-811071-3.00002-0

© 2018 Elsevier Inc. All rights reserved.

atomic lattice cells in any crystalline material, the smaller the size of a nanoparticle of this material can be with the well-defined material properties. The size of the smallest material element also determines the lower size limit of the well-defined material in any form. For this size of nanoparticles, a physical continuum can be defined.

A nanomaterial is usually defined as a material with at least one of its dimensions being less than 100 nm. Nanostructured materials consist of nanoscale structural elements comprised of nanoparticles, nanoscale crystals, and grains or nanostructures. Sometimes, a lower dimensional limit, i.e., 1 nm, is cited for the size limit of nanoparticles or a nanomaterial. However, a nanoparticle may have varying material properties for different sizes. In fact, material properties of nanomaterials are size-dependent; that is, distinct nanoparticles having different sizes may have material properties whose value varies with the size of these nanoparticles. This means that that the size of nanoparticles and nanostructures of about 1 nm has such variations in material properties. A homogenization criterion can be used to clearly define the lower size limit for the nanoparticles and nanostructures having well-defined material properties and uniquely defined mechanical properties. Later in this chapter, the size-invariant definition of a nanomaterial is presented with its uniquely defined material properties, which are independent of the size of nanomaterial powder particles. The size-invariant nanomaterials of the smallest dimensions are important for the computer industry and nanomedicine, where miniaturization and dose and response management is essential for the effectiveness of applications.

Understanding of nanoscale size effects on stress distribution can benefit from consideration of the length scales in the material structure of nanomaterials. The length scales in the atomic structure of carbon nanotubes, nanocomposites and carbon fiber-reinforced polymer matrix composites are illustrated in Fig. 2.1. In this schematic, the smallest length scale of 1.5 Å is associated with a carbon atom and the length of the C–C bond, which is involved in elastic interactions. The size of a carbon ring, a, is about 2.45 Å, while the diameter of the armchair (10,10) carbon nanotube is 13.6 Å. The effective thickness of carbon nanotube lattices may vary, depending on the type of loading and the size of their radius as will be shown later. The length of carbon nanotubes may vary between 100 nm and 20 cm and is much longer for the recently manufactured nanotubes. Therefore, long carbon nanotubes link micromechanical and nanoscale interactions in nanocomposites. The diameter of a carbon fiber is about 20 μm, and its length may vary between a few centimeters to meters. It should be pointed out that the thickness of the fiber–matrix interphase region around a carbon fiber in the polymer matrix composite is about 1–3 μm. The thickness of carbon fiber-reinforced polymer matrix composites may vary from 1 or 2 mm to 1 or 2 cm for automotive applications, and more for structural applications. The broad range of length scales between 1 cm and 1 Å requires robust analytical and multiscale computational methods, which are used to describe the complex material structures, from macroscale to nanoscale.

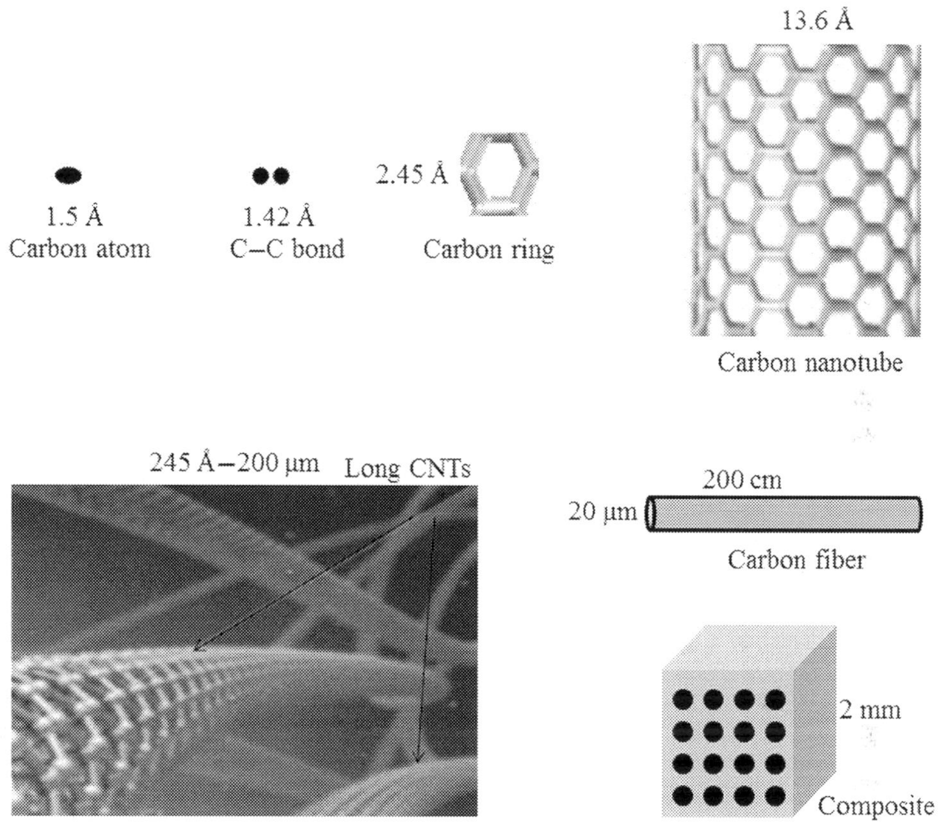

Figure 2.1
Schematic of the length scales for carbon materials: from a bonded carbon atom and the C–C bond length to a carbon ring, the diameter of a CNT, the length of CNTs, the diameter and the length of carbon fiber and the thickness of a carbon fiber polymer matrix composite. (Disclaimer: long CNTs can be toxic). *CNT*, Carbon nanotube.

2.1.1 Breakdown of Continuum Concepts and Limitations of Continuum Mechanics

The structure of nanomaterials and its length scales require appropriate methods of analysis to describe the mechanical behavior, deformation and interactions of material constituents, and their material structure. Atomic structure of carbon nanotubes requires the use of analytical and computational methods, which can predict the dynamics of individual atoms and the mechanical response of large groups of atoms in the lattice structure. Composite materials, which are reinforced by carbon nanotubes or carbon fibers, can be analyzed by the methods of continuum mechanics, with appropriate resolution of interfacial regions and carbon nanotubes. The framework of classical mechanics, which also includes micromechanics, is based on the fundamental assumption of the *continuum media* (see the diagram in Fig. 2.2). Continuum mechanics is based on macroscopic material properties, the

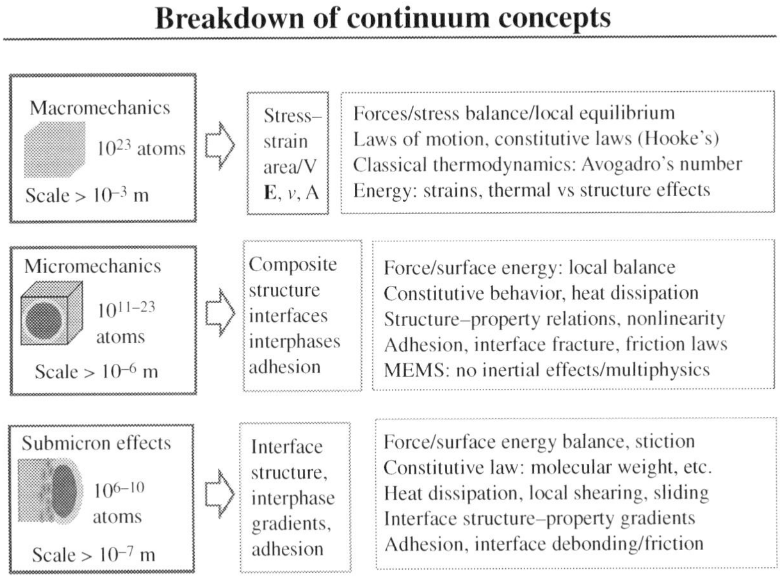

Figure 2.2
Schematic of the breakdown of continuum concepts of macromechanics and micromechanics with the emergence of submicron nanoscale effects. *© 2011 by Nanodesigns Consulting (reproduced with a permission).*

constitutive laws, and the conservation laws for the mass, energy, and momentum. Continuum concepts become more complex with submicron property gradients, interfacial structure, and interphases in composite materials, as well as interfacial adhesion and fracture.

Macromechanics of materials uses tensor variables such as stress and strain, which depend on small or large displacements and forces distributed over an area or volume of a continuum media (Fig. 2.2). The structural effects in macromechanical material behavior are influenced by the material constituents, which are continuous, homogeneous, and often isotropic. In composite materials, the microstructure can be anisotropic and inhomogeneous. This is when structure–property relations become important. In micromechanics, the role of interfacial regions and interfacial strength is critical for the effective load transfer across the fiber–matrix interface. The material structure of the fiber–matrix interface region may be complex, and represents an interphase with unique material properties and property gradients.

The property gradients at the fiber–matrix interface can be affected by the molecular weight of polymers used in the matrix, the structure of the fiber coating, and the local stoichiometry. The constitutive properties of local material structure can be sensitive to the molecular weight and the degree of entanglement of polymer networks. At the nanoscale, the discrete atomic lattices and the nanoscale molecular networks have to be homogenized (or averaged) over the representative structural elements in order to derive an equivalent continuum representation (see the diagram in Fig. 2.3). The direct application of the known homogenization approaches at the nanoscale is complicated by the dominance of the surface effects over the volumetric effects, and the need to accurately represent the discrete atomic structure at the length scale of its homogenization, or upscaling (as in the so-called coarse-grain representation of groups of atoms).

Assumptions of continuum mechanics about the existence of continuum media with billions of billions of atoms are no longer valid at the nanoscale (Fig. 2.4). In fact, the groups of 100 s or 1000 s atoms have to be analyzed as a part of a nanostructure, a grain, or a nanoparticle in a powder. Nanostructured materials may have macroscopic material properties, which may be somewhat different from their counterparts with a regular microstructure. However, small material elements may or may not have these macroscopic material properties, depending on its size. This phenomenon of size-dependence of material properties is unique to nanoscale and is quite different from the concepts of continuum mechanics and micromechanics (Fig. 2.4). The nanoscale phenomenon of size-dependence of material properties can be described in part by the nanoscale homogenization criterion and material parameters, which characterize this size dependence. Material properties of nanomaterials can be tailored at nanoscale by controlling their size, in particular the size of

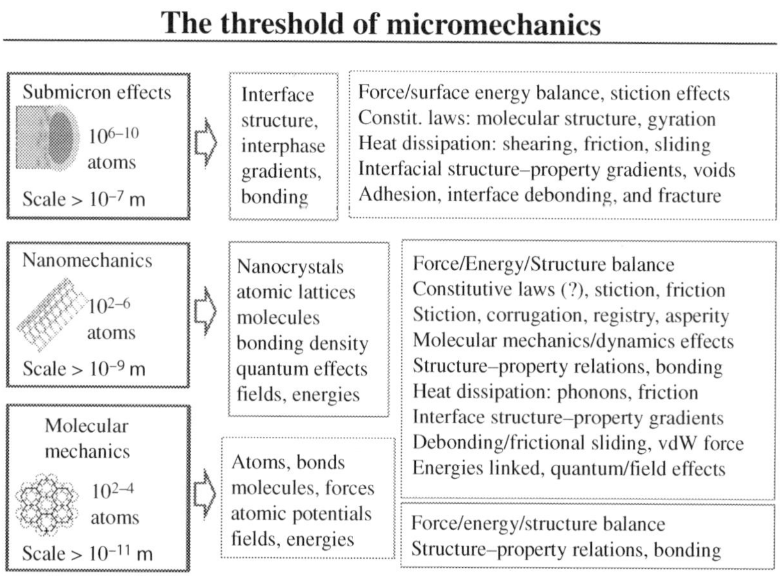

Figure 2.3
Schematic of the breakdown of continuum concepts of micromechanics with the emergence of nanoscale effects. © *2011 by Nanodesigns Consulting (reproduced with a permission).*

Continuum mechanics assumptions

List of assumptions:		*Current trends:*
- Solids contain at least 6 * 10^{23} atoms	=>	10^2–10^9 atoms
- Material element has macroscopic properties	=>	*Size effects*
- Solids/constituents: homogeneous, anisotropic	=>	*Multiscale* Discrete
- Separate phases (no phase transformations)	=>	*Interphases* Interfaces
- Motion is uniquely determined by forces	=>	*Probabilistic, statistical*
- Total energy = kinetic + potential energies	=>	*Quantum field* *Fermi energy*
- Only mechanical forces considered: (pressure, acceleration, stresses, and tractions)	=>	*Friction, fields* *atomic potentials* *interfaces, bonds*

Figure 2.4

Schematic of the breakdown of continuum assumptions of macromechanics and micromechanics with the emergence of nanoscale effects. © *2011 by Nanodesigns Consulting (reproduced with a permission).*

powder nanoparticles. Such ability to tailor material properties by design can be useful for nanoelectronics and pharmaceutical applications, as well as for nanomedicine.

Nanostructured materials have an inherently multiscale structure with discrete atomic lattices at nanoscale, different grain sizes, and texture at microscale, as well as surface coating and fiber–matrix interface in composites at microscale and macroscale. Distributions of grain sizes in the material result in probabilistic variations in average material properties at microscale and nanoscale. Variations in the distribution of reinforcing inclusions and fibers in composite materials and nonuniform dispersion of carbon nanotubes in nanocomposites also lead to local changes in material properties at microscale and nanoscale. Deviations in wettability of carbon fibers, variations in the processing conditions, and quality of material constituents forming polymer microstructure in composite materials also affect average material properties at microscale and nanoscale. Nanoscale variations in geometry of carbon nanotubes and the structure of nanoscale interfaces results in changes of electromechanical and thermo-elastic properties in multifunctional nanocomposites. Structural health monitoring in composite materials with embedded micro-electromechanical systems and sensors is also affected by local variations in material properties, while their functional performance depends on nanoscale effects. The size of transistors in nanoelectronic sensors available for multifunctional nanocomposites is now below 90, 44, and even 22 nm, with the new carbon nanotube-based 9-nm transistor designed by IBM. The thermo-mechanical response of nanoscale devices depends either on their discrete atomic structure or the structure of interfaces at the contact points. The mechanical behavior of carbon nanotube-based nanocomposites is clearly influenced by nanoscale phenomena, as well as by quantum effects in embedded nanodevices. Capabilities of analytical and computational methods have to correspond to the needs of describing such complex material systems, and to address challenges in the analysis of their physical properties and mechanical behavior.

2.1.2 Size Effects and the Structure–Property Relations for Nanomaterials

Material properties in nanomaterials, including nanocomposites, are affected by the local microstructure and nanoscale variations in the distribution of polymer chains, their molecular weight, and their entanglement. Dispersion of carbon nanotubes in polymer matrices and their local agglomeration also contribute to variability in material properties in nanocomposites. The size and form of such agglomerations of carbon nanotubes and the size of polymer entanglements affects local material properties in nanocomposites. The size of nanoscale fibers and the length of carbon nanotubes also influence the load-carrying ability of nanocomposites, as well as their elastic properties. The role of the length of carbon nanotubes in the mechanics of nanocomposites and its impact on the elastic moduli of nanocomposites will be discussed in greater details in the chapter on nanocomposite materials.

The typical length scales characterizing the material structure of carbon nanotubes, as compared to a simple polymeric structure of a crystalline polyethylene (PE) polymer, and the polymer matrix composites reinforced by the carbon fibers are shown in Table 2.1. The length scales associated with the structure of carbon nanotubes, which were discussed earlier, are presented in the first column. It should be noted that multiwall carbon nanotubes (MWCNTs) may have larger diameters, e.g., 50–100 nm, and much larger lengths, e.g., 5 cm–1 m, especially in the form of bundles (up to 5 m or more). Polymer chains also have important structural elements and the associated length scales (Table 2.1), which are important in multiscale modeling of nanocomposites. Carbon nanotube-based nanocomposites, or the nanotube-modified polymers, have unique structural length scales shown in the first two columns in Table 2.1, which are different from the hierarchical length scales associated with the microstructure of typical composites shown in the third column in Table 2.1. Therefore the multiscale modeling of nanocomposites with carbon nanotubes and with fiber-reinforced composite materials is different. Analysis of nanocomposites requires representative volume elements (RVEs) with discrete atomic

Table 2.1: Typical length scales of the material structure in composites

Carbon Nanotubes	Polymers	Polymer Composites
Carbon ring (2.46 Å)	(CH_2) zigzag (~3 Å)	Fiber/Matrix interphase (1–3 μm)
Diameter (~1–2 nm)	Crystalline unit (~7 Å)	Carbon fiber (20 μm)
Length (10 nm–5 cm)	Length of chains (~300 nm)	Fiber length (1–10 m)

structures, as well as RVEs at nanoscale and microscale. Analysis of carbon fiber composites requires a microscopic RVE, including a fiber, its coating, interface/interphase region, RVE at the ply level describing the size and volume fraction of fibers in the matrix, and a macroscopic RVE showing the orientation of fibers in the plies and the stacking sequence in the composite.

Material properties of the nanotube-based nanocomposites also depend on the structure–property relations and dependencies, which are different across the hierarchical length scales. In order to link a particular macroscopic material property with the nanoscale material characteristics or structural elements, an appropriate multiscale "tree" diagram is needed (Table 2.2). Dependence of the Young's modulus, E, on the material characteristics at different length scales is illustrated for the polymer composites based on the single-wall carbon nanotubes (SWCNTs) and MWCNTs, respectively, and on traditional carbon fibers. Elastic moduli of carbon nanotube-based nanocomposites depend on the volume or weight fraction of nanotubes, their length and orientation, and other properties listed in Table 2.2 for different length scales. Carbon fiber-reinforced composite materials also depend on their microstructure, as well as on nanoscale effects, which are not fully understood.

Selection of nanoscale RVEs for nanocomposite materials should take into account the nanoscale phenomenon of size-dependence of material properties, which was mentioned earlier. Nanoscale RVEs may have averaged material properties, which may be somewhat different from microscopic and macroscopic material properties. The nanoscale size-dependence of material properties means that small material elements or RVEs may or may not have macroscopic material properties. Moreover, the polymer interfacial region around each carbon nanotubes in nanocomposite may or may not have well-defined material properties. In fact, if material properties of the polymer interfacial region are uniquely defined, this region can be called the interphase (i.e., the separate distinct phase). The volume of RVEs and their material constituents have to be large enough. The nanoscale phenomenon of size-dependence of material properties is connected not only with the discrete atomic lattices of carbon nanotubes and graphene, but also with polymer networks

Table 2.2: Multiscale structure–property dependencies for Young's modulus

Length Scales	Nanocomposites	Polymer Composites
Macroscopic	Entanglement, CNT coil size	Volume fraction, E_m
Microscopic	CNT bundles, clustering	Carbon fiber (size, E)
Submicron	SWCNT vs MWCNT effects	Interphase/Interface
Nanoscale	CNT/Polymer interphase, polymer molecular weight	Crosslink density, interface structure
Atomic	CNT/Polymer bonding, molecular aromaticity	Polymer structure, molecular weight, etc.

MWCNT, multiwall carbon nanotube; *SWCNT*, single-wall carbon nanotube.

and agglomerations of polymer chains. This nanoscale phenomenon of size-dependence of material properties will be described later in this chapter, along with the nanoscale homogenization criterion and parameters, which characterize size-dependence at nanoscale.

2.2 Nanoscale Laws of Motion and Governing Equations

The motion of the atomic lattice structure of carbon nanotubes can be described by the system of dynamic equations of motion formulated for each carbon atom. The equations of motion for each atom and the acceleration, $\mathbf{a}_i$, of each carbon atom are derived from Newton's second law of motion:

$$\mathbf{F}_i = m_i \mathbf{a}_i, \tag{2.1}$$

where $\mathbf{F}_i$ is the force acting on an ith atom of mass, m_i. The movement of ith atom from time t with the time increment Δt can be described by changes of its position vector, $\mathbf{r}_i$:

$$\mathbf{r}_i(t + \Delta t) = \mathbf{r}_i(t) + \mathbf{v}_i(t)\Delta t + \left(\frac{1}{2}\right)\mathbf{a}_i(t)\Delta t^2, \tag{2.2}$$

which are approximated by a Taylor series (2.2) with the velocity $\mathbf{v}_i$ being the first time derivative of the position vector and the acceleration, $\mathbf{a}_i$, being its second time derivative. Changes in the velocity of atoms, $\mathbf{v}_i(t)$, with time are approximated also by a Taylor series with the velocity Verlet algorithm (Yakobson and Dumitrica, 2003):

$$\mathbf{v}_i(t + \Delta t) = \mathbf{v}_i(t) + \left(\frac{1}{2}\right)[\mathbf{a}_i(t + \Delta t) + \mathbf{a}_i(t)]\Delta t. \tag{2.3}$$

Acceleration, $\mathbf{a}_i$, of N atoms is determined from Newton's law (2.1) as force $\mathbf{F}_i$ divided by mass, m_i. The force, $\mathbf{F}_i$, acting on each atom can be evaluated as the gradient of a potential energy function, $V(\mathbf{r})$, which depends on interatomic interactions and the positions of each atom. Therefore Newton's second law (2.1) can be written as

$$\mathbf{a}_i(t) = -\left(\frac{1}{m_i}\right)\nabla_{\mathbf{r}_i} V(\mathbf{r}_1, \ldots, \mathbf{r}_N). \tag{2.4}$$

The acceleration, $\mathbf{a}_i$, of individual atoms can be evaluated as the first time derivative of their velocity, $\mathbf{v}_i$, or the second time derivative of their position vector, $\mathbf{r}_i$. Therefore the governing Eq. (2.4) results in the second order differential equation for the position vector, $\mathbf{r}_i$, of each atom. Newton's second law (2.1) provides the following governing equation for the motion of individual atoms:

$$\frac{d^2\mathbf{r}_i}{dt^2} = -\frac{1}{m_i}\nabla_{\mathbf{r}_i} V\left(\mathbf{r}_{i,}, \ldots, \mathbf{r}_N\right), \tag{2.5}$$

where the choice of the potential function, $V(\mathbf{r})$, determines the type of interatomic interactions. Vibrations of atoms in the lattice have the time scale of approximately 100 fs, therefore a smaller time step, Δt, is required, e.g., 2 fs (Frankland and Harik, 2003).

2.2.1 Conservation of Energy, Mass, and Momentum at Nanoscale

A system of N atoms forming a carbon nanotube or graphene lattice has the total internal energy, E_{total}, composed of the sum of the kinetic energies, T_i, of individual atoms and their potential energy, U_i:

$$E_{\text{total}} = \sum_{i}^{N} T_i + \sum_{i}^{N} U_i \tag{2.5}$$

where the kinetic energy, T_i, of each atom is $T_i = (m_i\ V_iV_i)/2$, and U_i is their potential energy. The law of conservation of energy is applicable at nanoscale to the entire system of atoms. The law of conservation of energy states that the total energy of a system of N atoms is conserved, that is,

$$E_{\text{total}} = \text{const.} \tag{2.6}$$

During molecular dynamics (MD) simulations or ab initio MD simulations, the total energy should be monitored in time. The computed value of the total energy should be constant. The law of conservation of mass is also applicable at nanoscale to the entire system of atoms. Mechanical behavior of nanoscale material systems does not involve chemical reactions. Therefore the sum of masses, m_i, of all atoms remains constant. The law of conservation of momentum, m_iV_i, is applicable at nanoscale to the interacting atoms and groups of interacting atoms.

2.2.2 Atomic Potentials for Interatomic Interactions

Interactions between individual atoms are determined by the potential energy of interatomic interactions and the atomic potentials, which describe these interactions. Atomic potentials can be empirical, tight binding, and based on first principles (Yakobson and Dumitrica, 2003). The many-body potential, $V(\mathbf{r})$, can be presented as a sum of pairwise terms with the energy of pair interactions, $\phi(r_{ij})$, depending on their relative distance r_{ij}:

$$V(r_1, \ldots, r_N) = \left(\frac{1}{2}\right) \sum_{i \neq j} \varphi(r_{ij}) \tag{2.7}$$

This relatively simple form of interatomic potential can be used to describe interactions of atoms in rarified gases, where atoms interact through van der Waals forces. In order to describe the bonding interactions between atoms, the atomic potential (2.7) can be modified for the covalent bonding in semiconductors (Tersoff, 1988) and for the bonding of carbon atoms in hydrocarbons (Brenner, 1990):

$$V = \frac{1}{2}\left[\sum_{i\neq j}\phi_{\mathrm{R}}(r_{ij}) + \sum_{i\neq j}B_{ij}\phi_{\mathrm{A}}(r_{ij})\right], \tag{2.8}$$

where the energy of pair interactions, $\phi(r_{ij})$, has attractive, $\phi_{\mathrm{A}}(r_{ij})$, and repulsive, $\phi_{\mathrm{R}}(r_{ij})$, terms. The weight function, B_{ij}, reduces the strength of the *ij*-bond with the *j*-atom, when it has other bonds. This function also describes the angular dependence with a cutoff that restricts this pair potential to the nearest neighbors. The Tersoff–Brenner potential for carbon atoms includes a switch-type function:

$$G(r) = \begin{cases} 1, r < r_1 \\ \frac{1}{2}\left\{1 + \cos\left[\frac{\pi(r - r_1)}{r_2 - r_1}\right]\right\}, r_1 \le r \le r_2, \\ 0, r > r_2 \end{cases} \tag{2.9}$$

which leaves the interatomic interactions unchanged for the distances between pairs smaller than the inner cutoff distance: $r_1 = 1.7$ Å. It is smoothed down to zero at the second cutoff distance: $r_2 = 2$ Å.

In MD simulations (Frankland and Harik, 2003), interatomic van der Waals interactions at the interface between a polymer matrix and a carbon nanotube was modeled with the Lennard-Jones potential (Lennard-Jones, 1924):

$$U = 4\varepsilon_{LJ}\left(\frac{\sigma_{\mathrm{LJ}}^{12}}{r^{12}} - \frac{\sigma_{\mathrm{LJ}}^{6}}{r^{6}}\right) \tag{2.10}$$

where U is the potential energy between a pair of atoms, r is the distance between a pair of atoms, $\varepsilon_{\mathrm{LJ}}$ is the potential well depth, and σ_{LJ} is the size of van der Waals spacing (Fig. 2.5). For the interaction between the carbon atoms of a carbon nanotube and the polymer $-CH_2-$ units, the potential was parameterized with $\varepsilon_{\mathrm{LJ}} = 0.4492$ kJ/mol and $\sigma_{\mathrm{LJ}} = 0.3825$ nm. The bond angle bending forces were modeled with a harmonic valence-angle potential having an equilibrium angle of 112.813° and a barrier of 520 kJ/mol. A torsional potential $\phi(\theta)$ of the form

$$\varphi(\theta) = A_1(1 + \cos\theta) + A_2(1 + \cos 2\theta) + A_3(1 + \cos 3\theta) \tag{2.11}$$

was used, where θ is the torsional angle around the CH_2-CH_2 bond and $A_1 = -18.087$, $A_2 = -4.88$, and $A_3 = 31.8$ kJ/mol. The Lennard-Jones potential was also used to describe nonbonding interactions between $-CH_2-$ units in either the same chain or between different chains. For these interactions, $\varepsilon_{\mathrm{LJ}} = 0.4742$ kJ/mol and $\sigma_{\mathrm{LJ}} = 0.428$ nm. The

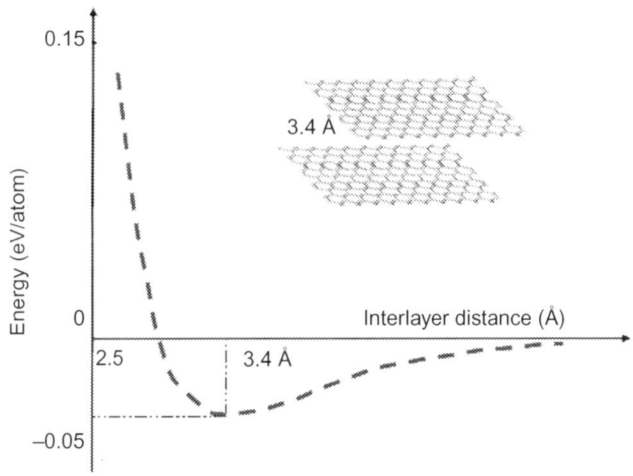

Figure 2.5
Schematics for the energy of interfacial interactions through van der Waals forces between the adjacent lattices (with the equilibrium distance of 3.4 Å). © *2011 by Nanodesigns Consulting (reproduced with a permission).*

carbon nanotube was modeled with a many-body bond-order potential (2.8) developed for carbon atoms (Brenner, 1990). This carbon potential is parametrized for C−C bonds of lengths up to 0.17 nm. MD simulations were carried out by using code named DL-POLY, which is a large-scale MD simulation package available from Daresbury Laboratory. This program was adapted to include the carbon potential for carbon nanotube. All simulations were carried out at 300 K, with a 2 fs time step.

2.2.3 Stress at Atomic Scale and Its Averaging for Nanomaterials

Atomic lattice structures can be subjected to various forces and the application of stress. Stress can be defined as the change in the internal energy, E, with respect to the strain per unit volume. Therefore the stress tensor, σ_{ij}, may have the following form for a linear-elastic material:

$$\sigma_{ij} = \frac{1}{V_0}\left(\frac{\partial E}{\partial \epsilon_{ij}}\right)_{S=\mathrm{const}} \tag{2.12}$$

where V_0 is the volume of a nanoscale material element, E is the total internal energy of the atomic lattice structure, and ε_{ij} is the strain tensor. The subscript S denotes that the entropy, S, of the strained system is constant during the deformation. During an elastic deformation of a nanostructured material, the internal energy of a nanoscale element can be equal to the strain energy of the deformed material element. In this case, Hooke's law may be derived from Eq.(2.12). Furthermore, if the strain energy is expressed in terms of the applied force acting over the surface area of a carbon nanotube (Frankland and Harik, 2003), then a more familiar form of stress such as force per unit area can be derived.

Atomic interactions during deformation of carbon nanotubes can produce dynamic stress on individual atoms, which can be evaluated by using energy contributions associated with the kinetic energy of nth atom, and the potential energy of its interactions with the m-atom. Stress in a nanoscale volume, V_0, of a group of atoms in the atomic lattice structure of a carbon nanotube or nanocomposite is given by

$$\sigma_{ij} = -\frac{1}{V_0}\sum_n \left(m_n v_i^n v_j^n + \sum_m F_i^{nm} r_j^{nm} \right), \tag{2.13}$$

where m_n is the mass of nth atom, v_n is its velocity with the i- and j-components, F^{nm} is the force of interactions between the nth atom and the mth atom with the i-components, and r^{nm} is the distance between the nth atom and the mth atom with the j-components. It is very important for proper nanoscale averaging of stress components that the size of the chosen nanoscale volume, V_0, is large enough. The criteria for the size of the nanoscale volume, V_0, for averaging will be presented with the criteria for nanoscale homogenization in the next section.

The independent variables in these equations are the spatial position of an atom and the time, which provide the basic length- and time-scales for the motion of such atomic systems, i.e., 0.1–10 Å (or 1 nm = 10^{-9} m) and 1 fs–1 ps (i.e., femto- or pico-second). At the macroscale of a few centimeters and at the time scale of a second or so, such atomic oscillations have no effect and make no difference, except in the sense of temperature: the higher the temperature, the higher the frequency of harmonic oscillations of atoms. In various instruments, thermal vibrations result in the so-called white noise. Vibrations of atomic lattices may have a deterministic, probabilistic and chaotic character, which may also involve various lattice distortions and lattice waves known as phonons.

2.2.4 Nanoscale Homogenization Criteria for Nanomaterials

The smallest length scale in the atomic structure of nanomaterials is represented by the length of the covalent bonds, which is different for distinct materials. This length scale is important for understanding the local origins and the propagation of the lattice waves, i.e., phonons, as well as the distribution of internal and external stresses. In the analysis of material properties of nanomaterials, the size of the smallest volume element in the atomic lattice is more important. When the atomic lattice cells in any crystalline material are of smaller volume, the smaller size of a nanoparticle of this material may have the well-defined material properties. Nanoparticles of nanomaterial powder achieve their macroscopic well-defined properties when their size reaches the next length scale associated with many smallest volume elements of the atomic lattice. A rigorous representation of this requirement is the so-called nanoscale homogenization criterion (Harik, 2001a,b, 2011, 2014).

The size of the smallest volume element also determines the lower size limit of the well-defined material. This connection between the material structure and its properties is reflected in the form of the nanoscale homogenization criterion, which can mathematically define the lower size limit for the nanomaterials with the well-defined material properties with the unique size-invariant values (Harik, 2001a,b). A criterion for the unique averaging (or *nanoscale homogenization*) of any material property over a volume element, V_0, of any material requires that the smallest dimension, L, of the volume element is noticeably greater than the size of the smallest element of the atomic lattice, a, in the direction of the smallest dimension (Harik, 2001a):

$$L \gg a \quad \text{or} \quad \frac{L}{a} \gg 1. \tag{2.14}$$

These inequalities require that the difference in sizes is roughly an order of magnitude, i.e., $L/a \approx 10$ or more. Nanoscale homogenization criteria (2.14) are applicable for the averaging of any physical property associated with any atomic lattice.

The definition of a nanomaterial has to include the upper limit of 100 nm for its smallest dimension, as well as the lower limit for the same dimension, which defines the size, when the well-defined material properties are reached, so their values are size-invariant, unique, and don't change with size. The lower limit for the dimensions of a nanomaterial is defined by the nanoscale homogenization criteria (2.14). Material properties of crystalline materials may depend on the orientation of their atomic lattice. However, the requirements of the homogenization criteria (2.14) have to be satisfied for any direction under consideration. The size of a nanomaterial in any direction has to reach the homogenization length, otherwise material properties of its atomic lattice will be size-dependent, ill-defined, and cannot be considered as uniquely defined.

In the analysis of mechanical deformation of nanomaterials, the nanoscale averaging of stress requires that the size of the smallest nanoscale volume, V_0, over which the stress is averaged, is large enough according to the criteria (2.14). The criteria (2.14) for the smallest size of the nanoscale volume, V_0, used for averaging of stress, are also the criteria for nanoscale homogenization of material properties and the definition of a physical continuum. Averaging of stress over a two-dimensional surface of an extended atomic lattice also has to satisfy the inequalities (2.14) for the dimensions of the surface area involved in the analysis and the size of the smallest element in the atomic structure of the surface.

Analysis of stresses during the sliding of two surfaces along the nanoscale interface has to satisfy criteria (2.14) with respect to the atomic lattices, in order to make sure that the laws of nanoscale friction are valid (see the next section). A surface area should also satisfy the nanoscale homogenization criteria (2.14) in order to be considered as a physical continuum during deformation. Moreover, the wave length, λ, of the deformation response of surface with an atomic structure has to be much larger than the size of the smallest element in the atomic lattice, a, i.e., $\lambda \gg a$, so that the lattice can be considered as a continuum surface on the local scale. It should be noted that the size requirements for nanoscale homogenization are also important for nanoscale scale characterization of various physical properties and physical parameters associated with surface interactions or interfacial interactions.

2.2.5 Nanoscale Friction Laws: An Analog of Newton's Friction Law

The nanoscale sliding of adjacent atomic lattices of two crystalline materials along an interface depends on the local lattice registry, the applied force, and other interfacial interactions mechanisms. The frictional sliding between two atomic lattices can be modeled on the length-scale level larger than a few atomic lattice cells. Multiple lattice cells have to be considered for the meaningful averaging of the lattice sliding (Harik, 2011). Dimensions of a surface element involved in the interfacial sliding, e.g., its length, L, should be sufficiently long, that is, $L/a > 10$, where a is the size of a surface lattice cell. At the length-scale larger than the length, L, a nanoscale friction can be analyzed as follows. The average shear stress, $\langle \tau_{xz} \rangle$, for the sliding interactions between two surfaces can be defined as $\langle \tau_{xz} \rangle \approx \langle f \rangle_{pull}/(A_{ss})$, where $\langle f \rangle_{pull}$ is the average force applied to the sliding surface, and A_{ss} is the common interfacial area during steady sliding between the two lattices separated by the van der Waals forces.

The average strain rate, $\langle \dot{\gamma} \rangle$, of interfacial sliding can be estimated as

$$\langle \dot{\gamma} \rangle \approx \frac{\partial \langle V_{sliding} \rangle}{\partial z} \tag{2.15}$$

where $\langle V_{sliding} \rangle$ is the average velocity of the surface lattice in the axial x-direction and z is the distance between the two adjacent surfaces. Then, a nanoscale analog of the Newton's law for the steady interfacial sliding can be written as (Harik, 2011)

$$\langle \tau_{xz} \rangle = \mu_{eff} \frac{\partial \langle V_{sliding} \rangle}{\partial z}, \tag{2.16}$$

where μ_{eff} is an effective viscosity for the interfacial sliding of the two surface lattices. As a result, an interfacial friction model for the entire sliding process can be developed for the total shear stress, $\langle \tau_{xz} \rangle_{pull}$, which, in addition to the sliding component $\langle \tau_{xz} \rangle$, includes an initial threshold pull stress, τ_0:

$$\langle \tau_{xz} \rangle_{pull} = \tau_0 + \mu_{eff} \frac{\partial \langle V_{sliding} \rangle}{\partial z} \tag{2.17}$$

After approximating the change in velocity across the interfacial separation, h_{vdW}, as the average sliding velocity, $\langle V_{sliding}\rangle$, Eq. (2.17) can be written as

$$\langle \tau_{xz}\rangle_{pull} = \tau_0 + \mu_{eff}\frac{\langle V_{sliding}\rangle}{h_{vdW}} \tag{2.18}$$

where the strain rate is approximated as $\left(\partial\langle V_{sliding}\rangle/\partial z\right) \approx \left(\langle V_{sliding}\rangle/h_{vdW}\right)$. Here, τ_0 is the threshold pull stress, and μ_{eff} is an effective viscosity for the interfacial sliding of the two atomic lattices. The threshold pull stress, τ_0, is affected by the initial commensurability of the two sliding lattices. "How [should] lattice commensurability effects impact sliding friction levels?" (Krim, 2002). The better initial commensurability of the adjacent lattices, the greater value of the threshold pull stress. The value of the effective viscosity for the interfacial sliding depends on the dynamic commensurability of the two atomic lattices during sliding as well as on other parameters including the strength of van der Waals interactions and the effective viscosity of the two layers of bonded electrons (Harik, 2011, 2014).

The dynamic commensurability contributes to nanoscale friction since "friction in the absence of wear must be due to strains building up in the sliding contact which were being released in the form of atomic vibrations" including phonons, according to Tabor (Krim, 2002). The nanoscale analogs of Newton's friction law (2.16) and (2.17) can be successfully used to describe the interfacial friction and sliding between two atomic lattices of graphene sheets (Chapter 3: Nanomechanics of Graphene Sheets and Atomic Registry Matrix Analysis). Moreover, the form (2.17) of the nanoscale analog of Newton's friction law can describe the role of initial commensurability between two graphene lattices through the threshold pull stress, τ_0, as well as the subsequent sliding between incommensurate lattices with some dynamic commensurability. These nanoscale friction laws can also describe the interfacial friction between a carbon nanotube and a polymer matrix and the adjacent lattices in the MWCNTs during telescoping sliding (see later chapters).

Nanoscale friction and tribological properties of carbon nanotubes have not been studied well, since these areas of research are quite challenging. For example, the application of the conventional definition of the coefficient of friction in nanoscale friction as a ratio of the lateral force to the normal force: $\sigma = F_l/F_n$, may need more investigation. Ni and Sinnott (2001) have discovered a paradox in the analysis of friction of the horizontally oriented carbon nanotubes and the application of the conventional definition of the coefficient of friction, σ. They have found that the coefficient of friction, σ, is about 30 for low normal loads, and 0.36 for the normal pressure of 13.7 GPa, as in the cases of surface damage and wear. These measurements indicate that there is a paradox in the analysis of friction of carbon nanotubes, because the sliding is easier on horizontal nanotube bundles, which have not been heavily compressed, as compared to the sliding on the compressed nanotube bundles. The aforementioned values of the friction coefficient, σ, seem to indicate the reserve trend and opposite characterization, which constitutes a paradox. However, this paradox by no means suggests that the definition of the coefficient of friction is not applicable for the analysis of nanoscale friction data and nanoscale friction itself.

2.2.6 Molecular Dynamics Simulations Computer Programs

Mechanical deformation of the molecular structure of nanostructured materials can be simulated by molecular mechanics computer codes (e.g., MM2 or MM3, TINKER), as well as by computer programs for MD simulations such as DL-POLY and Amber. For example, a molecular lattice structure of the crystalline part of the PE polymer with the $(CH_2)_n$ zigzag chains surrounding the armchair (10,10) carbon nanotube has been simulated by DL-POLY (Frankland and Harik, 2003). A chapter on carbon nanotubes in nanocomposites provides more details about the modeling of the molecular structure as well as the MD simulations. There are also computer programs that just allow building various molecular structures including the atomic lattices of nanostructures.

MD simulations of the polymer–nanotube interactions can be carried out with the DL-POLY package of subroutines written by W. Smith and T.R. Forester (Daresbury Laboratory, Warrington, United Kingdom, 1996). The aforementioned example of the molecular lattice structure of the crystalline part of PE polymer with the $(CH_2)_n$ zigzag chains and the (10,10) carbon nanotube includes numerous atoms ($\sim$880 carbon atoms in the armchair nanotube, and many more in the surrounding polymer chains). Such a complex MD simulation is computationally challenging; therefore a computer cluster has been used to carry out the numerical computations.

2.3 Continuum Models for Carbon Nanotubes

Continuum models have been used for the modeling of carbon nanotubes, as well as for the analysis of data of MD simulations (Yakobson et al., 1996). Continuum models of discrete atomic lattices have certain limitations in their ability to resolve local details of the discontinuous anisotropic material structures (Harik, 2001a, 2001b). Early predictions of the applicability of continuum models to the mechanics of carbon nanotubes have been quite optimistic: "A correspondence of atomistic model and macroscopic structural mechanics can be achieved with the proper choice of parameters of the continuum shell" (Smalley and Yakobson, 1998). Moreover, "with properly chosen parameters, the [shell] model provides a remarkably accurate 'roadmap' of nanotube behavior beyond Hooke's law." That is, "with properly chosen parameters, this analytical [shell] model predicts nanotube behavior not only at small deformations but also beyond the linear response." "This [hollow structure of fullerenes] should also manifest itself in dynamic properties of molecules, resembling so much the macroscopic objects known in continuum elasticity as *shells*." (Yakobson et al., 1996). "A less obvious property of CNTs is that the specific features of large nonlinear strain can be understood and predicted in terms of continuum theories." "Its relevance for a covalent-bonded system of only a few atoms in diameter was far from being obvious." (Yakobson and Dumitrica, 2003). Yakobson has been persistent advocate of the continuum shell models for carbon nanotubes, after such a model was successfully used to analyze the details of MD simulations (Yakobson et al., 1996). Other researchers have greatly extended and refined this area of research in the mechanics of carbon nanotubes, as shown by different reviews (Harik, 2011, 2014).

2.3.1 Continuum Shell Models for Carbon Nanotubes

The mechanical "evolution of the molecular structure can be put in the framework of continuum elasticity" (Yakobson and Dumitrica, 2003). "A curved sheet [of graphene] can be approximated by a uniform shell with only two elastic parameters: flexural rigidity D, and its in-plane stiffness, C." The material parameters of a two-dimensional cylindrical nanostructure are well-defined when the length of carbon nanotube and its circumference are large enough in order to satisfy the nanoscale homogenization criteria (2.14) (Harik, 2001a, 2001b). The elastic energy of a continuous shell representing a carbon nanotube with the well-defined material properties can be described by a surface integral of the quadratic form of local deformation (Timoshenko and Gere, 1961; Yakobson et al., 1996):

$$E = \frac{1}{2}\iint\left\{D\left[\left(\kappa_x+\kappa_y\right)^2 - 2(1-\nu)\left(\kappa_x\kappa_y - \kappa_{xy}^2\right)\right] + \frac{C}{(1-\nu^2)}\left[\left(\varepsilon_x+\varepsilon_y\right)^2 - 2(1-\nu)\left(\varepsilon_x\varepsilon_y - \varepsilon_{xy}^2\right)\right]\right\}ds,$$

where κ is the curvature variation, ε is the in-plane strain, and x and y are local coordinates. The dimensions of the area of surface integration have to satisfy the nanoscale homogenization criteria (2.14) as noted above. This representation of the elastic strain energy of a continuous shell can be adapted to a sheet of graphene in the structure of the armchair (10,10) carbon nanotube. The values of parameters D and C can be evaluated by using data from ab initio and semiempirical predictions for the strain energy of this carbon nanotube at small strains.

The in-plane stiffness, C, of a continuum shell can be evaluated for small strains as the second derivative of the total elastic strain energy with respect to the axial strain [$C = 56$ eV/atom $= 340$ J/m^2 (Yakobson et al., 1996)]. The strain energy is also related to the flexural rigidity, D ($D = 1.46$ eV), and the shell radius, R, as $D/(2R^2)$. The Poisson ratio ($\nu = 0.15$) is estimated by using the change of the diameter of the strained carbon nanotube during axial extension. This value of the Poisson ratio is supported by the experimental data for the elastic constants obtained for the single crystal graphite. The in-plane stiffness, C, of continuum shells depends on their wall thickness, h, and their elastic modulus, E, as $C = Eh$ (Timoshenko and Gere, 1961). The flexural rigidity, D, of continuum shells is determined by $D = Eh^3/12(1 - \nu^2)$ (Timoshenko and Gere, 1961). These two equations and the MD data for the shell parameters: C and D, yield the following estimates for the elastic modulus of the (10,10) carbon nanotube: $E_{NT} = 3.9$ TPa, and its effective thickness: $h_{NT} =$ 0.089 nm. Yakobson et al. (1996) have used these estimates and the linear stability analysis for continuum elastic shells (Timoshenko and Gere, 1961) to assess the buckling behavior of the (10,10) carbon nanotube under axial strain. The particular geometry of the (10,10) carbon nanotube is such that the material parameters of its two-dimensional cylindrical nanostructure are well-defined. That is, the length of the (10,10) carbon nanotube used in MD simulations and its circumference are large enough to satisfy the nanoscale homogenization criteria (2.14) (Harik, 2001a, 2001b). Therefore, there are carbon nanotubes for which this analysis has significant limitations.

Yakobson et al. (1996) have emphasized that the linear stability analysis for continuum shells is very effective for the investigation of different deformation modes of carbon nanotubes under axial compression. An axial perturbation of a cylindrical shell can be evaluated by its Fourier harmonics, with m azimuthal portions and n half waves along the axis of the shell, i.e., sines and cosines of arguments my/R and $n\pi x/L$. If the axial perturbation of a carbon nanotube results in a critical strain, ε_{cr}, the variation of the elastic strain energy vanishes for the most favorable shape disturbance (m, n). Under the critical strain, the cylindrical shell of a carbon nanotube becomes unstable and starts to deform in such a way, which results in the lower value of the strain energy by assuming the energetically favorable (m, n)-pattern. The critical strain, ε_{cr}, for a relatively short cylindrical shell occurs for the buckling pattern with $m = 2$ and $n \geq 1$. For a short shell or a corresponding carbon nanotube of the length, $L_{NT} = 6$ nm and diameter $d_{NT} = 1$ nm, the most favorable buckling pattern occurs for $m = 2$ and $n = 2, 3$. This deformation pattern includes the flattening of the shell in two directions perpendicular to each other. The axis of the buckling shell remains straight and does not buckle sideways. The transition states and minimum energy pathways for the collapse of carbon nanotubes can be also described with the Cauchy–Born rule for the deformation of atomic lattices (Zhang et al., 2006). The critical buckling strain, ε_{cr}, for cylindrical shells can be estimated as (Timoshenko and Gere, 1961)

$$\varepsilon_{cr} = 2\sqrt{D/C}/R = \left(1/\sqrt{3}\right)\left(1 - \nu^2\right)^{-1/2} h/R, \tag{2.19}$$

where the ratio of the shell thickness, h, to its radius, R, plays a critical role (Harik, 2001a, 2001b). Here, the thickness-to-radius ratio, h/R, actually determines for what type of carbon nanotubes the estimate (2.19) can apply to (Harik, 2011, 2014). It so happens that the radius of the (10,10) carbon nanotube is large enough to satisfy the so-called the thin shell criterion for carbon nanotubes (Harik, 2001a, 2001b).

The critical buckling strain, ε_{cr}, for carbon nanotubes can be estimated by the formula (2.19) in the form that emphasizes its dependence on the diameter, d_{NT}. The armchair (10,10) carbon nanotube buckles under the axial critical strain, ε_{cr}, which can be expressed as inversely proportional to the diameter, d_{NT} (Yakobson et al., 1996):

$$\varepsilon_{cr} = \frac{0.1nm}{d_{NT}}$$

This form of the formula (2.19) has to be used with the additional requirement for cylindrical shells when considering the smallness of the thickness-to-radius ratio, h/R (Harik, 2001a,b). Otherwise, the diameter of small carbon nanotubes can easily fall outside of the range of applicability of the critical strain formula (2.19) and all its forms.

The range of applicability of the classical mechanics of continuum shells for the discrete atomic structure of carbon nanotubes is an important topic in nanoscale mechanics (Harik, 2001a,b, 2011, 2014). In addition to the requirements to the value of the thickness-to-radius ratio, h/R (Harik, 2001a,b, 2002), one has to consider a long list of assumptions used in the derivation of the cylindrical shell theory, and especially, the thin shell theory (Timoshenko and Gere, 1961; Harik, 2001a, 2002, 2011). There are carbon nanotubes, which can be described by the thin shell theories in some cases (Harik, 2001a, 2002, 2011, 2014). However, there are groups of carbon nanotubes which cannot be treated as thin shells (Harik, 2001a,b, 2002). Therefore the thin shell theory cannot be used to investigate the mechanical behavior of such carbon nanotubes or the process of their buckling under an axial load.

The effective thickness of the atomic lattice of carbon nanotubes, h_{NT}, was estimated to be 0.06 nm for the cylindrical shell model (Yakobson et al., 1996). This estimate for the thickness of carbon nanotubes was later revised to 0.089 nm (Yakobson and Dumitrica, 2003). However, numerous researchers use the experimentally observed thickness of a graphene sheet in the graphite, i.e., 0.34 nm, for the thickness of carbon nanotubes with different diameters for various loading conditions. The later value is 3.5–5 times larger than the aforementioned estimates obtained with the shell model. This is a significant discrepancy which has not been addressed in the original investigations (Yakobson et al., 1996; Yakobson and Dumitrica, 2003). In fact, this discrepancy in the estimates of the effective thickness of the atomic lattice of carbon nanotubes, h_{NT}, can significantly change the range of applicability of the continuum shell theory to the mechanics of carbon nanotubes and other thin atomic shells, such as graphene (Harik, 2011, 2014). Discrepancy in the estimates of the effective thickness of carbon nanotubes results in the inconsistency in the evaluation of the Young's modulus of carbon nanotubes.

2.3.2 Continuum Beam Models for Carbon Nanotubes

Continuum beam models have been also used to describe the mechanical behavior of carbon nanotubes. The aforementioned case of the axial buckling of carbon nanotubes can be also described by the continuum beam models for long nanotubes (Harik, 2001a,b, 2002). The continuum beam theory is based on the Euler beam equation (Timoshenko and Gere, 1961). The Euler beam equation for the small deflection, $w(x)$, of a long elastic beam has the following form (Timoshenko and Gere, 1961):

$$EI\frac{\partial^2 w}{\partial x^2} + Pw = 0, \tag{2.20}$$

where E is the elastic modulus, I is the moment of inertia, and P is the load or the force per cross-sectional area, A. The second derivative of the deflection of the beam identifies the change in the curvature of the deflected beam along its length. It is physically understandable that the curvature of the deflected beam increases with the load, P, as well as with the magnitude of deflection, $w(x)$. The ranges of applicability of this equation to carbon nanotubes depend on a long list of assumptions (Harik, 2001a,b, 2002), which are discussed in details later in this book.

In order to use the continuum beam theory, one has to ensure that the geometry of carbon nanotubes allows for the definition of continuum media, or the equivalent continuum for the discrete atomic lattice structures (Harik, 2001a, 2001b). The definition of a continuum media is possible when the length of carbon nanotube satisfies the nanoscale homogenization criteria (2.14). Moreover, the value of the elastic modulus of carbon nanotubes is well-defined when the length, L_{NT}, is long enough to satisfy the same criteria for nanoscale homogenization. The linear stability analysis for continuum elastic beams (Timoshenko and Gere, 1961) can be also used to determine the critical buckling strain of long carbon nanotubes (Harik, 2001a,b, 2002).

The aspect ratio, d_{NT}/L_{NT}, of carbon nanotubes, or its equivalent, R_{NT}/L_{NT}, is the main nondimensional length scale parameter governing the nanoscale buckling and the buckling strain, ε_{cr}, of the high aspect ratio carbon nanotubes (Harik, 2001b, 2002):

$$\varepsilon_{cr} = 4\pi^2\frac{R_{NT}^2}{L_{NT}^2}\left[\left(1+\frac{h_{NT}}{2R_{NT}}\right)^2 - \left(1-\frac{h_{NT}}{2R_{NT}}\right)^2\right], \tag{2.21}$$

where h_{NT} is the equivalent thickness of the thick or thin lattice shells of carbon nanotubes. Note that the nanoscale buckling strain, ε_{cr}, also depends on the thickness-to-radius ratio, h_{NT}/R_{NT}, of high aspect ratio carbon nanotubes, but only in the second-order effects. The diameter-to-length aspect ratio, d_{NT}/L_{NT}, of carbon nanotubes is an important structural parameter in many applications, such as in the nanotube-based AFM probes for atomic force microscopes. This nondimensional ratio is critical for the analysis of any buckling process for long beam-like molecules.

It has been noted that the linear stability analysis (Timoshenko and Gere, 1961) for a long cylindrical shell, e.g., a carbon nanotube with $L_{NT} > 10$ nm, the buckling bifurcation is first attained for the buckling pattern with $m = 1$, $n = 1$ (Yakobson et al., 1996). A long carbon nanotube buckles sideways, while preserving its circular cross section for small strains. That is carbon nanotube buckles sideways as a whole cylindrical shell with the critical buckling strain, which is close to that for a nanoscale beam (Yakobson et al., 1996; Harik, 2001a, 2002):

$$\varepsilon_{cr} = \frac{\pi^2}{2}\left(\frac{d_{NT}}{L_{NT}}\right)^2 \tag{2.22}$$

or four times less for a carbon nanotube with hinged (or unclamped) ends. Here, we use the form of the formula for the critical buckling strain (Harik, 2001a, 2002) in order to emphasize the well-known importance of the aspect ratio, d/L, in the buckling of the beams (Timoshenko and Gere, 1961). In somewhat different approach which focuses on the shell model, Yakobson et al. (1996) have noted the importance of the ratio of the circumference of carbon nanotubes, πd, to their length, L, i.e., $\pi d/L$. Nevertheless, the long cylindrical shells initially buckle as long beams, and the key nondimensional parameter is their aspect ratio, d/L. The normalized circumference, $2\pi R_{NT}/a$, of carbon nanotubes, where a is the size of a carbon ring, is important in the evaluation of secondary effects in the axial buckling of long cylindrical lattices, as well as of the radial deformation of carbon nanotubes and the radial buckling of their lattices (Harik, 2011). It should be noted that the formula (2.22) for the critical buckling strain of carbon nanotubes is valid only for small axial strains, i.e., the change in their length, δL_{NT}, is such that $\delta L_{NT}/L_{NT} \ll 1$. This is an important requirement that has to be verified during numerical simulations.

2.4 Conclusions

In this chapter, the atomic structure of nanomaterials and its lengths scales are discussed, along with the breakdown of continuum concepts and limitations of continuum mechanics. Size effects in the material properties of nanomaterials are considered, along with the criteria for nanoscale homogenization. Nanoscale laws of motion and governing equations are reviewed, along with the laws for the conservation of energy, mass, and momentum at nanoscale. Atomic potentials for interatomic interactions are also reviewed, along with the formulae for the stress at atomic scale and the requirements for its proper averaging for nanomaterials. Nanoscale homogenization criteria are presented for nanomaterials. Nanoscale friction, an analog of Newton's friction law and a paradox of the conventional friction coefficient at nanoscale have been examined. The definition of stress for nanostructures is examined, as well as the limitations on the size of RVEs. Continuum beam and shell models for carbon nanotubes are presented, along with the ranges of their applicability for nanoscale buckling.

Sample Problems

1. Describe the length scales in the atomic structure of crystalline nanomaterials.
2. Formulate the nanoscale homogenization criteria for nanomaterials.
3. Prepare a viewgraph with the atomic structure of a nanomaterial.
4. Describe Newton's second law for the motion of atoms.
5. Describe the Taylor series for the position vector.
6. Describe the Lennard-Jones potential.
7. Describe an atomic potential for van der Waals interactions.
8. Describe the nanoscale analog of Newton's friction law.
9. Describe how the nanoscale homogenization criteria are used in the averaging of stresses at nanoscale level.
10. Describe the key nondimensional parameters for nanoscale buckling.
11. Describe a formula for the critical buckling strain of carbon nanotubes.
12. Describe how the nanoscale homogenization criteria can be connected to size effects in the material properties of nanomaterials.

References

Brenner, D.W., 1990. Empirical potential for hydrocarbons for use in simulating the chemical vapor deposition of diamond films. Phys. Rev. B 42, 9458–9471.

Frankland, S.J.V., Harik, V.M., 2003. Carbon nanotube pull-out from a polymer matrix. Surf. Sci. Lett. 525, L103.

Harik, V.M., 2001a. Ranges of applicability for the continuum beam model in the mechanics of carbon nanotubes and nanorods. Solid State Commun. 120 (7-8), 331–335.

Harik, V.M., 2001b. Ranges of applicability for the continuum beam model in the constitutive analysis of carbon nanotubes: nanotubes or nano-beams? NASA/CR-2001-211013. NASA Langley Research Center, Hampton, VA.

Harik, V.M., 2002. Mechanics of carbon nanotubes: applicability of the continuum-beam models. Comput. Mater. Sci. 24 (3), 328–342.

Harik, V.M., 2011. Mechanics of Carbon Nanotubes. Nanodesigns Press, Newark, DE.

Harik, V.M. (Ed.), 2014. Trends in Nanoscale Mechanics. Springer, Dordrecht.

Krim, J., 2002. Surface science and the atomic-scale origins of friction: what once was old is new again. Surf. Sci. 500, 741–758.

Lennard-Jones, J.E., 1924. The determination of molecular fields. I. From the variation of the viscosity of a gas with temperature. Proc. R. Soc. (London) 106A, 441.

Ni, B., Sinnott, S.B., 2001. Tribological properties of carbon nanotube bundles predicted from atomistic simulations. Surf. Sci. 487, 87–96.

Smalley, R.E., Yakobson, B.I., 1998. The future of fullerenes. Solid State Commun. 107 (11), 597–606.

Tersoff, J., 1988. New empirical approach for the structure and energy of covalent systems. Phys. Rev. B 37, 6991–7000.

Timoshenko, S.P., Gere, J.M., 1961. Theory of Elastic Stability. McGraw-Hill, New York, NY.

Yakobson, B.I., Dumitrica, T., 2003. Nanomechanics: physics between engineering and chemistry. In: Harik, V. M., Salas, M. (Eds.), Trends in Nanoscale Mechanics. Kluwer Academic Publishers, The Netherlands, pp. 3–33, Chapter 1.

Yakobson, B.I., Brabec, C.J., Bernholc, J., 1996. Nanomechanics of carbon tubes: Instabilities beyond the linear response. Phys. Rev. Lett. 76 (14), 2511–2514.

Zhang, S., Khare, R., Belytschko, T., Hsia, K.J., Mielke, S.L., Schatz, G.C., 2006. Transition states and minimum energy pathways for the collapse of carbon nanotubes. Phys. Rev. B 73, 075423.

CHAPTER 3

Nanomechanics of Graphene Sheets and Atomic Registry Matrix Analysis

Chapter Outline

3.1 Atomic Structure of Graphene Sheets and Its Lengths Scales

Graphene and graphene sheets are a part of commonly used pencils. Exfoliation of graphene sheets from the graphite has been challenging for a long time (Raza, 2011). Graphene has exceptional material and physical properties. Graphene sheets consist of a single layer of carbon atoms similar to carbon nanotubes. The atomic structure of graphene consists of carbon atoms arranged into a periodic hexagonal lattice of carbon rings (Fig. 3.1). The edge of graphene sheets or nanoribbons may have the familiar zigzag or armchair structure (Fig. 3.1), as well as other complex patterns with their special nomenclature (Raza, 2011). Control of edge patterns in graphene sheets requires special manufacturing or processing technologies.

The atomic structure of graphene is similar to the structure of graphite (Fig. 3.2), although it has the single layer structure, as opposed to many layers in graphite. Graphene platelets (Fig. 3.2) may consist of two or several layers of graphene. A schematic of the typical AAA

Mechanics of Carbon Nanotubes.
DOI: https://doi.org/10.1016/B978-0-12-811071-3.00003-2

© 2018 Elsevier Inc. All rights reserved.

Figure 3.1
Atomic lattice structure of a graphene sheet (or a graphene nanoribbon) with the zigzag edges on top and bottom.

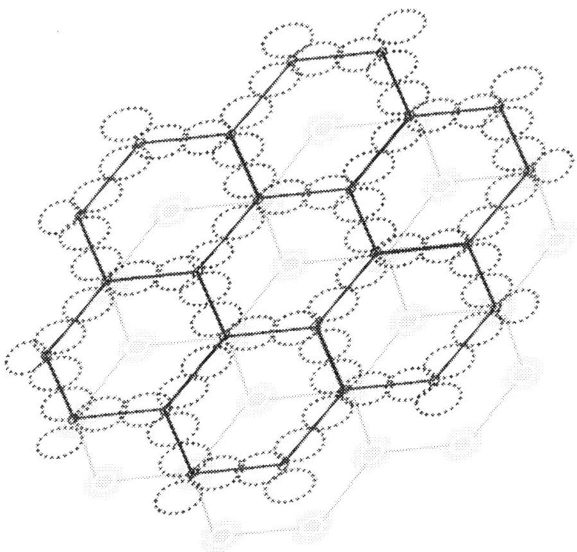

Figure 3.2
Schematic of the typical AAA stacking of the incommensurate graphene sheets in a two-layer graphene platelet with the carbon rings of the adjacent lattices positioned right above the opposite carbon ring and its C—C bonds of carbon atoms. © *2011 by Nanodesigns Consulting (reproduced with permission).*

stacking found in graphite is shown in Fig. 3.2, where the adjacent graphene sheets are incommensurate. Single graphene sheets have remarkable electronic properties, for example, the quantum Hall effect (Geim and Novoselov, 2007) and the band gap-width dependence (Raza, 2011). The band gap-width dependence will be described in more details later in this chapter. Graphene properties, such as electrical conductivity and Young's modulus, also depend on the minimum length, or the size of graphene sheets, i.e., the minimum number of carbon rings along its length, as was noted for carbon nanotubes and graphene (Harik, 2011, 2014). The minimum length of graphene sheets has to be long enough in order to satisfy the nanoscale homogenization criterion required for achieving the macroscopic value (Harik, 2011, 2014). A general formula for the dependence of material properties of graphene sheets or other materials on their size will be presented later.

Large graphene sheets can be used to form various tubular structures, since they tend to roll up. Carbon nanotubes can be considered as cylindrical structures made of the rolled-up graphene sheets. Two unit vectors, $\mathbf{a}_1$ and $\mathbf{a}_2$, defined at the origin (0,0) can be used to describe the orientation of carbon rings in the atomic lattice of graphene sheets (Fig. 3.3), as it was done for carbon nanotubes (Chapter 1: Nanotechnology of Carbon Nanotubes: Sensors, Transistors, and Nanocomposites). The width of a graphene sheet along the unit vector, $\mathbf{a}_1$, can be measured in the units of the carbon ring size, $a = 2.46$ Å, i.e., $L = 3a$ for the graphene flake shown in Fig. 3.3. It should be noted that the small graphene flake shown in Fig. 3.3 and the graphene nanoribbon shown in Fig. 3.1 do not have the material properties of large graphene sheets. Similarly, the graphene platelet shown in Fig. 3.2 does not have the macroscopic properties of graphite. The reason is the small size ($L = 3a$) of the graphene flake in Fig. 3.3 and the graphene platelet in Fig. 3.2, as well as the small width ($w \approx 5a$) of the graphene nanoribbon in Fig. 3.1.

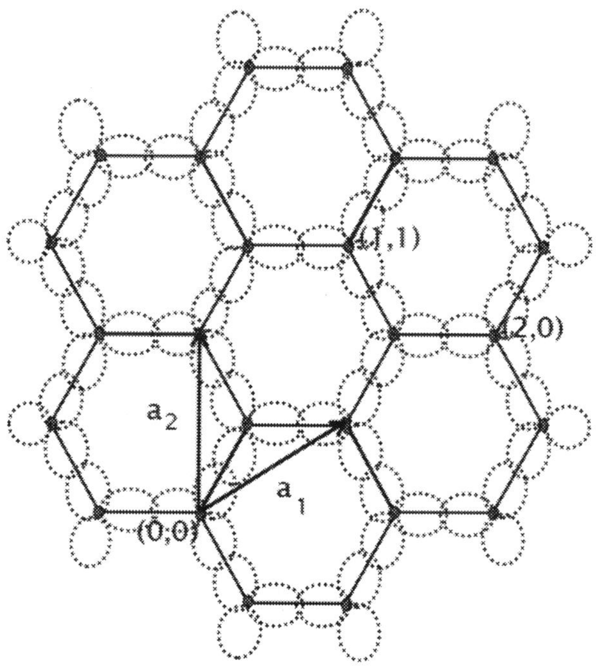

Figure 3.3
Schematic of the local atomic structure of graphene sheets made of the carbon rings of six carbon atoms with the σ-electrons along the C—C bonds of length, $l_{C-C} = 1.42$ Å. The (n,m) nomenclature is defined by the two unit vectors, $\mathbf{a}_1$ and $\mathbf{a}_2$, at the origin (0,0). © *2011 by Nanodesigns Consulting (reproduced with permission).*

The smallest length scale in the atomic structure of graphene is represented by the length of the C–C bond, $l_{C-C} \approx 1.42$ Å. This length scale is important for the propagation of the lattice waves, i.e., phonons, and for the distribution of stresses in the structure of graphene devices. However, in the analysis of material properties and atomic structure of graphene, the size of a carbon ring, a, is more important as the size of the smallest structural element. Therefore the next length scale in the structure of graphene sheets is measured not by the number of the C–C bonds, but by the number of carbon rings in the width or the length of graphene nanoribbons (Fig. 3.1). Graphene sheets and graphene platelets achieve their macroscopic properties when their size reaches the next length scale measured in at least 10 carbon rings. A rigorous representation of this requirement is the so-called nanoscale homogenization criterion (Harik, 2001a,b, 2011, 2014), which will be discussed below.

A criterion for the unique averaging (or homogenization) of any material property over an extended surface of any two-dimensional material, including the atomic structure of graphene, graphene nanoribbons, and platelets, requires that their size, L, is considerably greater than the size of their smallest structural element, a (Harik, 2001a):

$$L \gg a \quad \text{or} \quad \frac{L}{a} \gg 1. \tag{3.1}$$

This inequality means roughly an order of magnitude difference, i.e., $L/a \approx 10$ or more, where a is the size of a carbon ring: $a \approx 2.45$ Å. Nanoscale homogenization criteria (3.1) are applicable for the averaging of any physical property associated with a discrete atomic structure.

3.1.1 Deformed Fermi Cones of π-Electrons in Graphene

Atomic lattice structure of graphene and carbon nanotubes consists of carbon rings (Fig. 3.3), which are formed by the σ-electrons along the six covalent C–C bonds. Each carbon ring has an additional six covalent bonds associated with each carbon atom and the adjacent carbon rings. Each carbon atom also has a cone of π-electrons, a Fermi cone, with an axis perpendicular to the plane of a carbon ring (Fig. 3.4). Spatial arrangements of such complex atomic structure affect the intrinsic materials properties of graphene, as well as the physical response of graphene to the applied forces. Covalent bonds and their σ-electrons determine the potential energy of elastic and hyperelastic interactions, while π-electrons affect van der Waals interactions with adjacent atomic lattices.

A schematic of carbon rings is shown in Fig. 3.3, where carbon atoms form the sp^2 bonding with the σ-electrons along the six covalent bonds. The length, l_{C-C}, of the covalent C–C bonds in graphene is similar to graphite, i.e., $l_{C-C} = 1.42$ Å, as noted above. The approximate size, l_σ, of the spatial distribution of the σ-electrons is such that $l_\sigma = l_{C-C}/2 \approx 0.71$ Å, which is slightly larger than the approximate size of the interaction zone for the

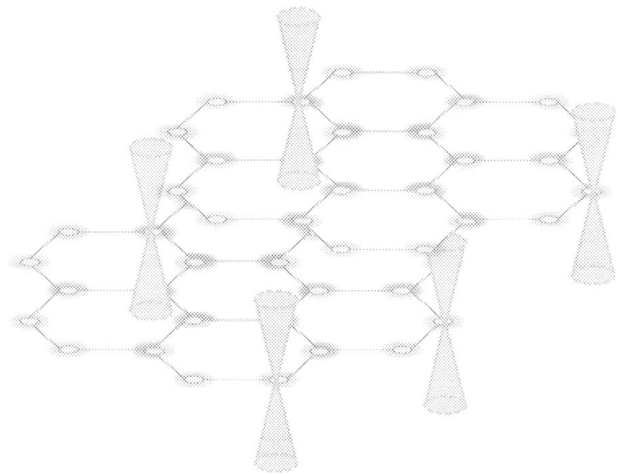

Figure 3.4
Schematic of two graphene atomic lattices with the schematically shown a few out-of-plane *Fermi cones* of the π-electrons above each carbon atom. © *2011 by Nanodesigns Consulting (reproduced with permission).*

out-of-plane π-electrons above the middle of the elastic C—C bonds. The π-electrons in the Fermi cones (Fig. 3.4) affect van der Waals interactions between adjacent atomic lattices in graphene platelets.

The spatially distributed electrons cover the entire surface area of the carbon rings, with the σ-electrons forming the sp^2-type bonding of the elastic C—C bonds, and the out-of-plane π-electrons distributed over the elastic C—C bonds and in the middle of the carbon ring (Fig. 3.4). The energy distribution of the out-of-plane π-electrons has its high point right above the midpoint of each C—C bond (the M-point of the so-called Brillouin zone[1] for the energy of states). In a carbon ring, it is the ΓK′K area of the carbon ring (Fig. 3.5). The highest value of the energy distribution of the out-of-plane π-electrons is in the center of a carbon ring, i.e., the Γ-point of the Brillouin zone, while its lowest zero-points are at the position of each carbon atom, i.e., the K-point of the Brillouin zone (Fig. 3.5). Distribution of the energy of states $E_G(k)$ for electrons in graphene can be described by

$$E_G(k) = \pm t\sqrt{1 + 4\cos\left(\frac{3k_x a_{cc}}{2}\right)\cos\left(\frac{\sqrt{3}k_y a_{cc}}{2}\right) + 4\cos^2\left(\frac{\sqrt{3}k_y a_{cc}}{2}\right)},$$

[1] The so-called Brillouin zone is used to schematically represent the energy dispersion relation for the energy of states of the graphene lattice vibration waves, i.e., *phonons*, and the energy of states of the oscillating electrons.

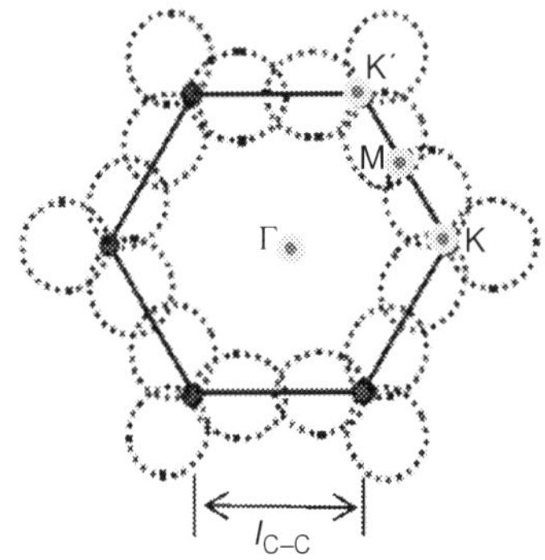

Figure 3.5
Schematic of the carbon ring of six carbon atoms with the schematically shown σ electrons ($l_\sigma \approx$ 0.71 Å) along six covalent C–C bonds with the length: $l_{C-C} = 1.42$ Å, and the approximate size of the spatial distribution of the σ-electrons, $l_\sigma \approx l_{C-C}/2$. The energy distribution of the out-of-plane π-electrons (not shown here) has its lowest zero-points at the position of each carbon atom, i.e., at the K-point, and its high points at the M-point and Γ-point of the Brillouin zone. *© 2011 by Nanodesigns Consulting (reproduced with permission).*

where k_x and k_y are the x- and y-components of the wave vector k, $a_{cc} = l_{C-C}$ is the length of the C–C bond, and t is the first nearest-neighbor tight binding parameter (Saito et al., 1998). The length of the covalent bonds, a_{cc} or l_{C-C}, is an important scaling parameter in the distribution of energy of electrons.

The model of a Fermi cone approximates the experimentally evaluated energy distribution of the π-electrons states near the point C (Fig. 3.6) above each carbon atom (after Bostwick et al., 2006). The energy upshift of 0.45 eV from the Fermi free energy to the Dirac energy of the fast π-electrons is marked by the C point in Fig. 3.6. The height of the Fermi cone over 18 C–C bonds around the six carbon atoms is approximately 1.65 eV, and the radius of the Fermi cone is about $(3/8)l_{C-C}/2$ toward the M-point of the Brillouin zone, and about $(1/3)l_{C-C}/2$ toward the Γ-point. Since there are three C–C bonds with the three M-points and the three carbon rings with the three Γ-points around each carbon atom (at the K-point), the Fermi cone is deformed along the three C–C bonds into a trilateral clover-type cone-deformation topology, i.e., the clover cone.

The length scale of $(3/8)l_{C-C}$ or $(3/4)l_{C-C}/2$ is the approximate size of the interaction zone (about $l_{C-C}/3$) for the out-of-plane π-electrons from the K-point and K′-point of the Brillouin zone above the middle of the elastic C–C bonds. The energy-to-length slope, E/L, of the Fermi cone is the gradient of the energy of states of about 6.20 eV/A or 2.2 eV per one-half of the half C–C bond, $l_{C-C}/2$, which is 1.65 eV over $(3/8)l_{C-C}/2$, and has the units of force. The experimental energy distribution of the π-electron states has the same energy-to-length slope, E/L, or the gradient of the π-electron density at the point B (the M-point of the Brillouin zone), and $(3/8)l_{C-C}/2$ away from the Γ-point (the point A).

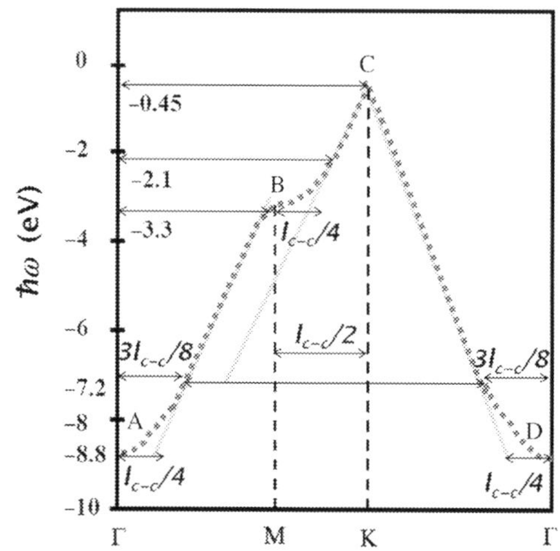

Figure 3.6
Approximate experimental energy distribution of the π-electron states as a function of the momentum along the principle directions of the carbon ring through the K, M, and Γ points of the Brillouin zone (analysis of experimental data of Bostwick et al., 2006). © *2011 by Nanodesigns Consulting (reproduced with permission).*

The radius of the concave top of the experimental energy distribution of the out-of-plane π-electrons states that are distributed over the center of the carbon ring is also about (3/8) l_{C-C} or $(3/4)l_{C-C}/2$ (see the point A and the point D in Fig. 3.6 at the Γ-point of the Brillouin zone). The energy-to-length slope, *E*/*L*, of the deformed Fermi cone $(3/4)l_{C-C}/2$ from the point D (i.e., the Γ-point) is about 9.4 eV/A or 1/150 in the units of one-half of the C–C bond, $l_{C-C}/2$, over 1 eV (i.e., the units of inverse Newton's, N^{-1}), which is 8.35 eV over $(5/8)l_{C-C}/2$, and has the units of force. The energy length of the deformed Fermi cone (the trilateral clover cone) on the three sides of the three Γ-points is 6.75 eV, which is much higher than 1.65 eV of the energy length on the three sides of the three M-points. The energy barrier for the π-electrons near-neighbor hopping, E_h, between the two bonded atoms is 2.82 eV between the K-point and K′-point along the C–C bond over the M-point of the Brillouin zone, as shown by Bostwick et al. (2006). The concave top of the π-electrons energy of states distribution is deformed into the sixfold dome by the 12 radius modulations pointing toward the six K- and the six M-points with the π/6 period of the π-electron density modulation around the center of the carbon ring, which is caused by the geometric symmetry of the Brillouin zones of the carbon ring (Fig. 3.7).

The energy states of the π-electrons distributed over the center of the carbon ring (the Γ-point) scale with the energy for the near-neighbor hopping, E_h, (the M-point) as 88/33 or 2.6(6) with respect to the Fermi free energy, E_F, and as 2.93 or ~3 with respect to the

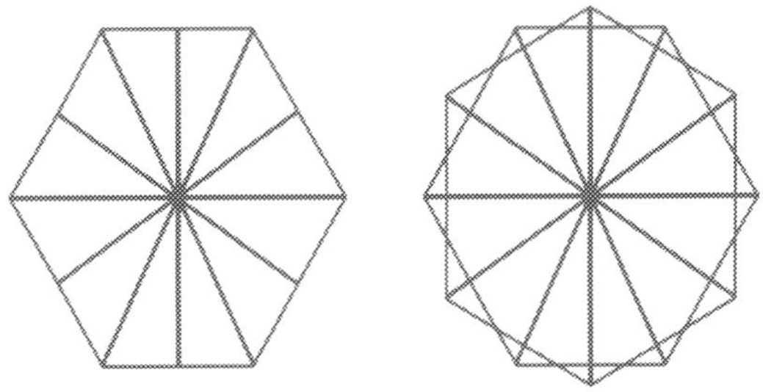

Figure 3.7
Schematic of the geometric symmetry of the six Brillouin zones within the carbon ring (left) and a schematic of the sixfold dome (or the three B-modulation dome) of the approximate experimental energy distribution of the π-electron states (i.e., the π-electrons density) with the 12 radial modulation toward the six K- and the six M points around the center of the carbon ring (right), which is the Γ point of all six Brillouin zones. *© 2011 by Nanodesigns Consulting (reproduced with permission).*

Dirac energy, E_D. The angular modulation of the π-electrons energy of states has lower values along the directions pointing toward the six M-points with the energy level of 3.3 eV (i.e., the sixfold dome with the angular period of π/6 for the π-electron density modulation around the center of the carbon ring). The energy distribution of the π-electrons has its lowest zero-points at the position of each carbon atom at the K-point of the Brillouin zone (Fig. 3.6).

3.1.2 Size Effects in the Material Properties of Graphene Sheets

Material properties of graphene depend on the orientation of carbon rings in its atomic structure, morphology of edges of graphene sheets, and the size of graphene nanoribbons. Graphene nanoribbons (Fig. 3.1) and graphene platelets (Fig. 3.2) do not have the macroscopic properties of graphene or graphite unless their size becomes large enough and reaches the "homogenization length scale." The nanoscale homogenization of graphene properties depends on the minimum length or the homogenization length, L_h, at which the size-invariant value is reached, i.e., the minimum number of carbon rings along its length ($L_h/a \gg 1$), as has been noted for the carbon nanotubes (Harik, 2001a) and graphene (Harik, 2011, 2014). If the length of graphene sheet is such that $L < L_h$, then all material properties of this graphene are size dependent. When material properties of atomic lattices are size dependent, then they are ill defined and cannot be considered as uniquely defined.

Mathematically defined criteria for obtaining the uniquely defined material properties of nanostructured graphene are given by inequalities (3.1). These inequalities can be used for nanoscale homogenization of any physical property of graphene nanoribbons and platelets.

The homogenization length, L_h, of graphene sheets has to be considerably greater than the size of their smallest structural element, a, i.e., $L_h \gg a$ (Harik, 2011, 2014). A graphene sheet can satisfy homogenization criteria (3.1) for the uniquely defined material properties in longitudinal dimension (as in Fig. 3.1) and be transversely smaller than the minimum length: $L_{\min} = L_h \approx 10a$, for the armchair configuration. The minimum length, $L_{\min} = L_h$, for the zigzag configuration is such that $L_{\min} \approx 20 l_{C-C}$.

Nanoscale homogenization of material properties is also important for electrical conductivity of nanostructured materials, and the conduction band gaps in semiconductors such as graphene or some carbon nanotubes. It is known that graphene sheets are semiconducting and their energy band gap, E_g, as a semiconductor depends on their width

$$E_g = \frac{1.65}{w}, \tag{3.2}$$

where w is the width measured in nanometers of a graphene sheet and 1.65 eV is a scaling constant for the band gap. The size-invariant value, E_g^h, for the energy band gap, E_g, of the zigzag graphene sheets is 0.21 eV at the homogenization width: $w_h \approx 20a_{zz}$, where the size, a_{zz}, of a carbon ring in the zigzag configuration is 2.84 Å, since $a_{zz} = 2a/\sqrt{3}$ (Raza, 2011). The energy band width, E_{bw}, of the midgap state for the zigzag graphene sheets reaches the size-invariant value for the energy band width: $E_{bw}^h = 0.21$ eV at the homogenization width: $w_h \approx 20a_{zz}$ with the scaling constant for the energy band width, $E_{bw} = 0.077$ eV (Raza, 2011).

Nanoscale homogenization criteria (3.1) can be also used to examine the stability of formation of nanostructured materials such as nanorods. It is interesting that such a property also requires at least 10 smallest structural elements for this material property to reach its invariant size-independent value, i.e., the homogenized macroscopic value. Researchers have noted that the stability of nanostructured $(AlO)_n$ nanorods increases along with their size (Harik, 2014). The available data indicates that the stability of such nanorods reaches its highest value as soon as their nanoscale structure becomes large enough to include at least 10 nanoscale structural elements (NSEs). This dependence can be characterized by a homogenization parameter $X_{10}(E)$. In the case of $(AlO)_n$ nanorods, the value of a parameter $X_{10}(E) = 0.99$ indicates that a material property (i.e., formation energy) reaches the 99% of its homogenized macroscopic value for the size of nanorods consisting of 10 structural elements. This example illustrates the intrinsic nanoscale size dependence of material properties below the critical size of the 10 structural elements or the homogenization length.

Different materials with the NSEs of varying intrinsic properties can be characterized by a unique $X_{10}(M)$ parameter for each material property, M. Hence, all material properties of a nanostructured material can be characterized by a set of homogenization parameters:

$X_{10}(M_1)$, $X_{10}(M_2)$, $X_{10}(M_3)$, or by $X_9(M_1)$, $X_{12}(M_2)$, $X_{11}(M_3)$ and so on. Ideally, each nanostructured material should be characterized by a pair of homogenization parameters: $[X_{10}(M_1), X_{11}(M_1)]$, $[X_{10}(M_2), X_{12}(M_2)]$, or $[X_{10}(M_3), X_{14}(M_3)]$ and so on, where the value of the second homogenization parameter must always be a unity, while the number of structural elements may vary for different materials and distinct material properties.

Nanostructured materials, which require less than 10 NSEs to reach the macroscopic value of their material properties, exhibit a strong intrinsic tendency to possess those properties. If the number of the minimum structural elements, which is required to reach the macroscopic value of material properties, is greater than 10, this nanostructured material exhibits a weak intrinsic tendency to possess those properties. In the above example, $(AlO)_n$ nanorods exhibit an intrinsic tendency for the stability of their formation, characterized by the formation energy, which is close to normal, since $X_{10}(E) = 0.99$. When any nanostructured material has the size below the critical size associated with the macroscopic properties, its material properties are highly dependent on the dimensions of such a material.

3.2 Nanoscale Friction and Sliding on Graphene

Graphene sheets or nanoribbons (Fig. 3.1) and other atomic lattices can slide on graphene in nanoscale devices. At the nanoscale level, the fundamental origins of sliding friction involve surface-to-surface interactions, such as the interlocking in graphene platelets (Fig. 3.2) and other registry effects between the adjacent graphene sheets (Fig. 3.4), as well as the sliding-induced excitation of the atomic lattice vibrations, i.e., phonons, with the interaction of phonons propagating along the sliding surfaces (Harik, 2011). Electrostatic interactions and interactions of electrons including but not limited to the $\pi-\pi$ bonding interactions, especially those involving aromatic ring structures, are also a part of nanoscale interfacial interactions. Some of these interfacial friction mechanisms have been studied by a quartz-crystal microbalance, scanning force microscopy, atomic force microscopy experiments, and theoretical modeling, including molecular dynamics (MD) simulations (Harik, 2014). The coupling of two or more interfacial interaction mechanisms such as strong registry effects, asperity interlocking, and a nonstandard stick–slip behavior with generation of lattice phonons is particularly challenging.

3.2.1 Atomic Registry of Graphene Lattice

The typical ABA stacking of the commensurate graphene sheets (Fig. 3.8) is composed of the adjacent graphene lattices arranged so that a carbon atom is positioned right over the center of the opposite carbon ring (i.e., the C-atom/C-ring registry or the O-registry). The three projections of its joined C–C bonds lie within the perimeter of the adjacent carbon ring in the O-registry (Fig. 3.8). The projection traces of the positions of the carbon atoms

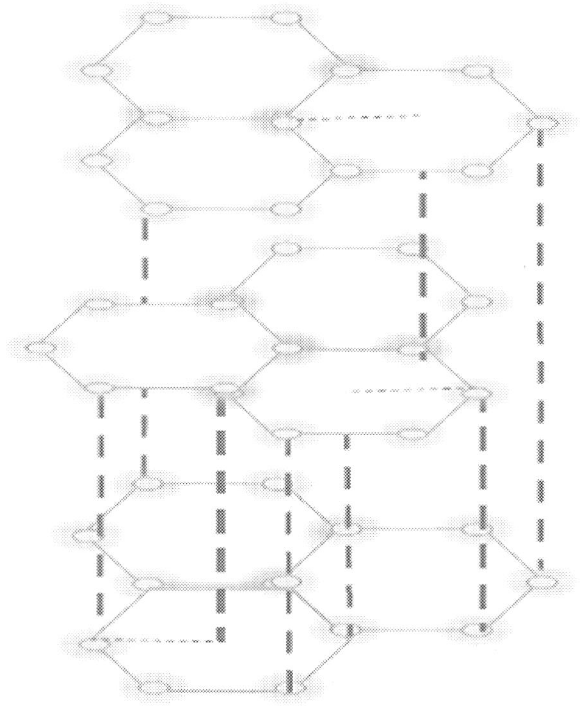

Figure 3.8
Schematic of the typical commensurate ABA stacking of the atomic lattices of the adjacent graphene sheets in graphene platelets with a top carbon atom and its three bonds in the O-registry with the opposite carbon ring as an asperity; hence, the O-registry. © *2011 by Nanodesigns Consulting (reproduced with permission).*

show that the three carbon atoms are positioned right above the carbon atoms of the opposite graphene sheet in the incommensurate positions, and the other three carbon atoms are positioned above the centers of carbon rings as asperities. The projections of the C–C bonds between the adjacent graphene lattices do not coincide, unlike in the case of the AAA stacking of the incommensurate graphene sheets (Fig. 3.2).

In the case of the commensurate graphene sheets, each graphene lattice in the ABA stacking is slightly corrugated, with the three carbon atoms positioned above the centers of the opposite carbon rings being closer to the opposite atomic lattice as asperities, and the three carbon atoms situated right above the carbon atoms of the opposite graphene sheet being a little bit further away from their counterparts. In case of the graphene sheets with the incommensurate stacking, all six carbon atoms are positioned right above their counterparts in the full ring–ring registry (or the O–O registry), with the opposite carbon rings being a little further from the opposite atomic lattice, as compared to the ABA stacking. The projections of the C–C bonds between the adjacent graphene lattices of the

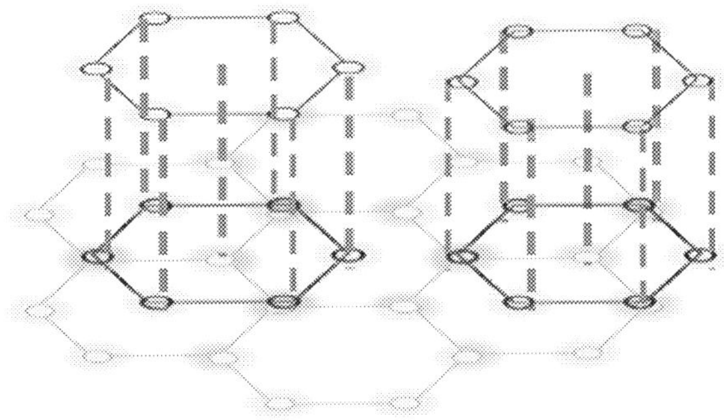

Figure 3.9
Details of the positions of carbon atoms in two representative carbon rings in the ABA stacking of the commensurate graphene sheets, where carbon atoms are positioned right above the center of a carbon ring and the shown projections of its joined C—C bonds lay within the perimeter of the opposite carbon ring in the O-registry. *© 2011 by Nanodesigns Consulting (reproduced with permission).*

AAA stacking coincide in the O—O registry, unlike in the case of the ABA stacking of the commensurate graphene sheets (Fig. 3.9).

Other registries between two graphene lattice sheets include

- the Φ- and the Θ-registry between a C—C bond and a carbon ring akin to the O-registry of an atom (Fig. 3.10);
- the Y-registry of a carbon atom and the opposite carbon atom with the three C—C bonds between the adjacent carbon rings in the O—O registry. The complete Y-registry is the symmetric rhombic < >-registry (Fig. 3.10C) of the ABA stacking (Fig. 3.9);
- the Ψ-registry of a carbon atom and the opposite C—C bond between the two adjacent carbon rings (Fig. 3.11A);
- the X- or the "+" registry of the two C—C bonds, i.e., the elastic C—C bond with the opposite C—C bond between the two adjacent carbon rings (Fig. 3.11C).

Note that the in-plane distance between the two carbon atoms and the nearest C—C bond in the Φ-registry (Fig. 3.10A) is slightly less than the Bohr radius,[2] $a_0 = 0.529$ Å. The typical van der Waals distance between two adjacent graphene sheets is 3.35 Å. The distance between two carbon atoms and the carbon atom on the two sides in the Θ-registry (Fig. 3.10B) is exactly one-half of the elastic C—C bond, $l_{C-C}/2$, i.e., 0.71 Å. The distance between these two atoms and the nearest C—C bond is exactly $a/2$ or 1.23 Å. This registry is quite stable. The O-registry and the companion complete Y-registry (Fig. 3.10C) are the most stable. The X-registry is unstable. The energy barrier for an electron to pass through

[2] The Bohr radius, $a_0 = 0.529$ Å, is the most probable distance between a proton and an electron in the hydrogen atom, according to Bohr's planetary model of an atom.

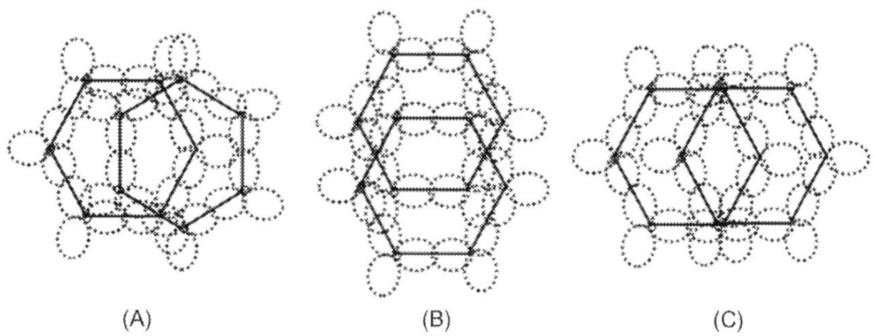

Figure 3.10
Schematics of the Φ-registry (A) and the Θ-registry (B) between the C—C bond of two carbon atoms and a carbon ring of six carbon atoms and the rhombic < >-registry (C), i.e., complete Y-registry and the incomplete companion Y-registry next to the Φ-registry (A) between a carbon atom and a carbon ring (i.e., the C-atom/C-ring registry). © *2011 by Nanodesigns Consulting (reproduced with permission).*

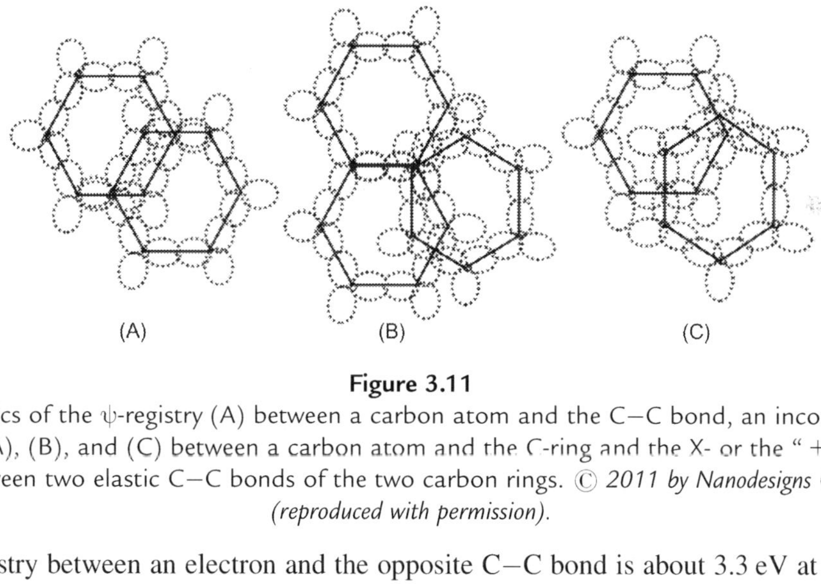

Figure 3.11
Schematics of the ψ-registry (A) between a carbon atom and the C—C bond, an incomplete Y-registry (A), (B), and (C) between a carbon atom and the C-ring and the X- or the " + " registry (C) between two elastic C—C bonds of the two carbon rings. © *2011 by Nanodesigns Consulting (reproduced with permission).*

the X-registry between an electron and the opposite C—C bond is about 3.3 eV at the M-point of the Brillouin zone for the energy of states in a carbon ring (see Fig. 3.5).[3] Energy barriers for different atomic registries are important for the analysis of surface—surface

[3] The onset of the Ψ-registry configuration (Fig. 3.11A) is associated with the need of electrons around a carbon atom to overcome the registry potential of the C—C bond and the associated Coulomb repulsion within the electron spatial exclusion (ESE) zone. Size of the ESE zone depends on the local atomic lattice configuration, the registry potential barriers, the nanoscale Coulomb repulsion proportional to $1/r^2$, and the nanoscale Pauli repulsion proportional to $1/r^{12}$. The combined effect results in the effect of spatial exclusion of electrons (SEE). The nanoscale Pauli repulsion for the electrons stems from the quantum Pauli principle for the identical particles with the spin 1/2 (fermions); the two identical particles cannot occupy the same energy state, as their combined wave function, ψ, is antisymmetric. The nanoscale Coulomb repulsion, the nanoscale Pauli repulsion, and the quantum Pauli principle for the electrons all affect precise dimensions of the ESE zone.

interactions between adjacent nanoscale surfaces. Transition between different atomic registries depends on local surface interactions within local energy barriers.

3.2.2 Atomic Registry Matrix Analysis

The atomic lattice structure of graphene sheets consists of carbon rings of six carbon atoms, which can be identified by six indices with the top left carbon atom being the first (Fig. 3.12), the top right carbon atom being the second in the top C–C bond, the third and the sixth carbon atoms situated below on the right and on the left, and the fourth and the fifth carbon atoms in the bottom C–C bond. The lattice registry of two adjacent carbon rings can be described as follows (Harik, 2011): When the two carbon rings with the corresponding atoms are directly on top of each other in the O–O registry of the AAA stacking, then the O–O lattice registry of their atoms can be described by the pairs of indices: (1,1), (2,2), (3,3), (4,4), (5,5), and (6,6). These pairs of indices can form a registry matrix for the O–O lattice registry with the presence of a carbon atom in a certain position of the lattice registry described by a pair of indices is indicated by a unity, 1. For the O–O lattice registry, the corresponding registry matrix R_{OO} is given by the matrix (3.3a), which indicates that two carbon atoms are present in the positions defined by the pairs of indices (1,1), (2,2), (3,3), (4,4), (5,5), and (6,6) in the same registry position on top of each other. The bond registry coefficient b for the C–C bonds is present in the registry positions defined by the pairs of indices: (1,2) and (2,1), (2,3) and (3,2), (3,4) and (4,3), (4,5) and (5,4), and (5,6) and (6,5).

All graphene lattice registries [e.g., the O–O registry, the C–O registry (Fig. 3.10C) and its companion complete Y-registry, and the Ψ-registry (Fig. 3.11A)] can be described by the corresponding atomic registry matrix, **R**, for two planar atomic lattices separated by the van der Waals spacing, h_{vdW}:

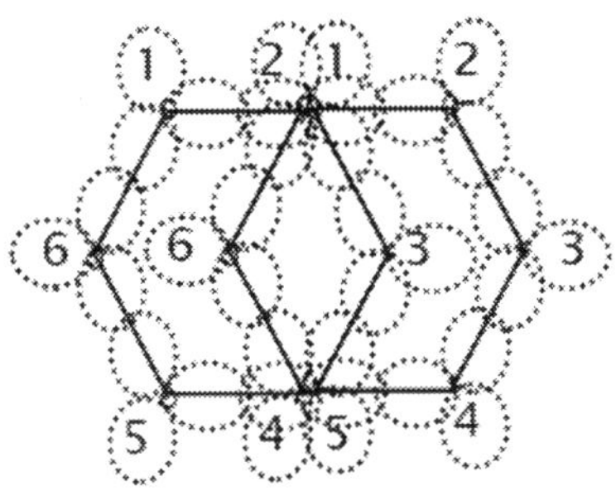

Figure 3.12
Schematic for the symmetric rhombic < >-registry described by the integer indices identifying carbon atoms in clockwise direction along the C–C bonds and the corresponding locations in a registry matrix (after Harik, 2011). © *2011 by Nanodesigns Consulting (reproduced with permission).*

$$\mathbf{R}_{\ddot{O}O} \equiv \begin{pmatrix} 2 & b & 0 & 0 & 0 & 0 \\ b & 2 & b & 0 & 0 & 0 \\ 0 & b & 2 & b & 0 & 0 \\ 0 & 0 & b & 2 & b & 0 \\ 0 & 0 & 0 & b & 2 & b \\ 0 & 0 & 0 & 0 & b & 2 \end{pmatrix}, \tag{3.3a}$$

for the O–O registry of the carbon rings in the adjacent atomic lattices in the AAA stacking;

$$\mathbf{R}_{C-O} \equiv \mathbf{R}_{Y-O} \equiv \begin{pmatrix} 1 & 0 & 0 & 0 & 0 & 0 \\ 1 & 1 & 0 & 0 & 0 & 0 \\ 0 & 0 & 1 & 0 & 0 & 0 \\ 0 & 0 & 0 & 1 & 1 & 0 \\ 0 & 0 & 0 & 0 & 1 & 0 \\ 1 & 0 & 0 & 0 & 0 & 1 \end{pmatrix}, \tag{3.3b}$$

for the C–O, Y-, or Y–O registry with a carbon atom over the center of the opposite carbon ring in the C–O registry; and

$$\mathbf{R}_{\Psi} \equiv \begin{pmatrix} 1+a_{\Psi}b_{\Psi} & b_{\Psi} & 0 & 0 & 0 & 0 \\ b_{\Psi} & 1 & 0 & 0 & 0 & 0 \\ 0 & 0 & 1+a_{\Psi}b_{\Psi} & 0 & 0 & 0 \\ 0 & 0 & 0 & 1 & b_{\Psi} & 0 \\ 0 & 0 & 0 & b_{\Psi} & 1+a_{\Psi}b_{\Psi} & 0 \\ 0 & 0 & 0 & 0 & 0 & 1 \end{pmatrix}, \tag{3.3c}$$

for the Ψ-registry, where b and b_{Ψ} are the bond overlap coefficients for the complete O–O registry and the Ψ-registry, for example; a_{Ψ} is the atom–bond overlap coefficient; the factors "1" and "2" correspond to the presence (probability density) of one or two carbon atoms.

The atomic registry matrix (3.3b) for the Y–O lattice registry of the atomic lattices of graphene sheets in the ABA stacking is formed as shown above. The lattice registry positions identified by the pairs of indices (1,1), (2,2), (3,3), (4,4), (5,5), and (6,6), and the corresponding terms in the registry matrix identified by the same pairs of indices describe the presence of one carbon atom by a unity, 1. In the Y–O lattice registry of the ABA stacking, the C–C bonds do not have overlapping registry. Therefore the registry matrix $\mathbf{R}_{YO}$ does not include the bond registry coefficient b. This registry matrix has a partial diagonal structure. A unity in the registry matrix position identified by the pair of indices (6,1) means that over the registry location of the sixth atom of a carbon ring in the bottom graphene sheets there is the first carbon atom from the top graphene sheet. Similarly, the pairs of indices (2,1) and (4,5) identify the pairs of carbon atoms in the lattice registry. The pairs of indices identify both the position in the registry matrix and the indices of carbon atoms from the two adjacent graphene sheets that are in that lattice registry (Harik, 2011). A sum of the registry matrix row elements gives the number of carbon atoms that are in the lattice registry positions. A single unity in a row means that there is no registry with that particular carbon atom, which is identified by a unity. The presence of the bond registry coefficient b in the registry matrix $\mathbf{R}_{OO}$ indicates that the C–C bonds have overlapping registry. The registry matrix $\mathbf{R}_{\psi}$ (3.3c) has the partial bond registry coefficient b_{υ}, and the atom–bond registry is described by the sum of unity and the atom–bond registry coefficient, a_{υ}. The registry matrix $\mathbf{R}_{\Theta}$ for the Θ-registry has the following form (Harik, 2011):

$$\mathbf{R}_{\Theta} \equiv \begin{pmatrix} 1 & 0 & 0 & 0 & 0 & b_{\Theta} \\ 0 & 1 & b_{\Theta} & 0 & 0 & 0 \\ 0 & b_{\Theta} & 1 & b_{\Theta} & 0 & 0 \\ 0 & 0 & b_{\Theta} & 1 & 0 & 0 \\ 0 & 0 & 0 & 0 & 1 & b_{\Theta} \\ b_{\Theta} & 0 & 0 & 0 & b_{\Theta} & 1 \end{pmatrix},$$

where the bond overlap coefficient b_{Θ} denotes the bond overlap registry between two carbon rings of the two adjacent graphene sheets.

The in-plane distance between all carbon atoms in the ψ-registry (Fig. 3.11A) along the C–C bonds is exactly one-half of the elastic C–C bond, $l_{C-C}/2$. The same is true for all atoms involved in the X- or the " + " registry between two C–C bonds (Fig. 3.11C). In the X-registry between two C–C bonds, two out of six carbon atoms are positioned in the incomplete Y-registry, i.e., the opposite carbon atoms of the two graphene sheets are located almost above each other in the incomplete incommensurate positions, as opposed to the O–O registry and the C–O registry, with the six and three carbon atoms in the incommensurate locations (i.e., the complete Y-registry). The full (six–six) and the partial incommensurate atom configurations of Y-registries are prone to sliding, as they are less stable than the three–three C–O registry and the companion complete Y-registry.

3.2.3 Matrix Analysis for the Interregistry Sliding on Graphene

Graphene sheets can easily slide off one another as in graphite pencils. Understanding of nanoscale details of the sliding process for individual graphene sheets and their parts can be achieved by examining the lattice registry of individual carbon rings and their sliding from one lattice registry to another. The interregistry sliding of a graphene sheet on the graphene surface between two lattice registries can be described by the interregistry matrices, which link one registry matrix with another matrix and its lattice registry through an operation of the matrix addition, e.g., $\mathbf{R}_{\Theta} + \Delta\mathbf{R}_{\Theta,OO} = \mathbf{R}_{OO}$, where the interregistry matrix $\Delta\mathbf{R}_{\Theta,OO}$ is given by

$$\Delta\mathbf{R}_{\Theta,\mathrm{OO}} \equiv \begin{pmatrix} 1 & b & 0 & 0 & 0 & -b_\Theta \\ b & 1 & b-b_\Theta & 0 & 0 & 0 \\ 0 & b-b_\Theta & 1 & b-b_\Theta & 0 & 0 \\ 0 & 0 & b-b_\Theta & 1 & b & 0 \\ 0 & 0 & 0 & b & 1 & b-b_\Theta \\ -b_\Theta & 0 & 0 & 0 & b-b_\Theta & 1 \end{pmatrix},$$

with the aforementioned bond registry coefficient b and the partial bond registry coefficient b_Θ (Harik, 2011). The O–O lattice registry of two carbon rings in the AAA stacking is easily perturbed into the Θ-registry through the interregistry sliding described by the interregistry matrix, $\Delta\mathbf{R}_{\Theta,\mathrm{OO}}$, and the interregistry sliding equation:

$$\mathbf{R}_{\mathrm{OO}} - \mathbf{R}_{\Theta,\mathrm{OO}} = \mathbf{R}_\Theta.$$

The Ψ-registry of two graphene sheets can be easily achieved from the O–O lattice registry through the interregistry sliding described by the interregistry matrix $\Delta\mathbf{R}_{\mathrm{OO},\Psi}$:

$$\Delta\mathbf{R}_{\mathrm{OO},\Psi} \equiv \begin{pmatrix} -1+a_\Psi b_\Psi & b_\Psi - b & 0 & 0 & 0 & b_\Psi \\ b_\Psi - b & -1 & -b & 0 & 0 & 0 \\ 0 & -b & -1+a_\Psi b_\Psi & -b & 0 & 0 \\ 0 & 0 & -b & -1 & b_\Psi - b & 0 \\ 0 & 0 & 0 & b_\Psi - b & -1+a_\Psi b_\Psi & -b \\ b_\Psi & 0 & 0 & 0 & -b & -1 \end{pmatrix}.$$

These interregistry sliding processes can be very useful in the design of the graphene-based electronic devices. Nanoscale interfacial interactions during deformation of nanostructured materials depend on local interregistry sliding, interfacial friction mechanisms, and nanoscale stiction. Dependence of nanoscale sliding on interregistry potential barriers and the applied forces can be examined by various nanoscale friction laws.

3.2.4 A Nanoscale Analog of Newton's Friction Law

The nanoscale sliding of two graphene sheets and surface cleaning from graphene flakes depends on the local lattice registry, the applied force, and other interfacial interactions mechanisms mentioned above. The frictional sliding between two graphene lattices can be modeled on the length-scale level larger than a few carbon rings in a particular registry configuration (Figs. 3.2 and 3.4). Multiple lengths of carbon rings have to be considered for the meaningful averaging of the graphene lattice sliding (Frankland and Harik, 2003). Dimensions of a graphene sheet, e.g., the length, L, should be sufficiently long, that is, $L/a > 10$, where a is the width of a carbon ring (Harik, 2011).

The relative sliding of two graphene sheets (Fig. 3.2) takes place from the energetically less favorable position to another with the applied forces that may allow a sliding sheet to overcome some registry-induced potential barriers. At the length-scale level larger than a carbon ring, a nanoscale friction can be analyzed as follows: the average shear stress, $\langle \tau_{xz} \rangle$, for the sliding interactions between two graphene sheets can be defined as $\langle \tau_{xz} \rangle \approx \langle f \rangle_{pull}/(A_{ss})$, where $\langle f \rangle_{pull}$ is the average force applied to the sliding graphene sheet, and A_{ss} is the common interfacial area during steady sliding between the two lattices separated by the van der Waals forces with the energy of interactions illustrated in Fig. 3.13.

The average strain rate, $\langle \dot{\gamma} \rangle$, of interfacial sliding can be estimated as $\langle \dot{\gamma} \rangle \approx \left(\partial \langle V_{sliding} \rangle / \partial z\right)$, where $\langle V_{sliding} \rangle$ is the average velocity of the graphene sheet in the axial x-direction and z is the distance between the two graphene sheets. Then, a nanoscale analog of Newton's law for the steady interfacial sliding can be written as (Harik, 2011):

$$\langle \tau_{xz} \rangle = \mu_{eff} \left(\frac{\partial \langle V_{sliding} \rangle}{\partial z} \right), \tag{3.4}$$

where μ_{eff} is an effective viscosity for the graphene–graphene interfacial sliding. As a result, an interfacial friction model for the entire pulling process can be developed for the total shear stress, $\langle \tau_{xz} \rangle_{pull}$, which, in addition to the sliding component $\langle \tau_{xz} \rangle$, includes an initial threshold pulling stress, τ_0:

$$\langle \tau_{xz} \rangle_{pull} = \tau_0 + \mu_{eff} \frac{\partial \langle V_{sliding} \rangle}{\partial z}. \tag{3.5}$$

After approximating the change in velocity across the interfacial separation, h_{vdW}, as the average sliding velocity, $\langle V_{sliding} \rangle$, Eq. (3.5) can be written as

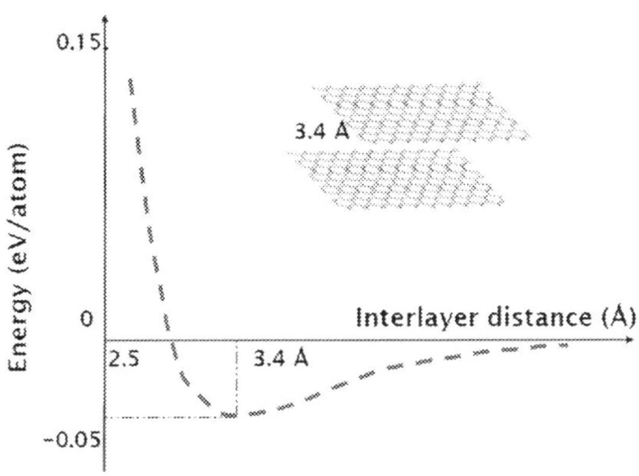

Figure 3.13
The energy of interfacial interactions associated with van der Waals forces between the adjacent graphene lattices (with the graphene-based equilibrium distance of 3.4 Å).

$$\langle \tau_{xz} \rangle_{\text{pull}} = \tau_0 + \mu_{\text{eff}} \frac{\langle V_{\text{sliding}} \rangle}{h_{\text{vdW}}}, \tag{3.6}$$

where the strain rate is $\left(\partial\langle V_{\text{sliding}}\rangle/\partial z\right) \approx \left(\langle V_{\text{sliding}}\rangle/h_{\text{vdW}}\right)$.

In order to complete the description of the interface model, the effective viscosity, μ_{eff}, should be described. The interfacial sliding velocity is, on average, linearly related to the average applied force, $\langle f \rangle$ (Harik, 2011, 2014), and the slope, χ_{eff}, can be determined. The resulting force–velocity dependence

$$\langle f \rangle = \chi_{\text{eff}} \langle V_{\text{sliding}} \rangle, \tag{3.7}$$

is a nanoscale analog of Newton's friction law.[4] The applied force, $\langle f \rangle$, in Eq. (3.7) is related to the shear stress in Eq. (3.4) by the force balance: $\langle f \rangle = \langle \tau_{xz} \rangle A_{\text{ss}}$. Therefore, the viscosity coefficient, χ_{eff}, can be related to the effective viscosity μ_{eff} as follows:

$$\mu_{\text{eff}} = \frac{1}{w} \frac{h_{\text{vdW}}}{L} \chi_{\text{eff}}. \tag{3.8}$$

The effective viscosity of the spatially distributed electrons, μ_{eff}, has the $1/w$ dependence, which is similar to some other property relations for graphene sheets. The viscosity of electrons is higher for higher densities of π-electrons and the associated greater values of the van der Waals separation, h_{vdW}. Then, the interfacial friction model for the entire pulling process is

$$\langle f \rangle_{\text{pull}} = f_0 + \chi_{\text{eff}} \langle V_{\text{sliding}} \rangle, \tag{3.9}$$

where f_0 is the critical force.

3.2.5 Intrinsic Viscosity of the Spatially Distributed Electrons

Evaluation of the effective interfacial viscosity of the graphene–graphene interface and other surface–surface interfaces are important for the understanding of nanoscale stiction and the dominant influence of the surface interactions at the nanoscale. The intrinsic sliding viscosity of the spatially distributed electrons can be estimated by using the MD simulations data for the force–velocity dependence at the low temperature of 10K and the above nanoscale analog of Newton's friction law (Frankland and Harik, 2003; Harik, 2011, 2014). According to this methodology, the effective viscosity coefficient χ_{eff} is estimated to be about 5 nN ps/Å, and the intrinsic sliding viscosity of the spatially distributed electrons, μ_e, is predicted (Harik, 2011) to be on the order of 1 cP (i.e., centipoise). It should be noted

[4] This fundamental research was partially supported by the Princeton-based NASA-funded URETI Institute (http://bimat.org) for the Bio-inspired Nanostructured Multifunctional Materials (award No. NCC-1-02037). Nanodesigns Consulting was formed in 2004 in order to assist URETI.

that the intrinsic viscosity of the spatially distributed electrons depends on the atomic structure of nanoscale interfaces.

3.3 Deformation of Graphene

At the nanoscale level, oscillations of the graphene lattice are inherent in most physical settings due to thermodynamic motion of carbon atoms. Such oscillations of graphene are associated with the constant thermal vibrations of the atomic lattice and stochastic appearance and scattering of small low-energy phonons [i.e., border value (BV) phonons]. The nature of these BV phonons of frequency, ω_i, and vector, k_i, is associated with the stochastic thermal vibrations of the atomic lattice, exchanges between different vibrational energies, and the random small-scale stochastic order (Harik, 2011). Waves in the atomic lattice of graphene, i.e., phonons, may occur in transverse and longitudinal directions through the deformation of C–C bonds out-of-plane of graphene and in-plane of graphene. However, graphene sheets may also deform as two-dimensional material with much larger deformation waves, having a short wave length and a large wave length (Fig. 3.14).

Deformation waves in graphene may have noticeable amplitude, especially, when the wave length of the deformation pattern is small (Fig. 3.14). The same amplitude may be not as pronounced, when the wave length is larger. It can be easily observed from the deformation pattern that, when the wave length, λ, of the deformation response is much larger than the size of a carbon ring, a, i.e., $\lambda \gg a$, then the lattice structure can behave as a continuum surface at the local scale comparable with the size of a single carbon ring. This is an interesting manifestation of the nanoscale homogenization criterion (3.1) in action during a

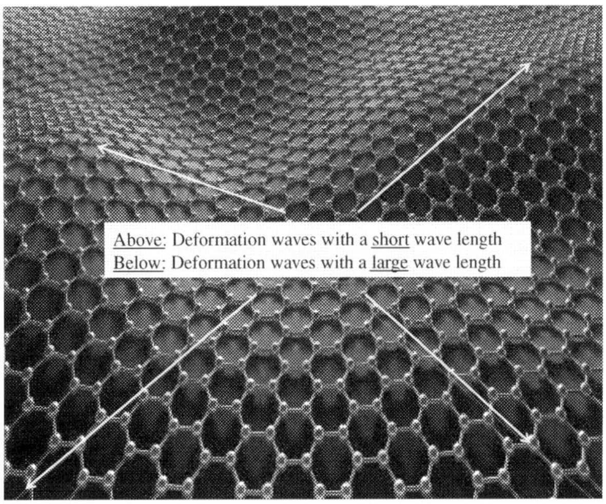

Figure 3.14
Deformation waves in graphene with a short wave length (top) and a long wave length (bottom) (after KTS Design).

deformation response. In other words, when the wave length, λ, of the deformation response satisfies the homogenization criterion (3.1) as noted above, the atomic lattice of graphene behaves as a homogenized material.

It is important to note that graphene lattice can approximate the deformation of continuum surfaces as long as there are enough carbon rings to approximate the waves along the deforming surface (Fig. 3.15). Here, the aforementioned homogenization criterion can be relaxed as far as the approximation of the deformation response takes place. However, one cannot claim that the local material properties of graphene lattice are the same as those of the large-size graphene. This is a fundamental difference between the nanoscale properties of the size-dependent discrete atomic lattices and continuum two-dimensional materials. Graphene sheets may have the well-defined deformation patterns such as circular waves with the well-defined valleys and ridges (Fig. 3.15). Such surface deformation patterns could be observed in nanoscale experiments and nanoscale manufacturing. However, when the size of nanostructures is reduced these deformation patterns start to break down, and only loosely approximate the continuum-like deformation response (Harik, 2014).

Thickness of graphene is the basic geometric parameter describing the shape and size of any graphene lattice in graphene nanoribbons and graphene sheets. This parameter is critical in any nanoscale design involving graphene sheets. The value of effective thickness of graphene is also important for the evaluation of its elastic modulus, which is a key material parameter (Huang et al., 2006). Therefore, any discrepancy in the estimates of graphene thickness affects the predicted value of the elastic modulus of graphene, which may result in potential errors and problems with nanoscale designs. The thickness of discrete atomic lattice of graphene can be evaluated by using different methods involving atomic potentials (Huang et al., 2006) or experiments. The effective thickness of graphene

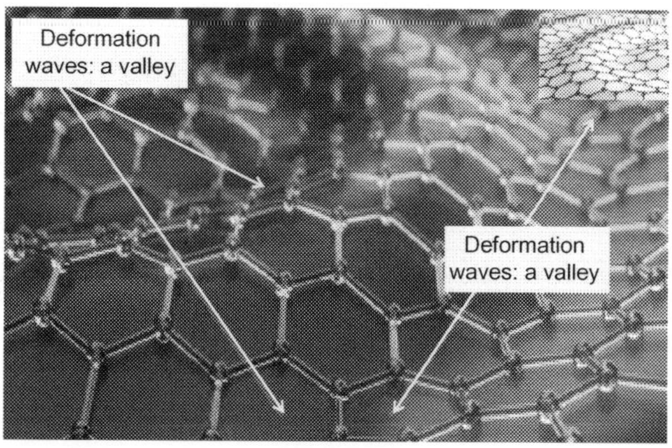

Figure 3.15
Circular deformation waves in graphene with a short wave length and the well-defined valleys and ridges.

can be the same as the effective thickness of graphene sheets in the graphite, that is, 3.4 Å, when it is subjected to transverse loading. However, when the loading is in-plane, then the effective thickness of graphene sheets can be considerably smaller (Huang et al., 2006).

3.4 Conclusion

In this chapter, the basic structure of graphene sheets has been examined, along with its effects on the interfacial sliding between the adjacent graphene layers. A nanoscale analog of Newton's friction law has been presented to describe the nanoscale sliding between the adjacent graphene sheets. The registry effects between adjacent graphene lattices have been analyzed by the atomic registry matrices, which are used to describe the atomic registry in graphene stacking, and to develop a new registry matrix analysis for the sliding of graphene sheets at the atomistic level. The new registry matrix analysis is very useful for the design of nanoscale devices.

It has been also shown that interfacial sliding of graphene sheets depends on the interfacial registry potentials and the previously explained "effect of the spatial exclusion of electrons" at the interface of two graphene sheets, which can be viewed as the nanoscale analog of Pauli's exclusion principle. Understanding of the nanoscale sliding phenomena between graphene sheets and other surfaces is critical for improving manufacturing technology for the graphene-based nanodevices.

Sample Problems

1. Describe all of the length scales in the atomic structure of graphene sheets.
2. Formulate the nanoscale homogenization criteria.
3. Prepare a viewgraph with the atomic structure of graphene.
4. Describe the main length scale characteristics in the distribution of energy of states for electrons over a carbon ring in graphene.
5. Describe the location of M-point, K-point, and Γ-point of the Brillouin zone for a carbon ring, as well as the ΓKK̍ area of a carbon ring.
6. Describe the O–O atomic registry between adjacent graphene lattices in the AAA stacking.
7. Prepare a viewgraph for the O–O atomic registry between adjacent graphene lattices in the AAA stacking.
8. Describe the atomic registry matrix for the O–O atomic registry between adjacent graphene lattices.
9. Describe the C–O (or Y–O) atomic registry between adjacent graphene lattices in the ABA stacking.

10. Prepare a viewgraph for the C—O (or Y—O) atomic registry between adjacent graphene lattices in the ABA stacking.
11. Describe the atomic registry matrix for the C—O (or Y—O) atomic registry between adjacent graphene lattices.
12. Describe the nanoscale analog of Newton's friction law.
13. Describe how the nanoscale homogenization criteria can be connected to the averaging of stresses at the nanoscale level.
14. Describe how the nanoscale homogenization criteria can be connected to the wave length of the deformation response of a graphene sheet.

References

Bostwick, A., Ohta, T., Seyller, T., Horn, K., Rotenberg, E., 2006. Nature 3, 36.

Frankland, S.J.V., Harik, V.M., 2003. Carbon nanotube pull-out from a polymer matrix. Surf. Sci. Lett. 525, L103.

Geim, A.K., Novoselov, K.S., 2007. The rise of graphene. Nature 6, 183.

Harik, V.M., 2001a. Ranges of applicability for the continuum beam model in the mechanics of carbon nanotubes and nanorods. Solid State Commun. 120 (7—8), 331—335.

Harik, V.M., 2001b. Ranges of applicability for the continuum beam model in the constitutive analysis of carbon nanotubes: nanotubes or nano-beams? NASA/CR-2001-211013. NASA Langley Research Center, Hampton, VA.

Harik, V.M., 2011. Mechanics of Carbon Nanotubes. Nanodesigns Press, Nanodesigns Consulting, Newark, DE.

Harik, V.M. (Ed.), 2014. Trends in Nanoscale Mechanics. Springer, Dordrecht.

Huang, Y., Wu, J., Hwang, K.C., 2006. Thickness of graphene and single-wall carbon nanotubes. Phys. Rev. B 74 (24), 24541.

Raza, H., 2011. Graphene Nanoelectronics: Metrology, Synthesis, Properties and Applications. Springer, Dordrecht.

Saito, R., Dresselhaus, G., Dresselhaus, M.S., 1998. Physical Properties of Carbon Nanotubes. Imperial College Press, London.

Further Reading

Xu, L., Ma, T.-B., Hu, Y.-Z., Wang, H., 2011. Nanotechnology 22, 285708.

CHAPTER 4

Classification of Carbon Nanotubes: 20 Classes of Atomic Structures

Chapter Outline

4.1 Atomic Structure of Carbon Nanotubes

The atomic structure of carbon nanotubes has been analyzed since the discovery of multiwall carbon nanotubes (MWCNTs) in 1991 by Iijima (Iijima, 1991). Their hollow, tubular structure was made of the multiple layers of carbon lattice. The MWCNTs or the "helical microtubules" discovered by Iijima belong to the class of the thick lattice shells, because of their large internal radius and a few layers of carbon atoms. The atomic structure of single-wall carbon nanotubes (SWCNTs) is formed by one layer of carbon

Mechanics of Carbon Nanotubes.
DOI: https://doi.org/10.1016/B978-0-12-811071-3.00004-4
© 2018 Elsevier Inc. All rights reserved.

atoms, or a "single-shell" (Iijima and Ichihashi, 1993). It means that the wall of SWCNTs is composed of a one-atom-thick layer of carbon lattice. Therefore, these hollow cylindrical nanostructures are likely to belong to the class of the thin lattice shells. However, since the discovered SWCNTs are of low diameters, they may not necessarily behave as shells or the thin shells. Therefore, rigorous criteria are needed in order to clearly define when an atomic lattice shell can be considered a thin shell. Moreover, variations in the atomic structure of carbon nanotubes produced by different methods show significant deviations in their geometry, which requires a more refined analysis of their structural properties. Such analysis is needed for the understanding of mechanical properties of carbon nanotubes, as well as their behavior in various nanotechnology applications. Rigorous criteria for the analysis of atomic lattices of carbon nanotubes have to be sensitive to their atomic structure and the length scales associated with the formation of their lattice configurations.

4.1.1 Length Scales in the Atomic Structure of Carbon Nanotubes

Atomic structure of carbon nanotubes (Fig. 4.1) involves several length scales associated with the structural characteristics of their atomic lattice. The key length-scale parameters that describe the atomic structure of lattices of carbon nanotubes and influence their global and local deformation include the length of the covalent σ-bonds, l_{C-C}, 1.41–1.44 Å, the size of a carbon ring, a = 2.46 Å, the radius, R_{NT}, of lattices (~0.2–2 nm, for SWCNTs, and ~35–50 to 80 nm for the MWCNTs) or their diameter, d_{NT}, and their length, L_{NT}, varying between a few nanometers and 100 s of microns (even a few centimeters or more,

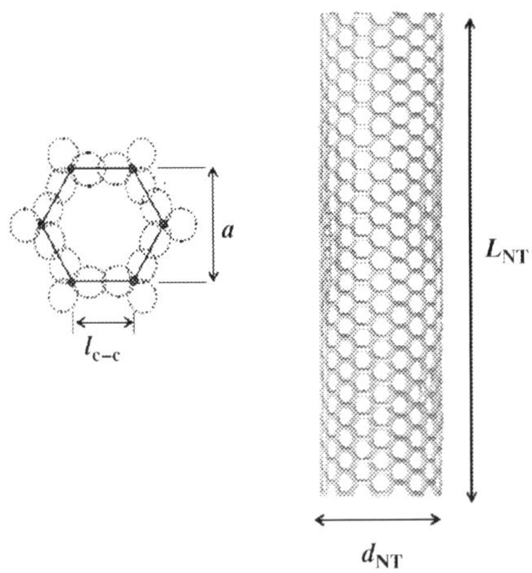

Figure 4.1

Schematics of the length scales in the atomic lattice structure of a carbon nanotube showing a carbon ring with the length of the C–C bond, l_{C-C} = 1.41–1.42 Å, and the size of 2.46 Å, and the lattice structure of an armchair (10,10) carbon nanotube of diameter, d_{NT} = 13.6 Å, and the length, L_{NT} = 52.5 Å, which represent four length scales.

Table 4.1). The approximate thickness, h_{NT}, of the atomic lattice of SWCNTs is also a key geometric parameter for describing the structure and geometry of SWCNTs. Thickness of MWCNTs depends on the number of layers or walls in their structure.

The length scales in the atomic structure of carbon nanotubes shown in Table 4.1 may vary for different geometries of SWCNTs and under different types of loading. For example, in SWCNTs, the ranges of values of the C–C bond length, l_{C-C}, are associated with the bond stretching in the atomic lattices of small radii. The carbon ring size, a, may vary under the tensile and compressive loading, and with lattice corrugation. The effective thickness, h_{NT}, of SWCNTs depends on the type of loading (Harik, 2011). The number of length scales characterizing the atomic structure of SWCNTs and the long list of geometric parameters that describe them: l_{C-C}, a, h_{NT}, R_{NT}, d_{NT}, and L_{NT}, pose a challenge for the quantitative analysis of structural properties of carbon nanotubes. Variations in the diameter, d_{NT}, and the length, L_{NT}, of carbon nanotubes result in a wide variety of atomic lattice structures with unique structural properties.

Another quantity that may describe the length scales is the number of carbon atoms involved in a crystalline lattice or an amorphous atomic cluster. Material properties of structured or unstructured agglomerates of atoms are highly sensitive to their size and the number of atoms involved. It should be noted: material properties of any nanomaterial, including carbon nanotubes consisting of a few dozens of atoms or so, are different from the macroscopic material properties of the same nanomaterial consisting of the same type of atoms but having much larger number of atoms.

This is a consequence of the nanoscale phenomenon associated with the size-dependence of the material properties of nanomaterials. This size-dependence is addressed by the nanoscale homogenization criteria (Harik, 2001a,b, 2002), which require that the length of any homogenized carbon nanotube, L_h, be much larger than the size of its smallest structural element, a, a carbon ring: $L_h \gg a$, or $L_h/a > 10$. These homogenization criteria, which were discussed earlier, are very important for the uniqueness of the material properties of all carbon nanotubes considered here in the classification. These criteria define

Table 4.1: Length scales in the structure of single-wall carbon nanotubes

Parameters	Typical Values	Ranges of Values[a]
C–C bond length, l_{C-C}	1.41–1.42 Å	1.41–1.62 Å
Carbon ring size, a	2.46 Å	~2.22–2.70 Å
Effective thickness, h_{NT}	0.66–3.4 Å	Load dependent
Diameter, d_{NT}	~1 nm	4–20 Å
Length, L_{NT}	~20–80 nm	AFM/NEMS
Length, L_{NT}	100 nm–1 μm–3 cm	Nanocomposites

AFM, atomic force microscope.
[a]Ranges of values may vary under tensile and compressive loading.

the size-invariance of the material properties in the framework of the existing length scales. Therefore, the homogenization criteria can be formulated as the criteria for the size-invariance of material properties of any carbon nanotube. That is, material properties of any nanomaterial, including carbon nanotubes, are size-invariant, or independent of the size, when their size, L_h, is much larger than the size of their smallest structural element, a, associated with one of the smaller length scale.

Therefore, when the size or the length of any carbon nanotube, L, is such that $L > L_h$, where L_h is the homogenization length, then this nanomaterial is completely homogenized, so any addition of any number of the same atoms does not change its material properties. The homogenization criteria also ensure that the material properties are well-defined, as opposed to being ill-defined. These criteria ensure that the value of material properties of carbon nanotubes is stable with respect to the small changes associated with the addition of any number of atoms to a sample. This is important for building a proper atomic model in the molecular dynamics (MD), and ab initio simulations of any nanostructure, and any carbon nanotube. As a result, the homogenization criteria provide rigorous criteria for ensuring that any sample of any nanomaterial, however small in size, represents the bulk properties of that nanomaterial. Consequently, the homogenization length, L_h, is the smallest size and length of any carbon nanotube, which is considered here.

4.1.2 Effect of Chirality of Carbon Nanotubes on Their Structural Properties

Carbon nanotubes may have different material properties depending on the orientation of hexagonal carbon rings in their atomic lattice (Fig. 4.2). The armchair SWCNTs are metallic and more flexible (Fig. 4.1), while the zigzag SWCNTs are semiconductive and

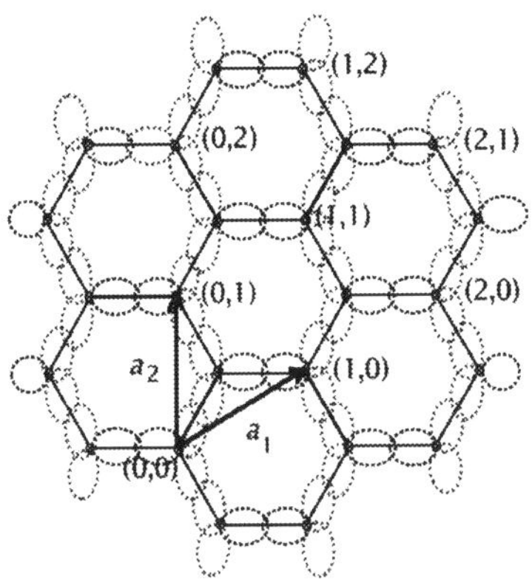

Figure 4.2
Schematics describing the chirality of carbon nanotubes by using a coordinate system with the unit vectors $\mathbf{a}_1$ and $\mathbf{a}_2$, as well as a pair of integers (n, m) (after Fujita et al., 1992). © *2011 by Nanodesigns Consulting (reproduced with a permission).*

more brittle (Dresselhaus et al., 1996). Consequently, the stiffness and strength of SWCNTs are also influenced by their chirality, i.e., the orientation of their atomic lattice structure. Chirality of carbon nanotubes can affect the ability of their atomic lattices to undergo shear deformation. The structure of zigzag SWCNTs is such that they are not only more brittle, and they are not susceptible to the shear strain.

Carbon nanotubes are called chiral, and hence, susceptible to a shearing deformation when their atomic lattice has some orientation other than zigzag. Chirality is associated with a skewed orientation of hexagonal carbon rings. Chirality of carbon nanotubes and, hence, their ability to undergo a shear deformation can be characterized by an angle specifying a particular orientation of hexagonal carbon rings in the atomic structure of SWCNTs. Fujita et al. (1992) have introduced a description of chirality in the atomic structure of carbon nanotubes by using a coordinate system with the unit vectors $\mathbf{a}_1$ and $\mathbf{a}_2$ (Fig. 4.2) and integers m and n identifying the components of the so-called chirality vector, $\mathbf{C}_h$. The chirality of carbon nanotubes is typically described by a pair of integer numbers: (n, m), e.g., a zigzag edge is denoted by the pair $(n, 0)$, while an armchair edge is denoted as (n, n). A study of shearing deformation in carbon nanotubes leads to the use of the Timoshenko beam model (Timoshenko and Gere, 1961), which can describe deformation of beams with internal shear.

4.1.3 Geometry of Carbon Nanotubes: Beams or Shells

Geometry of SWCNTs has a significant effect on their elastic and structural properties. Overall geometry of the atomic structure of SWCNTs can be characterized by two key geometric parameters, their diameter, and their length. It has been noted that changes in the diameter of carbon nanotubes result in noticeable changes in their radial flexibility (Ruoff et al., 1993). It is not surprising, since the diameter of SWCNTs determines one of the key length scales in their atomic structure. Carbon nanotubes with large diameters are flexible and deform radially as shells, while those with small diameters have higher radial stiffness. Curvature of carbon nanotubes, which is inversely proportional to their radius, increases the stiffness of carbon nanotubes and their elastic modulus. Analysis of radial dependence of elastic properties of carbon nanotubes has been carried out by many researchers (Harik, 2014). However, the fact that carbon nanotubes could undergo structural transformation into a separate class of nanocrystals was allusive until the scaling analysis of their atomic structure (Harik, 2001a, 2002).

The length of SWCNTs can also influence their mechanical properties and mechanical behavior (Harik, 2001a,b). The length of SWCNTs also determines one of the key length scales in their atomic structure. Long SWCNTs behave like beams for different values of their diameters (Harik, 2001a, 2002). Therefore long carbon nanotubes can be considered as nanoscale beams or "nanobeams" (Wong et al., 1997). In fact, in vibration experiments long

SWCNTs have been treated as nanoscale beams for the analysis of experimental data (Treacy et al., 1996; Wong et al., 1997). As a result, one could ask: Can carbon nanotubes be considered as the single-wall nanoscale shells or nanoscale beams? In fact, carbon nanotubes can behave as both nanoscale shells for large diameters, and as nanoscale beams, for large lengths, after their radial collapse. Therefore, large diameters of SWCNTs determine their shell-like behavior, and their large length controls their beam-like behavior. However, these are rather qualitative considerations, which are based, in part, on the experimental data and computational experiments. Quantitative criteria are needed for the analysis of structural properties of complex atomic lattices of carbon nanotubes for a wide range of geometric parameters and different length scales (Harik, 2001a, 2002).

4.1.4 Effective Thickness Paradox in the Analysis of Carbon Nanotubes

Thickness of the atomic lattice of carbon nanotubes is the basic geometric parameter describing their shape and size as lattice shells. This parameter is critical for the evaluation of Young's modulus of SWCNTs, which affects their mechanical behavior. The lattice thickness of SWCNTs also influences their structural behavior. Therefore, any discrepancy in the estimates of SWCNT thickness can affect the evaluation of structural behavior of SWCNTs. For example, thickness estimates can alter the evaluation of SWCNTs as either thin or thick shells, and thus change the predictions of their structural response. Theoretical modeling of carbon nanotube crystals has indicated that the thickness of SWCNT lattices seem to be the same as the thickness of graphene sheets in the graphite, i.e., 3.4 Å (Ruoff et al., 1993). MD modeling of Yakobson et al. (1996) and their continuum shell model have shown that the effective thickness of the (10,10) carbon nanotube is about 0.66 Å. The estimate of 0.66 Å for the effective SWCNT thickness is five times smaller than the earlier predicted thickness of nanotube lattices and the thickness of graphene sheets. This discrepancy in the evaluation of the thickness of carbon nanotubes constitutes the so-called thickness paradox for SWCNTs (Harik, 2011, 2014). It has not been noticed, however, that the higher thickness estimate corresponds to the transverse loading through van der Waals forces, while the lower estimate is associated with the axial loading. This load dependency of the thickness estimates is an important phenomenon associated not only with carbon nanotubes, but also with other atomic lattices (e.g., graphene).

In experiments with vibrating nanotubes, a hollow cylinder of thickness 3.4 Å was used for nanotube geometry, and a continuum beam model was utilized to predict Young's modulus of nanotubes as "nanobeams" (Wong et al., 1997). The vibration experiments demonstrate that carbon nanotubes vibrate similarly to the high aspect ratio beams. The estimate of the thickness of nanotubes affects the evaluation of the vibration frequency, which is a structural response. In the analysis of applicability of the continuum beam model for the modeling of SWCNTs, a new estimate for the effective thickness, i.e., 0.72 Å, was presented (Harik, 2001a, 2002) and adjusted later to 0.71 Å, or 1/2 l_{C-C} (Harik, 2011, 2014). This estimate for the thickness of SWCNTs has further exposed the existence of the thickness paradox with the thickness of 3.4 Å. These thickness estimates can also change the predictions for the vibration and the structural response of SWCNTs as either thin or thick shells. The thickness paradox has to be taken into account in the refined analysis of structural properties and of mechanical behavior of carbon nanotubes.

4.1.5 Scaling Analysis of the Atomic Structure of Carbon Nanotubes

Scaling analysis of the atomic lattice structure of carbon nanotubes involves the evaluation of relations between the key geometric parameters, which characterize the length scales of the structure (Harik, 2001a,b). Dimensional analysis allows one to reduce the number of key parameters by identifying one or two nondimensional parameters, which determine the global and local deformation response of SWCNTs (Harik, 2001a,b). The length scales characterizing the atomic structure of SWCNTs are associated with the geometric parameters that describe them, i.e., the length, l_{C-C}, of the C–C bond, the size of a carbon ring, a, the thickness, h_{NT}, of SWCNTs, their radius, R_{NT}, their diameter, d_{NT}, and their length, L_{NT}. Here, the radius and diameter of SWCNTs are not independent parameters, i.e., $d_{NT} = 2R_{NT}$. Similarly, the length, l_{C-C}, of the C–C bond, and the size of a carbon ring, a, are connected by the hexagonal geometry. Therefore, the number of nondimensional parameters based on independent geometric parameters will be lower.

The ratio of two adjacent length scales in the atomic structure of SWCNTs, which are associated with their transverse dimension, i.e., their diameter, d_{NT}, and their longitudinal dimension, i.e., their length, L_{NT}, provides the first nondimensional parameter, d_{NT}/L_{NT}. This nondimensional parameter is the aspect ratio of SWCNTs that governs the mechanical response of long SWCNTs. Variations in the diameter, d_{NT}, and the length, L_{NT}, of carbon nanotubes result in a wide variety of atomic lattice structures with unique structural properties. When the aspect ratio, d_{NT}/L_{NT}, of SWCNTs is unity, then SWCNTs start behaving as cylindrical shells, as opposed to their beam-like behavior for large and moderate lengths. For very large or infinite lengths, SWCNTs become flexible as threads or nanoscale fibers, which are prone to spontaneous Euler buckling, due to defects and environmental forces.

The ratio of two other adjacent length scales in the structure of SWCNTs, which are connected with their radial dimension, i.e., their radius, R_{NT}, and their thickness, h_{NT}, provides the second nondimensional parameter, h_{NT}/R_{NT}. This is the thickness-to-radius ratio that governs the radial deformation response of SWCNTs, as well as MWCNTs. When the thickness-to-radius ratio, h_{NT}/R_{NT}, is unity, then the thickness of MWCNTs is equal to their radius, and the internal radius vanishes with an increasing number of lattice walls. When the radius, R_{NT}, is very large, then SWCNTs become very flexible and start to deform under van der Waals forces, or even collapse as membranes.

The ratio of two adjacent length scales in the atomic lattice structure of SWCNTs, which characterizes their transverse dimension, i.e., their radius, R_{NT}, and the size of the smallest structural element, i.e., a carbon ring, a, yields the third nondimensional parameter, R_{NT}/a. When the normalized radius, R_{NT}/a, of SWCNTs reaches unity, then the curvature effects become noticeable in their impact on the elastic modulus of SWCNTs. For this radius, van der Waals forces inside of SWCNTs also contribute to the internal stiffening of SWCNTs. Furthermore, when the normalized radius, R_{NT}/a, falls below unity, then a structural transformation of SWCNTs takes place, which will be discussed later.

The aforementioned nondimensional parameters, which describe the relations between the adjacent length scales in the atomic structure of SWCNTs, can be used for the analysis of quantitative trends in the role of geometric parameters of SWCNTs on their structural properties and behavior. Moreover, these key nondimensional parameters can be used to examine how geometry of SWCNTs controls their structural properties as various changes in their geometric parameters take place. A more detailed scaling analysis can also provide rigorous criteria for the analysis of atomic lattices of carbon nanotubes, which will be sensitive to their atomic structure and the length scales associated with different lattice configurations.

4.2 Classification of Carbon Nanotubes into Different Classes

Carbon nanotubes can behave as cylindrical shells and as hollow beams. However, variations in the diameter of SWCNTs and their length result in a variety of cylindrical atomic lattices, with a wide range of structural properties and even distinct material properties. Scaling analysis of the atomic structure of carbon nanotubes provides the key nondimensional parameters, which can describe the relations between the geometry of SWCNTs and their structural properties. These nondimensional parameters can also demonstrate how the length scales in the atomic lattice of SWCNTs can influence their mechanical behavior, and even their material properties. The three independent nondimensional parameters, which have been derived by the scaling analysis of SWCNTs, can be also used to introduce a classification of SWCNTs and MWCNTs into different classes with distinct structural and material properties. A wide variety of atomic lattice structures forming SWCNTs and MWCNTs can be represented by 20 different classes of carbon nanotubes. Classification of carbon nanotubes is very important for nanoscale designs and nanotechnology applications, as well as for understanding of the basics of nanoscale mechanics.

4.2.1 Ten Classes of the Beam-Like Carbon Nanotubes

Carbon nanotubes with the elongated atomic lattice structures can behave as nanoscale beams, or even as long threads or nanoscale fibers, when their length is large enough. When their aspect ratio, d_{NT}/L_{NT}, is high (Fig. 4.3), carbon nanotubes behave as nanobeams, nanoscale beams, or the high aspect ratio cylindrical shells with either one or multiple shells. Their beam-like behavior persists for large and moderate lengths. For very large or infinite lengths, SWCNTs become flexible as long threads or nanoscale fibers, which are prone to spontaneous Euler buckling due to environmental forces, as was mentioned earlier. The aspect ratio, d_{NT}/L_{NT}, of SWCNTs and MWCNTs is the key nondimensional parameter that governs the mechanical behavior of long carbon nanotubes. The rigorous criteria that define the 10 classes of long carbon nanotubes are given by the following inequalities (Harik, 2001a,b, 2002):

$$\frac{d_{NT}}{L_{NT}} < \frac{1}{10} \quad \text{or} \quad \frac{d_{NT}}{L_{NT}} \ll 1, \tag{4.1}$$

where the later inequality means at least an order of magnitude difference. According to these criteria, any SWCNT or MWCNT with the atomic structure that satisfies the inequalities (4.1) behaves as beams during nanoscale buckling, bending, and deflection for small strains. The nanoscale beam criteria (4.1) also define the ranges of diameters of SWCNTs, their maximum value for the considered lengths, as well as the minimum length

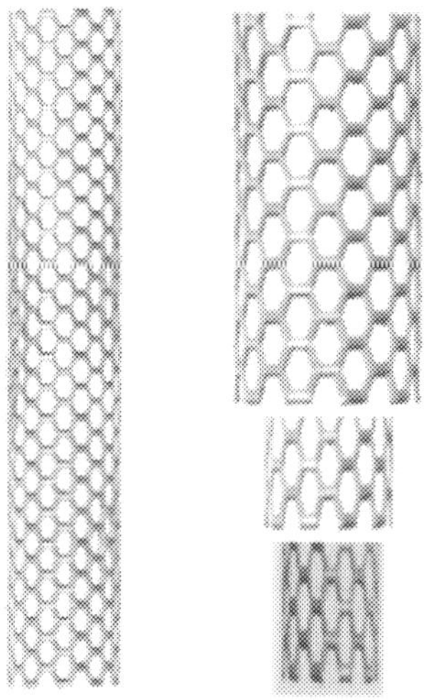

Figure 4.3
Atomic lattice structure of a thin lattice shell of the single-wall (10,10) carbon nanotube (left) with the length: L_{NT} = 52.5 Å, and the diameter: d_{NT} = 13.6 Å, as well as a schematics of the thick lattice structure of a small multiwall carbon nanotube (right). © *2011 by Nanodesigns Consulting (reproduced with a permission).*

of MWCNTs required for the given diameters. As a result of considerable variations in values of the dimensions associated with the atomic structure of SWCNTs and MWCNTs, one can define the following classes of nanoscale beams (Harik, 2011):

the high aspect ratio thin SWCNT shells (Fig. 4.3),
the high aspect ratio thick SWCNT shells,
the extra-long thin and thick SWCNT fibers,
the high aspect ratio thin MWCNT shells,
the high aspect ratio thick MWCNT shells (Fig. 4.3),
the extra-long thin and thick MWCNT fibers,
the high aspect ratio MWCNT beams, and
the extra-long MWCNT beams.

Notice that all 10 classes of nanoscale beams have the high aspect ratios, d_{NT}/L_{NT}, which satisfy the criteria (4.1) for the structure of carbon nanotubes. The extra-long SWCNT fibers and MWCNT fibers have very long lengths, which are greater than the Euler buckling length for the spontaneous buckling from environmental forces, i.e., $L_{NT} > L_E$. The normalized length of the high aspect ratio shells, $L_E = L_{NT}/d_{NT}$, which is associated with the onset of the Euler buckling, can be used to characterize not only the ambient forces acting on the lattice shell, but also the degree of defects in the suspended carbon nanotubes. This evaluation can be carried out by comparing the onset of the Euler buckling and the value, L_E, with that of the high quality lattice shells. The number of defective lattice shells, as well as the extent of lattice defects, can be estimated by direct experimental observation involving variations in the ambient forces, e.g., a degree of ultrasound sonication.

Evaluation of mechanical behavior of a particular class of long beam-like nanotubes can be done with a few high aspect ratio shells. According to the laws of similitude (Harik, 2011), the mechanical behavior of the entire class of the high aspect ratio thin SWCNT shells is similar. Furthermore, if their aspect ratio, d_{NT}/L_{NT}, is the same or close, then the values of their critical buckling strain, ε_{cr}, will be close. The same is true for all nanostructures belonging to each class of the high aspect ratio beam-like shells. This qualitative extrapolation of the analysis of one nanoscale shell to the entire class of nanostructures, based on the laws of similitude, greatly reduces the amount of analysis needed. Robustness of such qualitative extrapolation may depend on the type of analysis and characteristics of the typical atomic structure of a particular class of nanostructures.

The two classes of thin SWCNTs and thin MWCNTs are defined by the following criteria (Harik, 2001a,b, 2002):

$$\frac{h_{NT}}{R_{NT}} < \frac{1}{10} \quad \text{or} \quad \frac{h_{NT}}{R_{NT}} \ll 1, \tag{4.2}$$

where the estimates for the effective thickness, h_{NT}, of SWCNTs and the ranges of values for their radius, R_{NT}, have been discussed earlier. Since the thickness of MWCNTs may include several layers of graphene lattice, the values of their radius should be much higher. This is usually true in the case of the structure of MWCNTs.

The atomic structure of the thick SWCNTs and thick MWCNTs (Fig. 4.3) satisfy the following criterion for the thickness-to-radius ratio, h_{NT}/R_{NT} (Harik, 2001a,b, 2002):

$$\frac{h_{NT}}{R_{NT}} > \frac{1}{10}, \tag{4.3a}$$

where the thickness, h_{NT}, of MWCNTs may be considerable with the number of internal walls. Hence, the thickness-to-radius ratio may vary significantly. On the other hand, the radius of SWCNTs may be small; therefore the thick SWCNTs are those with the small values of their radius. It should be noticed that when the number of walls in the structure of MWCNTs is such that their thickness-to-radius ratio, h_{NT}/R_{NT}, satisfies the following criterion:

$$\frac{h_{NT}}{R_{NT}} > \frac{1}{3}, \tag{4.3b}$$

then these high aspect ratio MWCNTs are called MWCNT beams. The fraction of "1/3" in the inequality (4.3b) is approximate. However, the key requirement for the cross-section of MWCNT beams is that they are not prone to the thick shell buckling under the axial load. That is, their structure deforms as that of nanoscale beams regardless of their aspect ratios. The aforementioned extra-long fibers are being utilized for nanocomposite applications. The high aspect ratio SWCNTs and MWCNTs are being used for various nanoscale probes, including those for atomic force microscopes (AFM).

4.2.2 Two Classes of Thin Lattice Shells of Carbon Nanotubes

Carbon nanotubes may have large diameters of their atomic lattice shells, which behave as thin cylindrical shells in both cases of SWCNTs and MWCNTs. The atomic structure of SWCNTs is formed by one layer of carbon atoms (Fig. 4.1) or a "single-shell" of graphene, as noted by Iijima and Ichihashi (1993). It is not surprising that these hollow cylindrical nanostructures are likely to belong to the class of the thin lattice shells (Harik, 2001a,b, 2002). It has been noticed that the larger values of the diameter of carbon nanotubes result in noticeable increases in their radial flexibility (Ruoff et al., 1993). This is understandable, since the diameter of SWCNTs determines one of the key length scales in their atomic structure. Carbon nanotubes with large diameters are flexible and deform radially, as shells or even as membranes. Cylindrical membranes can deform and even collapse under the influence of van der Waals forces. Carbon nanotubes with large values of radius, R_{NT}, have small curvature, i.e., $1/R_{NT} \ll 1$, and atomic lattices with elastic properties similar to the graphene sheets.

The two classes of thin SWCNTs and thin MWCNTs are defined by the criteria (4.2) for the thin nanoscale shells (Harik, 2001a,b, 2002), i.e., their thickness-to-radius ratio, h_{NT}/R_{NT}, has to be small enough. When the radius, R_{NT}, is very large, then the normalized thickness, h_{NT}/R_{NT}, of SWCNTs is very small. As a result, the very thin lattice shells become very flexible and start to deform, even under van der Waals forces, or to collapse as membranes. Thickness of MWCNTs may include several layers of graphene lattice; therefore, these relatively thin shells are much stiffer than their counterparts with a single shell. However, under the axial loading they still respond as thin cylindrical shells, since their radius is much higher than that of thin SWCNTs. Thin MWCNT shells may have an axial buckling response with the some interfacial sliding between the adjacent layers.

The radial buckling of MWCNTs can be similar to that of SWCNTs, when their thickness-to-radius ratio, h_{NT}/R_{NT}, is identical. Moreover, according to the laws of similitude (Harik, 2011), the mechanical behavior of the entire class of thin SWCNT shells is similar. Furthermore, if their thickness-to-radius ratio, h_{NT}/R_{NT}, is the same or close then the values of their critical buckling strain, ε_{cr}, will be close. The chirality of such thin lattice shells has to be the same or almost identical. According to the same laws of similitude (Harik, 2011), the structural behavior of the entire class of thin MWCNT shells is similar. Moreover, the laws of similitude extend the qualitative applicability of the analysis results obtained for one atomic lattice structure to the entire class of nanostructures.

The role of chirality of the atomic lattice structures can be taken into account in the criteria (4.2) for the thin nanoscale shells (Harik, 2001a,b, 2002) by expressing the radius, R_{cr}, as:

$$R_{NT} = \frac{a}{2\pi}\sqrt{n^2 + m^2 + nm},$$

where the (n, m) pair defines the so-called chirality vector. The criteria (4.2) for the class of thin lattices shells can be written as follows (Harik, 2011, 2014):

$$h_{NT} \ll \frac{a}{2\pi}\sqrt{n^2 + m^2 + nm}, \tag{4.4a}$$

or

$$\frac{h_{NT}}{R_{NT}} = \frac{2\pi h_{NT}}{l_{C-C}\sqrt{3(n^2 + m^2 + nm)}} \ll 1. \tag{4.4b}$$

The advantage of this form is evident when the radius of the atomic lattices is hard to uniquely evaluate with the carbon rings being deformed in the (n, m) lattice. Carbon nanotubes and other cylindrical nanoscale shells, which satisfy the scaling criteria, (4.4a,b), deform as thin lattice shells. Therefore, the macroscopic scaling condition (4.2) for the class of the thin lattice shells is now defined by its lattice-based nanoscale analog (4.4a,b) without using the explicit value of the radius, R_{NT}.

Since the effective thickness, h_{NT}, of SWCNTs can be estimated as a half of the C–C bond length, 0.72 Å (Harik, 2001b, 2002), the presence of the carbon ring size, a, the effective thickness, h_{NT}, and the C–C bond length, l_{C-C}, in the new thin lattice-shell conditions (4.4a,b) can be avoided by presenting them in the following form, which is solely based on the (n, m) nomenclature for the atomic lattices:

$$\frac{\pi}{\sqrt{3(n^2 + m^2 + nm)}} \ll 1 \tag{4.5}$$

As a result, the macroscopic criteria (4.2) for the class of thin lattice shells is now presented in its lattice-based form (4.5) without using the explicit values of h_{NT} and R_{NT}, as in the ratio: h_{NT}/R_{NT}. An example of the thin lattice shell is presented in Fig. 4.1, where the armchair (10,10) SWCNT lattice is shown. This SWCNT has the radius: $R_{NT} = 6.8$ Å, which satisfies the thin lattice-shell conditions (4.2). It is so happens that the armchair (10,10) SWCNT lattice is one of the smallest armchair nanotube, which fulfills this condition for the class of thin lattice shells (Harik, 2001a, 2002).

Carbon nanotubes of larger radii: $R_{NT} > 6$ Å, or $R_{NT}/a > 2$, have the radius, which is larger than that of the SWCNT nanobeams examined below. Another difference is the greater number of the circumscribed carbon rings along their circumference (e.g., more than 12), which corresponds to the SWCNT lattice shells larger than the (12,0) zigzag carbon nanotube with the normalized circumference:

$$2\pi\frac{R_{NT}}{a} = \sqrt{n^2 + m^2 + nm},$$

e.g., the (15,0) zigzag lattice shell. Carbon nanotubes with many circumscribed carbon rings along their circumference behave as distinct cylindrical lattice shells. Their structural properties are quite different from those of the SWCNT nanobeams. For example, SWCNT lattice shells of large diameters, $d_{NT} > 12$ Å, have much smaller intratubular van der Waals forces and small curvature effects. The highly concentrated intratubular van der Waals forces and significant curvature effects in SWCNT nanobeams represent volumetric, as opposed to the surface effects.

4.2.3 Two Classes of Thick Lattice Shells of Carbon Nanotubes

Carbon nanotubes can behave as thick lattice shells, when their atomic structure is characterized by the relatively large thickness, h_{NT}, as compared to their radius, R_{NT}. The length scales of the atomic structure of such thick SWCNT shells and thick MWCNT shells have to satisfy the criterion (4.3a). It means that their thickness-to-radius ratio, h_{NT}/R_{NT}, is such that $h_{NT}/R_{NT} > 1/10$ (Harik, 2001a,b, 2002). Thickness, h_{NT}, of MWCNT shells may be considerable with the high number of internal walls. However, the number of walls in the structure of MWCNT thick shells is such that they still can deform as shells, because their thickness-to-radius ratio, h_{NT}/R_{NT}, does satisfy the criterion (4.3b), i.e., $h_{NT}/R_{NT} < 1/3$. Therefore, these MWCNTs are not MWCNT beams.

The class of thick SWCNT shells satisfies the criterion (4.3a) for the radius, R_{NT}, of carbon nanotubes and the effective thickness, h_{NT}, in the following form (Harik, 2011, 2014):

$$\frac{h_{NT}}{R_{NT}} = \frac{2\pi h_{NT}}{a\sqrt{n^2 + m^2 + nm}} > \frac{1}{10}, \tag{4.6}$$

which takes into account the (n, m) chirality of SWCNT lattice structure. Carbon nanotubes of small radii, $R_{NT} < 6$ Å or $R_{NT}/a < 2$, satisfy this condition for the class of the thick SWCNT lattice shells. These radii are close to the radii of SWCNT nanobeam*s* (Harik, 2001a, 2002), which have smaller number of the circumscribed carbon rings along their circumference and much greater curvature. The difference between the class of thick SWCNT shells and the class of SWCNT nanobeams is that these cylindrical SWCNT shells can still deform as thick shells, and their thickness-to-radius ratio, h_{NT}/R_{NT}, does satisfy the criterion (4.3b) for nanobeams, i.e., $h_{NT}/R_{NT} < 1/3$. Moreover, thick SWCNT shells have very small van der Waals forces inside their hollow structure as compared to the hollow SWCNT nanobeams.

4.2.4 A Class of Carbon Nanotubes with Small Radius: Nanobeams

Carbon nanotubes with small radius have an atomic lattice structure which has unique material and structural properties. The length scales associated with the atomic structure of small SWCNTs merge at the level of the two smallest length scales, i.e., the radius, R_{NT}, and, the size, a, of a carbon ring. It means that the two smallest length scales satisfy the following condition: $R_{NT}/a \approx 1$, for the normalized radius (Harik, 2001a,b, 2002). The length scales associated with the radius, R_{NT}, and the size, a, of the carbon ring are closely related to the effects of circumferential curvature, $1/R_{NT}$, and the ability of a carbon ring to fit along the idealized circumference. For these SWCNT lattices of small radii, i.e., below ~6 Å, the curvature effects are much more significant (Fig. 4.4). As a result, the effective thickness, h_{NT}, of these SWCNT lattices is larger than those carbon nanotubes (i.e., thick and thin SWCNT shells) that have larger values of the radius and better approximated circumferences.

In fact, these nanostructures are quite unique, since their atomic structure undergoes a degeneration of their cylindrical symmetry, when their radius is such that $R_{NT}/a \approx 1$. This means that the well-known cylindrical shape of SWCNTs is replaced by the less helical shapes of polygonal SWCNT nanobeams (Fig. 4.4). The number of carbon rings in their polygonal circumference is smaller, hence they cannot smoothly approximate cylindrical shells. In fact, the hexagonal SWCNT nanobeam has a unique symmetric cross-section (Fig. 4.4) with the carbon rings coinciding with the faces of the hexagon along the circumference. Therefore, degeneration of the cylindrical symmetry is associated with the emergence of polygonal symmetries. Moreover, the degeneration of the cylindrical symmetry in SWCNT nanobeams manifests itself in their circumferential buckling of the carbon rings (Figs. 4.5 and 4.6), due to their small number along the idealized circumference. The circumferential buckling of SWCNT nanobeams is a part of the curvature effects (Figs. 4.5 and 4.6).

Figure 4.4
Atomic lattice structure of the zigzag (6,0) SWCNT nanobeam with the carbon rings coinciding with the faces of the circumferential hexagon. *SWCNT*, single-wall carbon nanotube. © *2011 by Nanodesigns Consulting (reproduced with a permission).*

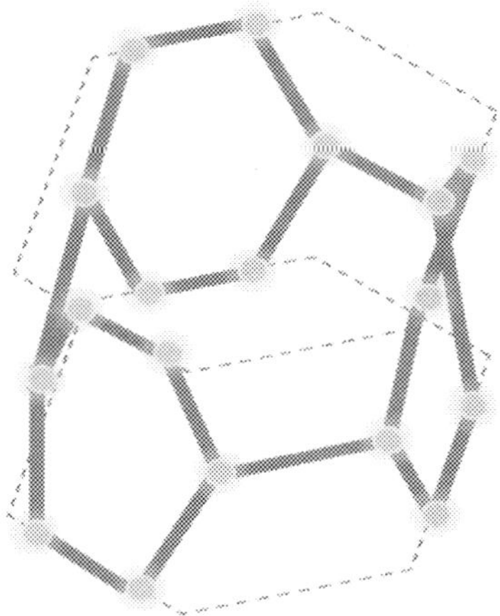

Figure 4.5
Atomic lattice structure of the (3,3) SWCNT nanobeam with the corrugated carbon rings due to the unique circumferential buckling with some of the C–C bonds positioned along the six hexagon faces. *SWCNT*, single-wall carbon nanotube. © *2011 by Nanodesigns Consulting (reproduced with a permission).*

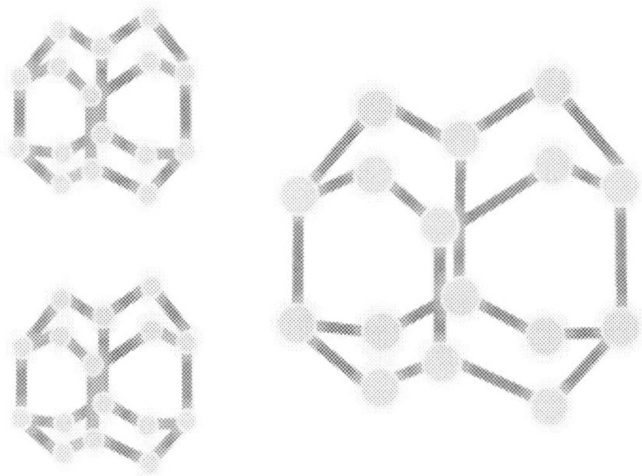

Figure 4.6
Atomic lattice structure of the (4,0) SWCNT nanobeam with the four buckled carbon rings spanning the faces of the size, $a/2$, of the circumferential octagon. *SWCNT*, single-wall carbon nanotube. © *2011 by Nanodesigns Consulting (reproduced with a permission).*

Another difference between the class of thick SWCNT nanobeams and the class of thick SWCNT shells is that these polygonal SWCNTs deform as thick hollow beams, and not as thick shells. When their aspect ratio is small or high, they buckle as beams under axial loads. These hollow nanobeams are, in fact, SWCNT nanocrystals (Harik, 2011, 2014). Moreover, the cross-section of SWCNT nanobeams can be corrugated (Fig. 4.5) and their thickness-to-radius ratio, h_{NT}/R_{NT}, can satisfy the criterion (4.3b) for nanobeams, i.e., $h_{NT}/R_{NT} > 1/3$. As a result, the hollow space inside of the thick SWCNT nanobeams is smaller (Fig. 4.6) than that of the thick SWCNT shells; consequently, van der Waals forces inside the hollow structure of nanobeams are noticeable as compared to these forces inside any of the hollow SWCNT shells.

The effective internal radius of the circumscribed inner circle in SWCNT nanobeams is smaller, and their effective lattice thickness, h_{NT}, is greater due to the curvature effects ($\sim 1/R_{NT}$) and the buckling of carbon rings. In summary, the unique lattice properties of SWCNT nanobeams are characterized by noticeable curvature effects, circumferential buckling of carbon rings (Figs. 4.5 and 4.6), the increased effective lattice thickness, and greater intratubular van der Waals forces which make the class of SWCNT nanobeams quite different from the other lattice shells. This difference is evident in their characterization as SWCNT nanobeam*s* with the unique crystalline lattice structure. It is not surprising that SWCNT nanobeams have material properties (e.g., Poisson ratio and Young's modulus) that are different from the material properties of other SWCNT lattice shells with larger values of their radius. The elastic C–C bonds of in nanobeams are stretched due to the curvature effects, the structural buckling of carbon rings, and the intratubular van der Waals forces.

Structural properties of SWCNT nanobeams are similar to those of the short MWCNT nanobeams with large and small radii. The short MWCNT nanobeams of small and large radii are such that their thickness-to-radius ratio, h_{NT}/R_{NT}, can easily satisfy the criterion (4.3b) for nanobeams, i.e., $h_{NT}/R_{NT} > 1/3$ (Fig. 4.3). Moreover, their multiwall thickness is such that they are not susceptible to the thick shell buckling under axial loads. When the radius of MWCNT nanobeams is small, their thickness may be such that $h_{NT}/R_{NT} \approx 1$, and their cross-section may be closer to that of nanoscale rods, as opposed to nanotubes. Therefore, both short MWCNT nanobeams of large and small radii form their own distinct classes of multiwall nanostructures: thick MWCNT nanobeams and thin MWCNT nanobeams.

4.2.5 A Class of Unique Carbon-Nanotube Nanocrystal Rods

Carbon nanotubes with very small radius have distinct atomic lattice structures, which have exceptionally unique material properties. In these atomic lattices, the length scales describing the geometric proportions between different parts of their atomic structure merge even further at the level of the smallest length scales. The radius, R_{NT}, of SWCNT nanobeams can become noticeably smaller than the size, a, of a carbon ring beyond the aforementioned level of $R_{NT}/a \approx 1$. It means that the two length scales, the radius, R_{NT}, and the length of the C–C bond, l_{C-C}, merge and satisfy the following condition: $R_{NT}/l_{C-C} \approx 1$, for the renormalized radius (Harik, 2001a,b, 2002, 2011). For these rather unique nanostructures, the atomic structure of SWCNTs undergoes the second degeneration of their hollow cylindrical structure, when their radius becomes comparable with the size of the C–C bonds.

The degeneration of the cylindrical symmetry means that the well-known hollowness and the cylindrical shape of SWCNTs are replaced by SWCNT nanocrystal rods with polygonal lattice (Fig. 4.7) and the cross-section made of the buckled carbon rings (Fig. 4.8). Their polygonal cross-section has a smaller circumference and the internal hollow space that merges into the regular interatomic lattice space (Fig. 4.7). Therefore, SWCNTs undergo the "topological transformation," from the hollow cylindrical SWCNT shells into the hollow polygonal SWCNT nanobeams and into SWCNT nanocrystal rods without the internal hollow space and cylindrical symmetry (Harik, 2001a,b, 2002, 2011). In fact, SWCNTs undergo through the breaking of the symmetry of $sp3$ bonds in their atomic structure (Stojkovic et al., 2001). The topology of the cross-sectional area of SWCNT nanocrystal rods is different from the topology of the well-known cylindrical SWCNTs, such as the (10,10) carbon nanotube. A topological transition in the atomic structure of a material means that the material properties change as well, into the unique class of SWCNT nanocrystal rods.

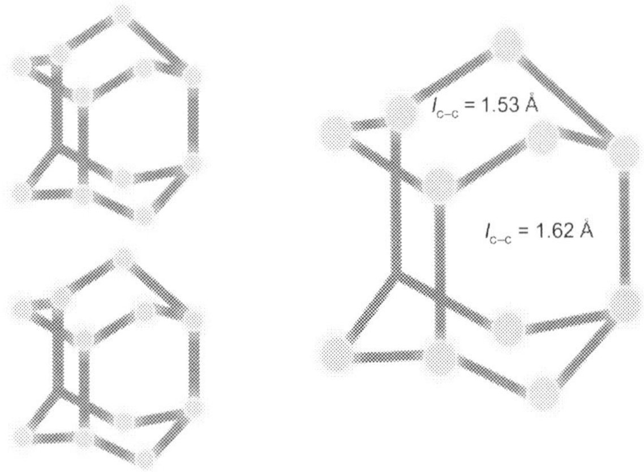

Figure 4.7
An illustration of the atomic lattice structure of the (3,0) SWCNT nanocrystal rod with the length of vertical elongated C—C bonds, l_{C-C} = 1.62 Å, and the adjacent stretched C—C bonds of 1.53 Å determined by Stojkovic et al. (2001). *SWCNT*, single-wall carbon nanotube. *© 2011 by Nanodesigns Consulting (reproduced with a permission).*

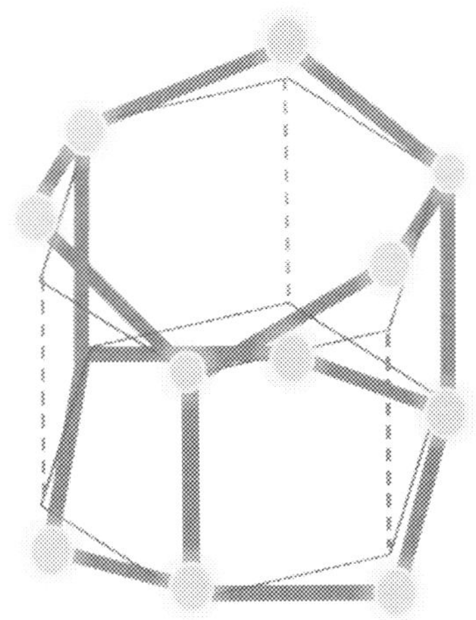

Figure 4.8
Atomic lattice structure of the zigzag (3,0) SWCNT nanocrystal rod with the halves of the buckled carbon rings coinciding with the faces of the circumferential hexagon. *SWCNT*, single-wall carbon nanotube. *© 2011 by Nanodesigns Consulting (reproduced with a permission).*

Another difference between the class of SWCNT nanocrystal rods and the class of thick SWCNT nanobeams is that the polygonal SWCNT nanobeams deform as thick hollow beams, while SWCNT nanocrystal rods deform as rods without the traditional hollowness of SWCNTs. The cross-section of SWCNT nanocrystal rods is such that their thickness-to-radius ratio, h_{NT}/R_{NT}, satisfies a criterion for nanobeams, i.e., $h_{NT}/R_{NT} > 1/3$. As a result, the hollow space inside of SWCNT nanocrystals is not only smaller than that of SWCNT nanobeams, but it vanishes into the normal interatomic space of crystalline lattice structures (Fig. 4.9). Consequently, van der Waals forces inside SWCNT nanocrystals reach their highest value, as compared to these forces inside any of the hollow SWCNT shells and SWCNT nanobeams. In SWCNT nanocrystal rods, the curvature effects become very significant with the pronounced carbon ring buckling (Fig. 4.8) and the noticeable degree of bond stretching (Fig. 4.7). Both of these structural characteristics contribute to the uniqueness of the material and structural properties of these smallest carbon nanotubes, i.e., SWCNT nanocrystals (Harik, 2011, 2014).

Atomic lattice structure of the armchair nanocrystal nanobeams (Fig. 4.5) and the armchair nanocrystal rods (Fig. 4.9) is quite different from the lattices of the aforementioned zigzag nanocrystal nanobeams (Figs. 4.4 and 4.6) and zigzag nanocrystal rods (Fig. 4.7) as well as the armchair nanotubes (Fig. 4.1). The armchair nanobeams (Fig. 4.5) and nanocrystal rods (Fig. 4.9) have corrugated lattice structure with the distinct structural buckling of the carbon rings. The atomic lattice of the (2,2) SWCNT nanocrystal rod is shown in Fig. 4.9 with its corrugation. The lattice structure of the armchair (3,3) SWCNT nanobeam (Fig. 4.5) has some similarity with the atomic lattice of the (2,2) SWCNT nanocrystal rod (Fig. 4.9), including some longitudinal corrugation (i.e., a form of the circumferential buckling of the

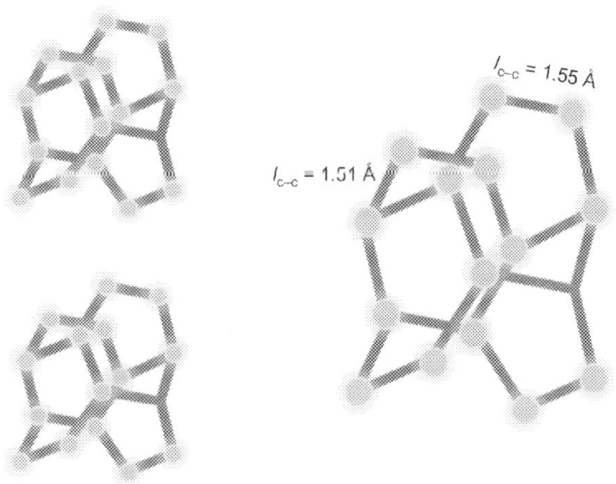

Figure 4.9
Atomic lattice structure of the (2,2) nanocrystal rods with the length of the C–C bonds, l_{C-C} = 1.55 Å, for the stretched horizontal bonds, and l_{C-C} = 1.51 Å, for the adjacent C–C bonds (an illustration of the lattice model of Stojkovic et al., 2001). © *2011 by Nanodesigns Consulting (reproduced with a permission).*

carbon rings). Circumferential buckling of carbon rings in the zigzag nanocrystal nanobeams (Figs. 4.4 and 4.6) and zigzag nanocrystal rods (Fig. 4.7) has polygonal symmetry along the circumference and no longitudinal corrugation. The circumferential buckling in SWCNT nanocrystals is a part of the curvature effects at the onset of the topological transition in their atomic structure.

The carbon rings do not fit well along the idealized circumference of the (3,0) nanocrystal (Fig. 4.7) and the (6,0) nanobeam (Fig. 4.4), since the curvature effects ($\sim 1/R_{NT}$) are greater with the smaller effective internal radius. Their effective lattice thickness, h_{NT}, is also greater, because the extra distance between the circumscribed chords of the atomic lattice and the idealized circumference. The effective lattice thickness for these SWCNT nanobeams and SWCNT nanocrystals also increases, because the length of their covalent bonds has been stretched, i.e., $h_{NT} \approx l_{C-C}/2$. The (3,0) and (6,0) lattices have unique atomic lattice structures, since their faces form a hexagon so that the hexagon face size coincides with the length of a half and the full carbon ring, i.e., $a/2$ and a. The atomic structure of (3,0) nanocrystal is formed with the circumferential buckling discussed above. The structure of nanocrystal rods is intrinsically connected with their material properties of atomistic beams. This becomes evident when one recalls that carbon nanotubes with the high aspect ratios, d_{NT}/L_{NT}, have distinct beam-like mechanical properties. Furthermore, when the atomic lattice of the (2,2) nanocrystals (Fig. 4.9) satisfies the homogenization criterion: $L_{NT} \gg a$, for the material property averaging, their length is such that the aspect ratio becomes high: $d_{NT}/L_{NT} < 1/10$. The same is true for the (3,0) nanocrystals (Fig. 4.7), the diameter of which equals the size of a carbon ring, $a = 2.46$ Å.

The size effects associated with the length, L_{NT}, of SWCNTs persist in all long nanocrystal rods, since their length can influence their global deformation through the value of their aspect ratio, d_{NT}/L_{NT}. In fact, the aspect ratio, d_{NT}/L_{NT}, becomes critical during the degeneration SWCNT nanocrystal rods into the thin molecules with the vanishing diameters. The length, L_{NT}, or the size of such SWCNT nanocrystal rods is also important for their homogenization length, L_{NT}/a, i.e., the minimum homogenization length, L_H, when their material properties become independent of their size or their length. The varying size of their cross-sectional area has also an effect on their atomic structure. The number of carbon rings in their small polygonal circumference is very limited; hence, they cannot even roughly approximate the symmetric polygonal circumference of nanobeams. Therefore, the degeneration of the cylindrical symmetry in SWCNT nanocrystal rods manifests itself in the corrugated circumferential buckling of the carbon rings. Degeneration of SWCNT nanocrystal rods into thin rod-like molecules is associated with a topological transformation of their cross-section into just one carbon ring, or one C–C bond.

4.2.6 A Summary of the Classification of Carbon Nanotubes

Carbon nanotubes may have different geometries that result in distinct mechanical behavior during various deformation processes. The seemingly continuous dependence of their elastic modulus on their radius has been investigated by many researchers. However, the unique characteristics of the atomic lattices of different groups of carbon nanotubes have not been carefully investigated and fully appreciated until recently. The first classification of carbon nanotubes (Table 4.2) was introduced, however, along with the discovery of the uniqueness of structural and material properties of the class of SWCNT nanobeams (Harik, 2001a,b, 2002). Around the same time, the uniqueness of the atomic lattice of the smallest nanotubes was independently underscored by the demonstration of the "breaking the symmetry of sp3 bonds in the tubular geometries" (Stojkovic et al., 2001). It is the uniqueness of the structural and material properties of different atomic lattices of distinct groups of carbon nanotubes, which requires their classification into appropriate classes of unique nanostructures.

Table 4.2: Classification of single-wall carbon nanotubes and classification criteria[a]

Classes of Nanotubes	Separation Criteria	Material Characteristics
Class Ia: High aspect ratio beams and the beam-like shells	$d_{NT}/L_{NT} \leq 1/10$ h_{NT}/R_{NT} varies	Some curvature, small internal van der Waals forces, beams, $l_{C-C} \geq 1.42$ Å
Class Ib: X-long nanotube fibers (thin and thick shells)	$L_{NT} > L_E$ (Euler buckling length for spontaneous buckling)	Buckling, some curvature, small internal van der Waals forces, fibers, $l_{C-C} \geq 1.42$ Å
Class IIa: Thin lattice shells	$h_{NT}/R_{NT} < 1/10$ $d_{NT}/L_{NT} \geq 1/10$	Low curvature, vanishing internal van der Waals forces, shells, $l_{C-C} \approx 1.41$ Å
Class IIb: Thick lattice shells	$h_{NT}/R_{NT} > 1/10$ $d_{NT}/L_{NT} \geq 1/10$	Some curvature, small internal van der Waals forces, shells, $l_{C-C} \geq 1.42$ Å
Class IIIa: Nanobeams	$R_{NT}/a \leq 1$, $h_{NT}/R_{NT} > 1/10$	High curvature effects, large internal van der Waals forces, distorted covalent bonds, $l_{C-C} >$ 1.42 Å, nanobeams or nanocrystals
Class IIIb: Nanocrystals	$R_{NT}/l_{C-C} \approx 1$, $h_{NT}/R_{NT} > 1/10$	

Harik V. M., *Mechanics of Carbon Nanotubes and Graphene*, Lecture notes: a short course, Nanodesigns Consulting, Nanodesigns Press, Wilmington, Delaware, 2015. Classification of carbon nanotubes is based on the methodology originated by the author at NASA Langley Research Center (Hampton, Virginia) in order to characterize the unique behavior of different carbon nanotubes: the thin and thick lattice shells and the long high aspect ratio nanotubes (Harik, 2001a,b, 2002). The discovery of the existence of carbon nanotube nanocrystals came as a result of a thorough scaling analysis of the atomic lattice structures, which pointed out their unique material properties that differ from the material characteristics of other classes of carbon nanotubes (Harik, 2001a,b, 2002). Crespi's group (Stojkovic et al., 2001) has independently demonstrated the "breakdown of the sp^3 symmetry" of the covalent bonds in tubular geometries of small carbon nanotubes. The scaling analysis of the atomic lattices of carbon nanotubes has demonstrated that *the breakdown of the cylindrical shell symmetry for carbon nanotubes* occurs at the atomic scale limit for the applicability of continuum shell models, when the normalized radius satisfies the following condition: $R_{NT}/a \approx 1$ (Harik, 2001a,b, 2002).

[a]© 2015 by Nanodesigns Consulting (reproduced with a permission).

Scaling analysis of the atomic lattice structure of carbon nanotubes has described the unique relations between different length scales in their nanostructure. Carbon nanotubes can behave as the high aspect ratio shells or beams (Table 4.2), as well as thin or thick cylindrical shells, in the case of SWCNTs and MWCNTs. A collapse of the adjacent length scales associated with the SWCNT radius, R_{NT}, and the size, a, of the carbon ring, which is the smallest structural element in the SWCNT lattice, results into the formation of a distinct class of the atomic lattices, i.e., the class of SWCNT nanobeams with the lattice radius, $R_{NT} \approx a = 2.46$ Å. The further collapse of the adjacent length scales associated with the SWCNT radius, R_{NT}, and the length, l_{C-C}, of the C–C bond, which is the smallest element in the lattice, results into the formation of a distinct class of the atomic lattices of SWCNTs, i.e., the SWCNT nanocrystal rods.

A summary of the separation criteria derived by the scaling analysis is presented in Table 4.2, along with unique characteristics of material properties associated with different classes of carbon nanotubes. The updated 2001 classification of carbon nanotubes includes six classes of atomic lattices of single SWCNTs (Harik, 2001a,b, 2002):

- the class of the high aspect ratio nanotubes (Class Ia), which behave similarly to the high aspect ratio hollow beams with the thin or thick lattice shells,
- the class of the extra-long high aspect ratio SWCNT fibers (Class Ib), which are prone to the spontaneous Euler buckling and flexible as threads,
- the class of thin lattice shells (Class IIa), which have the mechanical properties similar to the cylindrical thin shells of hyperelastic materials,
- the class of thick lattice shells (Class IIb), which behave similarly to the thick shells,
- the class of SWCNT nanobeams (Class IIIa), which have polygonal symmetry with the material properties distinct from the carbon nanotubes with cylindrical shapes,
- the class of thin SWCNT nanocrystals (Class IIIb), which have polygonal shapes and unique material properties distinct from the other classes of carbon nanotubes.

The two classes of the high aspect ratio carbon nanotubes can be expanded in order to include both SWCNTs and MWCNTs. Similarly, the two classes of the thin and thick lattice shells can be expanded in order to describe the atomic lattices of MWCNTs as well. The class of nanobeams can also include MWCNT nanobeams of small and large radii.

The modified classification of carbon nanotubes includes 20 classes of nanoscale beams and shells (Fig. 4.10), i.e., the six classes of the long and extra-long beams, thin and thick high aspect ratio MWCNT shells (i.e., Classes I_m, I_{mx}, I_{ms}, I_{msx}, I_{mb}, and I_{mbx}), where the subscript "m" stands for the multiwall nanostructure, the subscript "x" identifies the extra-long nanostructure, the subscript "s" indicates for the thin shell nanostructure, and the subscript "b" stands for the bulky thick wall, beam-like nanostructure. The combination of these subscripts identifies various classes of MWCNT shells and beams. The four classes of the long and extra-long thin and thick high aspect ratio SWCNT shells are represented by

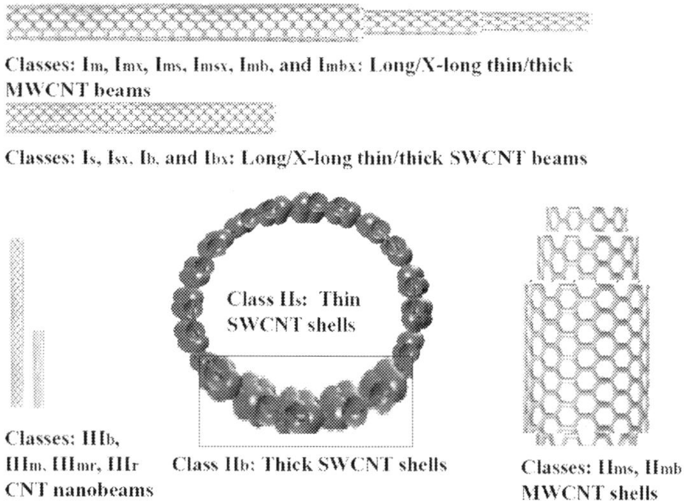

Figure 4.10

The modified classification of carbon nanotubes into 20 classes: the six classes of *the long and extra-long beams, thin and thick high aspect ratio MWCNT shells* (Classes: I_m, I_{mx}, I_{ms}, I_{msx}, I_{mb}, and I_{mbx}), four classes of *the long and extra-long thin and thick high aspect ratio SWCNT shells* (Classes: I_s, I_{sx}, I_{sb}, and I_{sbx}), *the thin and thick SWCNT shells* (Classes: II_s and II_b), *the thick and thin MWCNT shells* (Classes: II_{ms} and II_{mb}), and the six classes of *SWCNT nanobeams, short MWCNT nanobeams of large and small radii, and SWCNT nanocrystal rods*, which may have small radii and moderate or high aspect ratios (Classes: III_b, III_{bx}, III_{mb}, III_{mr}, III_r, and III_{rx}). *MWCNT*, multiwall carbon nanotubes; *SWCNT*, single-wall carbon nanotube.

the beam-like classes of shells: I_s, I_{sx}, I_{sb}, and I_{sbx}, where the meaning of the subscripts is the same as described above. The two classes of the thin and thick SWCNT shells include the Classes: II_s and II_b. The two classes of the thick and thin MWCNT shells are identified as the Classes: II_{ms} and II_{mb}. The six classes of nanobeams are represented by the classes of SWCNT nanobeams, short MWCNT nanobeams, and SWCNT nanocrystal rods, which have small radii and moderate or high aspect ratios (i.e., Classes: III_b, III_{bx}, III_{mb}, III_r, and III_{rx}). The subscript "b" identifies the beam-like nanostructure, while the subscript "r" identifies the rod-like nanostructure (Fig. 4.10).

The classification of carbon nanotubes is an important part of mechanics of carbon nanotubes and its applications. Classification of atomic lattices of SWCNTs and MWCNTs has important implications for the understanding of their mechanical behavior during deformation. Distinct classes of carbon nanotubes behave differently under the same loading conditions. Selection of carbon nanotubes for specific nanoscale designs, nanodevices, and nanoscale probes depends on the understanding of their structural and material properties. Criteria for the purification of carbon nanotubes into different groups of geometries can be developed on the basis of the classification of carbon nanotubes.

4.3 Material Maps for Carbon Nanotubes

Carbon nanotubes may have different geometries of their atomic lattice structures, which have distinct material properties. Geometric parameters characterizing the atomic structure of carbon nanotubes and the length scales of their lattices can be used to construct material maps for different classes of carbon nanotubes. The ranges of values for the key geometric parameters can be examined in the framework of criteria for different classes of carbon nanotubes. The boundaries of the material map can be identified by using the nanoscale homogenization criteria for the appropriate geometric parameters. The detailed analysis of the ratios of geometric parameters associated with the key nondimensional parameters can be utilized to refine the local boundaries for different classes of carbon nanotubes in the material maps.

4.3.1 A Material Map for the Beam-Like Classes of Carbon Nanotubes

The material properties of SWCNTs and MWCNTs become unique and independent of their size or the number of atomic layers, when their atomic structure satisfies the nanoscale criteria for the unique averaging (or *homogenization*) of the material properties over a volume of atomic lattice, or a surface of a lattice shell, i.e., $L_{NT} \gg a$, or $L_{NT}/a \gg 1$. Here, the length, L_{NT}, should be considerably greater than the size of a carbon ring, $a = 2.46$ Å, which is the smallest structural element in the lattice. The inequality in the homogenization criteria means roughly an order of magnitude difference, i.e., $L/a \approx 10$. Therefore, this relation defines the lower boundary in the material map for SWCNT nanobeams and shells (Fig. 4.11). A similar boundary can be defined for MWCNT beams and shells. The lower boundary of the material map is associated with the lower values of the radius of carbon nanotubes, which are represented by the atomic structures of the classes of SWCNT nanobeams and nanocrystal rods. Classification of carbon nanotubes (Fig. 4.10) identifies four appropriate classes of nanobeams and nanocrystal rods, with the parameters represented in the material map (Fig. 4.11). The atomic lattices of nanobeams have small radii and do not have the atomic structure of cylindrical shells. The class of thin nanocrystal rods has the atomic structure with the highly distorted bonds and considerable internal van der Waals forces. These nanocrystal rods are prone to cleavage (Stojkovic et al., 2001).

A material map (Fig. 4.11) also describes the class of the high aspect ratio carbon nanotubes, which may include up to 10 subclasses of atomic lattices (i.e., the six subclasses of MWCNT beams and shells, as well as the four subclasses of SWCNT shells, as shown in Fig. 4.10). The key length scale parameters that describe the dimensions of atomic lattices have to be reflected in the material map. For example, the radius, R_{NT}, of lattices: $\sim 0.2-2$ nm, for SWCNTs, and ~ 35 nm, for MWCNTs, and their length, L_{NT}, varies between few nanometers and 100 s of microns, or even a few centimeters. Now, the length of carbon nanotubes and their bundles may reach meters. Note that the material properties

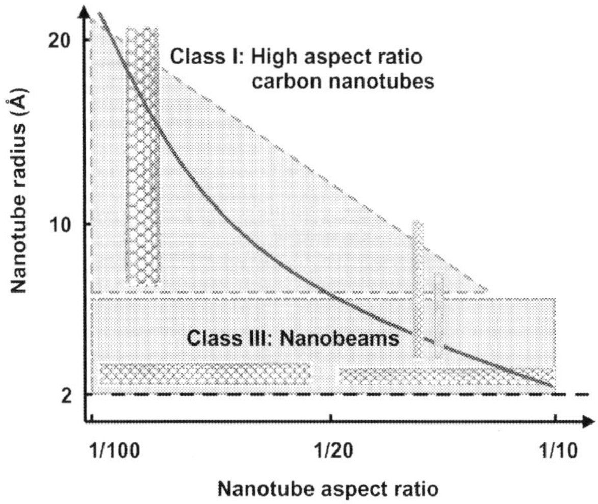

Figure 4.11
A material map for the two classes of the high aspect ratio carbon nanotubes: *long and extra-long thin and thick shells* (Class I with up to 10 subclasses) and the class of nanobeams and nanocrystals *of* small radii, i.e., Class II with up to five subclasses. The thick line [$\sim 1/(10d_{NT}/L_{NT})$] represents a boundary for the optimal material structure of the high aspect ratio lattices. © *2011 by Nanodesigns Consulting (reproduced with a permission).*

of thin MWCNT nanobeams will not be uniquely defined in the transverse direction, because the addition of an extra layer of atoms in the transverse radial directions would continue to slightly change the average or the bulk properties of MWCNT *nanobeams* until its transverse size satisfies the inequality: $d_{NT} \gg a$, or a similar inequality: $d_{NT} \gg l_{C-C}$, where l_{C-C} is the length of the covalent bond.

It should be noted that the material map (Fig. 4.11) can be used to examine the material parameters associated with the formation of unique classes of carbon nanotubes in the range of parameters along the lower boundary of the map. For example, the unique atomic structures of SWCNT nanobeams and nanocrystal rods are formed when the radius, R_{NT}, becomes comparable with the size, a, of the carbon ring, which is the smallest structural element in the lattice. It has been shown earlier that such a collapse of the adjacent atomic length scales results into the formation of a distinct class of the atomic lattices, the nanobeams with the lattice radius, $R_{NT} \approx a = 2.46$ Å. This is the value that defines the neighborhood of the lower line in the material map (Fig. 4.11). The class of carbon nanotube nanocrystals is defined by this limit in their atomic structure. For example, the (3,0) nanocrystal (Fig. 4.7) is just one example of carbon nanotube nanocrystals. The (6,0) nanobeam (Fig. 4.4) represents the atomic structure with the radius of about one carbon ring.

Ranges of applicability of the nanoscale analogs of the Euler formula for the axial buckling of the high aspect ratio carbon nanotubes are defined by the carbon ring size, a, on one side of the model applicability map (Fig. 4.11) and by the values of the lattice aspect ratio, i.e., the so-called shell-beam transition border line $\sim 1/(10d_{NT}/L_{NT})$, beyond which geometric parameters define the class of thin SWCNT shells (Harik, 2001a,b, 2002, 2011, 2014). The key nondimensional parameters that govern the materials map for the beam-like carbon nanotubes (i.e., the class of the high aspect ratio lattices and the class of nanobeams), include the aspect ratio, d_{NT}/L_{NT}, the homogenization ratio: L_{NT}/a, and the normalized radius, R_{NT}/a (Harik, 2001a,b, 2002). Examples of the unique atomic lattices, for which the beam models can be used, can be summarized by the following list of SWCNT nanobeams and nanocrystal rods:

- the (2,2) nanocrystal rods,
- the (4,0) nanobeams,
- the (3,3) nanobeams,
- the (6,0) nanobeams,
- the (3,0) nanocrystal rods.

The atomic lattice structure of the SWCNT nanobeam*s* has characteristic bond stretching due to the curvature, the small radii, and the structural buckling and corrugation of carbon rings. These unique carbon nanotube-nanobeams have higher elastic moduli (Stojkovic et al., 2001), since their curvature is inversely proportional to their radius. As a result, the higher values of curvature increase the stiffness of atomic structures and their elastic modulus. These changes in the material properties have to be reflected in the models used for these nanostructures.

The effect of high curvature in atomic lattices is so significant that these nanostructures become a new type of nanomaterial, the so-called SWCNT nanocrystals. The radius of SWCNT nanocrystals is smaller than the size of a carbon ring (Harik, 2001a,b, 2002). The uniqueness of these carbon nanotubes is associated with their unique material properties, which are quite different from other SWCNTs of larger radii. SWCNT crystals do not have very high strain-to-failure shown by most carbon nanotubes, which should be reflected in the constitutive models for the beam theories used for these nanostructures. Note that for the lengths greater than the spontaneous Euler buckling length, $L_E \leq L_{NT}$, nanobeams behave like nanowires.

Material maps for the classes of nanobeams and the beam-like nanocrystals indicate that the unique material properties of their lattices have the mechanical properties and the deformation response of the beam-like structures. However, appropriate constitutive models have to be used to address the uniqueness of their atomic lattices and the associated material properties. Ranges of applicability for the equivalent-continuum beam models (Harik, 2001a,b, 2002) span two different groups of geometric parameters (Fig. 4.11), which define two different classes of atomic lattices with up to 10 subclasses with small and large values of radii. It has been shown that these carbon nanotubes have similar buckling behavior, although other mechanical properties (e.g., radial buckling) diverge due to different structural characteristics.

4.3.2 A Material Map for All Single-Wall Carbon Nanotubes

The material map shown in Fig. 4.11 can be expanded to include all SWCNTs. Estimates for the effective thickness of carbon nanotubes can be used to refine the radial boundaries in the material map for the thin and thick lattice shells of the high aspect ratios (Table 4.3). It should be noted that the estimate of 0.66 Å does not apply to the effective thickness of SWCNT nanobeams and SWCNT nanocrystals, which have higher curvature and atomic lattice shells that are of higher thickness. Another estimate for their thickness, i.e., $l_{C-C}/2$, is more applicable for the classes of nanobeams and nanocrystals under the axial loading, because of its form involving a fraction of the covalent bond. According to this estimate, the effective nanotube thickness increases as the length of the covalent bond increases. Estimates for the effective thickness of SWCNTs determine the ranges of values for the radius of carbon nanotubes, which can be considered thin shells or thick shells (Table 4.3).

The aspect ratio, d_{NT}/L_{NT}, in the material map (Fig. 4.11) represents the length scale or size effects associated with the length, L_{NT}, of SWCNTs. The length scale of nanostructures influences their material properties and mechanical behavior (Table 4.3). The material map (Fig. 4.11) can be refined in order to reflect not only the length scale of atomic lattices characterized by their aspect ratio, d_{NT}/L_{NT}, but also the homogenization length, L_{NT}/a, or the minimum homogenization length, L_H. The material map shown in Fig. 4.11 and the length sales described in Table 4.4 indicate that SWCNT shells and nanobeams have some similarities. At the length scale level associated with the radius, R_{NT}, of the diameter, d_{NT}, the material properties of carbon nanotubes are influenced not only by their chirality, but also by the thickness-to-radius ratio, h_{NT}/R_{NT}. For the nanobeams, the curvature effects and the degree of bond stretching are also important for the onset of circumferential buckling. These effects can be reflected in the refined material map as well. A refined material map (Fig. 4.12) for the atomic lattices of carbon nanotubes belonging to four different classes: the high aspect ratio shells (i.e., Class I with up to 10 subclasses), thin shells (i.e., Class IIa), thick shells (i.e., Class IIb), nanobeams, and nanocrystal rods of small radii (i.e., Class

Table 4.3: Classification of thin and thick carbon nanotubes

	Class IIa	Class IIb
Estimates of Thickness, h_{NT}	**Thin Shells[a] (R_{NT})**	**Thick Shells (R_{NT})**
Graphene analog: 3.4 Å (van der Waals spacing)	37 Å–...	2–34 Å
MD shell analog: 0.66 Å (axial buckling)	7.2 Å–...	2–6.6 Å
Half of the C–C bond length, l_{C-C}: 0.72 Å	7.9 Å–...	2–7.2 Å

MD, molecular dynamics.

[a]A transitional range of 10% is assumed.

Table 4.4: The length scales in SWCNT shells and nanobeams

Length Scales	SWCNT Lattice Shells	SWCNT Nanobeams
The length, L_{NT}	Aspect ratio, d_{NT}/L_{NT} Homogenization ratio: L_{NT}/a	Aspect ratio, d_{NT}/L_{NT} Homogenization ratio: L_{NT}/a
The radius, R_{NT}	Chirality: armchair/zigzag the thickness-to-radius ratio, h_{NT}/R_{NT}, curvature	Chirality: armchair/zigzag cross-sectional area and the thickness, h_{NT}
Carbon ring, a	Chirality: armchair/zigzag Lattice defects	Chirality: armchair/zigzag l_{C-C} bond stretch, curvature
C–C bond	Effective thickness, h_{NT}	Degree of the l_{C-C} bond stretch, corrugation

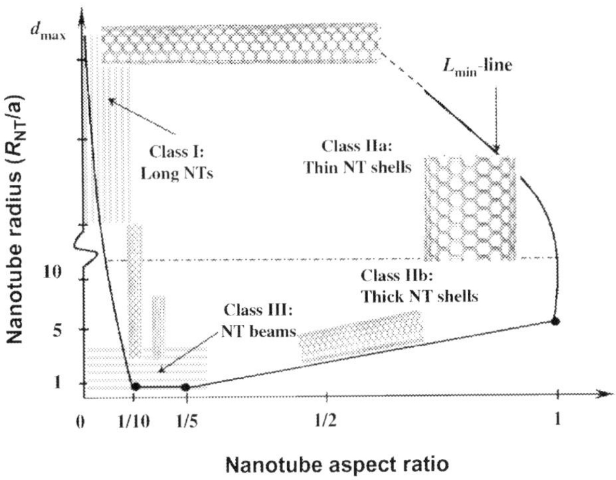

Figure 4.12

A material map for the atomic lattices of carbon nanotubes belonging to four different classes: *the high aspect ratio shells* (i.e., Class I with up to 10 subclasses), *thin shells* (i.e., Class IIa with two subclasses), *thick shells* (i.e., Class IIb with two subclasses), *nanobeams* and *nanocrystal rods* of small radii (i.e., Class III with up to five subclasses). © *2011 by Nanodesigns Consulting (reproduced with a permission).*

III). An expanded material map can include two classes of the thin shells (i.e., Class IIa for SWCNT shells) and a similar class for MWCNT shells with the higher values of radius, two classes of the thick shells (i.e., Class IIb for SWCNT shells), and a similar class for the thick MWCNT shells with the higher values of radius. Similarly, the class of nanobeams can be expanded to include SWCNT nanobeams (i.e., Class III) as well as up five subclasses (see Fig. 4.10).

Material maps for the classes of beam-like nanocrystal rods and nanobeams indicate that the material properties of the corresponding atomic lattices have the mechanical properties and the associated deformation response of the beam-like structures. Therefore, for the ranges of geometric parameters specified in the material maps (Figs. 4.11 and 4.12), various macroscopic beam models can be extended to the nanoscale mechanics for the nanoscale beam-like lattices under appropriate conditions. Ranges of applicability for the equivalent-continuum beam models (Harik, 2001a,b, 2002) span different groups of geometric parameters that define distinct classes and subclasses of atomic lattices with small and large values of radii. For example, it has been shown that these carbon nanotubes have the same buckling behavior, although other mechanical behaviors and properties (e.g., transverse stiffness) may diverge due to different structural characteristics.

4.4 Parametric Maps for the Applicability of Models for Carbon Nanotubes

The material map (Fig. 4.11) describes the ranges of geometric parameters, which are associated with different classes of the beam-like carbon nanotubes. This material map can be also used as a parametric map for the applicability of continuum beam models to different classes of SWCNTs and MWCNTs. Similarly, a parametric map for the applicability of continuum shell models can identify the ranges of geometric parameters (Fig. 4.12), when these models can be applied for the atomic lattices of carbon nanotubes. The ranges of applicability of the continuum beam and continuum shell models are shown to depend on the estimates of the effective thickness of the carbon nanotube lattices (Harik, 2001a,b, 2002). According to these parametric maps, carbon nanotubes can be modeled as either hollow beams or flexible shells (Harik, 2001b, 2002, 2011, 2014). These parametric maps are important, because there are clear distinctions between the two different types of deformation behavior associated with various classes of carbon nanotubes, i.e., the beam-like or the shell-like behavior. Variations in the geometry of carbon nanotubes as well as in their mechanical behavior require the use of classification of carbon nanotubes with respect to their geometric parameters associated with their atomic lattice structure. The effective thickness estimates for carbon nanotubes (Table 4.3) are important for identifying the ranges of applicability of different models in the parametric maps. The 2001 classification of carbon nanotubes (Harik, 2001a,b, 2002) is useful for the design of the carbon nanotube based probes for the atomic force microscopy, i.e., the AFM probes.[1]

The beam models can be used for the classes of long lattice shells with the high aspect ratios, $d_{NT}/L_{NT} < 1/10$, and considerable surface effects proportional to their surface area: $\pi d_{NT} L_{NT}$ or $a\sqrt{n^2 + m^2 + nm}\, L_{NT}$. The expanded material map (Fig. 4.12) and the thick line [$\sim 1/(10 d_{NT}/L_{NT})$] represent a boundary for the optimal material structure of the high aspect ratio lattices:

$$\frac{R_{NT}}{a} = \frac{1}{10 d_{NT}/L_{NT}},$$

[1] Nanodesigns Standards, v. 1, n. 3, Nanodesigns Consulting, Nanodesigns Press, Wilmington, Delaware, 2015.

where R_{NT}/a is the normalized radius, $R_{NT}/a \geq 1$. The optimal aspect ratio for the long *lattice shells* is such that $d_{NT}/L_{NT} < 1/20$.

Material properties of the thin and thick carbon nanotubes depend on their atomic structure, characterized by the structure–property relations illustrated by the length scales in Table 4.4, and the key nondimensional parameters that govern the corresponding regions in the material map (Fig. 4.12). The (n, m) pair of parameters describe the chirality of the lattices and their susceptibility to shearing deformation. The length-scale-based parameters that describe the atomic structure of the thin and thick lattices and affect their global and local deformation include the thickness-to-radius ratio, h_{NT}/R_{NT}, the ratio of the shell circumference to the size of a carbon ring, a, or the normalized radius, R_{NT}/a. These nondimensional parameters have to be reflected in the models and formulae describing their deformation.

The class of thin shells starts from the normalized radius, R_{NT}/a, 2.9 (or 3.2), and 15, depending on the estimates of the effective lattice thickness, h_{NT} (Table 4.3), or the complexity of the applied loads. The class of thick C_{NT} shells starts with the atomic lattices of the normalized radius, R_{NT}/a, 2.7 (or 2.9) and down toward the normalized radius, $R_{NT}/a \approx 1$. Below this critical value of the normalized radius, $R_{NT}/a = 1$, the Class III of SWCNT nanobeams and nanocrystals with small radii extends (Fig. 4.11). The material map (Fig. 4.12) for four classes of carbon nanotubes is bounded by the L_{min}-line corresponding to the homogenization ratio: $L_{NT}/a \geq 10$ or so. The material map is the result of the scaling analysis of atomic structures, therefore it can serve as the parametric map for the models, which are appropriate for different carbon nanotubes. The differences in the structure of carbon nanotubes and their mechanical behavior are reflected in the classification of the carbon nanotube lattices. The four distinct classes of SWCNT lattices: thin shells, thick shells, long shells (i.e., the high aspect ratio lattices), and the beam-like nanocrystals and nanobeams require appropriate constitutive models and deformation models. For the ranges of the atomic lattice parameters specified in the material map (Fig. 4.12) for the lattice shells, various equivalent-continuum shell models can be extended to the nanoscale mechanics under appropriate conditions. Discrepancies in the strain predictions for nanoscale strain failures, ε_{cr} of the (n, m) lattices, which are predicted by various continuum shell models, can be partially avoided by using the formulae with the sensitivity to the chirality of SWCNT lattices. In the case of nanoscale buckling, the formulae for critical buckling strain, ε_{cr}, which are sensitive to the chiral lattice structure of carbon nanotubes, should be used (Harik, 2011, 2014).

4.5 Conclusions

Atomic structure of carbon nanotubes, their geometry, and the length scales characterizing the structure of atomic lattices have been examined in the context of their influence on material properties. Relations between different geometric parameters describing the geometric and structural properties of carbon nanotubes, as well as the length scales of their atomic lattices, have been analyzed by using the scaling analysis and nondimensional parameters. Classification of carbon nanotubes based on the scaling analysis of their atomic lattice structures provides the foundation for the nanoscale analysis of different types of lattice shells in the mechanics of carbon nanotubes. The discovery of the uniqueness of carbon nanotubes with very small radii, along with their distinct material properties, has provided an additional need for the understanding of a wide variety of atomic lattice structures associated with these nanostructures.

Classification of carbon nanotubes identifies 20 classes of distinct atomic lattice structures with unique structural and material properties. Criteria for the separation of different classes of carbon nanotubes are presented in the framework of the scaling analysis of their atomic structures. Classification of carbon nanotubes answers not only when the well-known cylindrical shells behave as beams or shells, but also which atomic lattices represent the actual nanoscale beams with unique material properties. Moreover, classification of carbon nanotubes describes the process of degeneration of cylindrical symmetry from the nanoscale shells to nanoscale rods through the polygonal structure of nanobeams and the circumferential buckling of carbon rings. Carbon nanotubes may, indeed, behave as nanoscale beams or beam-like shells, with the 10 classes of their atomic lattices displaying the deformation response of elongated structures. The classes of thin and thick cylindrical shells also have unique structural properties for both single wall and MWCNTs.

The classification of carbon nanotubes into different classes has also allowed the developing of material maps for similar groups of these nanostructures. Parametric maps for the applicability of different models to distinct atomic structures of carbon nanotubes have also been presented. The applicability of various models within a particular class of carbon nanotubes is governed by the laws of geometric similitude. The laws of geometric similitude stipulate that different lattice shells belonging to one class of carbon nanotubes have similar mechanical behavior. When the values of the nondimensional length-scale-based parameters are the same for such lattice shells, the magnitude of mechanical deformation response may be identical, as in the formulae for the critical buckling strain.

Sample Problems

1. List approximate dimensions of different SWCNTs and MWCNTs.
2. List approximate dimensions of the atomic lattices for different SWCNTs and MWCNTs.
3. Draw a schematic representation of different known geometries and sizes of SWCNTs .
4. Make a Powerpoint viewgraph with a schematic representation of different known geometries and sizes of SWCNTs.

5. Make a Powerpoint viewgraph with a schematic representation of atomic structure of any zigzag and armchair carbon nanotube.
6. Draw a schematic representation of atomic structure of any armchair SWCNT along the description of its chirality with a coordinate system and unit vectors.
7. The normalized circumference of the zigzag carbon nanotube lattice shells is given by $2\pi(R_{NT}/a) = n$. Show that the radii of the (12,0) and (18,0) nanotube shells are such that $R_{NT}/a \approx 2$ (or $R_{NT} < 2a$) and $R_{NT}/a \approx 3$ (or $R_{NT} < 3a$), respectively.
8. Find a few examples of the normalized radius, R_{NT}/a, and the aspect ratio of carbon nanotube lattice shells along the lower boundary of the material map in Fig. 4.11.
9. What is the separation criterion for the class of thin lattice shells of carbon nanotubes?
10. What is the separation criterion for the class of thick lattice shells of carbon nanotubes?
11. What is the separation criterion for the class of the high aspect ratio lattice shells of carbon nanotubes?
12. What are the separation criteria for the class of carbon nanotube nanobeams?
13. What are the separation criteria for the class of carbon nanotube nanocrystal rods?
14. Use the separation criteria for different classes of carbon nanotubes, the (n, m) lattice nomenclature, and the formula for the radius to show which armchair and zigzag lattice shells belong to the class of the thick shells.
15. Use the separation criteria for different classes of carbon nanotubes, the (n, m) lattice nomenclature, and the formula for the radius to show which armchair and zigzag lattice shells belong to the class of the thin shells.

References

Dresselhaus, M.S., Dresselhaus, G., Eklund, P.C., 1996. Science of Fullerenes and Carbon Nanotubes. Academic Press, San Diego.

Fujita, M., Saito, R., Dresselhaus, G., Dresselhaus, M.S., 1992. Formation of general fullerenes by their projection on a honeycomb lattice. Phys. Rev. B. 45, 13834–13836.

Harik, V.M., 2001a. Ranges of applicability for the continuum beam model in the mechanics of carbon nanotubes and nanorods. Solid State Commun. 120 (7–8), 331–335.

Harik, V.M., 2001b. Ranges of applicability for the continuum beam model in the constitutive analysis of carbon nanotubes: nanotubes or nano-beams? NASA/CR-2001-211013. NASA Langley Research Center, Hampton, VA.

Harik, V.M., 2002. Mechanics of carbon nanotubes: applicability of the continuum-beam models. Comput. Mater. Sci. 24 (3), 328–342.

Harik, V.M., 2011. Mechanics of Carbon Nanotubes. Nanodesigns Press, Nanodesigns Consulting, Newark, DE.

Harik, V.M. (Ed.), 2014. Trends in Nanoscale Mechanics. Springer, Dordrecht.

Iijima, S.S., 1991. Helical microtubules of graphitic carbon. Nature 354, 56.

Iijima, S.S., Ichihashi, T.T., 1993. Single-shell carbon nanotubes of low diameters. Nature 363, 603.

Ruoff, R.S., Tersoff, J., Lorents, D.C., Subramoney, S., Chan, B., 1993. Radial deformation of carbon nanotubes by van der Waals' forces. Nature 364, 514–516.

Stojkovic, D., Zhang, P., Crespi, V.H., 2001. Smallest nanotubes: breaking the symmetry of *sp3* bonds in tubular geometries. Phys. Rev. Lett. 87 (12), 125502.

Timoshenko, S.P., Gere, J.M., 1961. Theory of Elastic Stability. McGraw-Hill, New York, NY.
Treacy, M.M.J., Ebbesen, T.W., Gibson, J.M., 1996. Exceptionally high Young's modulus observed for individual carbon nanotubes. Nature 381, 680.
Wong, E.W., Sheehan, P.E., Lieber, C.M., 1997. Nanobeam mechanics: elasticity, strength, and toughness of nanorods and nanotubes. Science 277, 1971.
Yakobson, B.I., Brabec, C.J., Bernholc, J., 1996. Nanomechanics of carbon tubes: instabilities beyond linear response. Phys. Rev. Lett. 76, 2511.

Further Reading

Harik, V.M., 2015. Mechanics of Carbon Nanotubes and Graphene, Lecture Notes: A Short Course, Nanodesigns Consulting. Nanodesigns Press, Wilmington, DE.

CHAPTER 5

Mechanics of the High Aspect Ratio Carbon Nanotubes: Beam Models, Nanoscale Buckling, Nanobeams and Phonons

Chapter Outline

5.1 Unique Properties of Nanoscale Beams and Carbon Nanotubes

Carbon nanotubes with elongated atomic lattice structures can behave as nanoscale beams when their aspect ratio, d_{NT}/L_{NT}, is high, or when they acquire the polygonal cross-section of hollow nanobeams or thin nanoscale rods (Fig. 5.1). The high aspect ratio cylindrical shells of carbon nanotubes have different structural properties when they have just one or multiple shells. The thickness of a single or multiple shells is also important for their mechanical response. The longitudinal and circumferential size of nanoscale lattice beams

Mechanics of Carbon Nanotubes.
DOI: https://doi.org/10.1016/B978-0-12-811071-3.00005-6

© 2018 Elsevier Inc. All rights reserved.

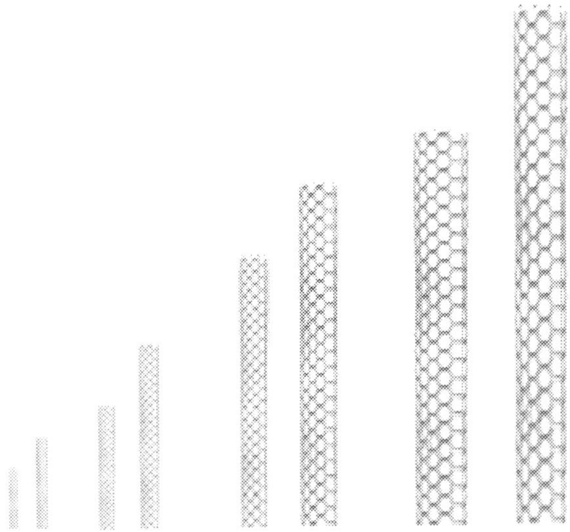

Figure 5.1
Examples of the beam-like structures of carbon nanotubes with the high aspect ratios. © *2011 by Nanodesigns Consulting (reprinted with a permission).*

affects the mechanical response of lattice cells, which can display smoother two-dimensional response with larger dimensions. The aspect ratio, d_{NT}/L_{NT}, of the diameter, d_{NT}, of single wall carbon nanotubes (SWCNTs) and multiwall carbon nanotubes (MWCNTs), over their length, L_{NT}, is the key nondimensional parameter that governs the mechanical behavior of any long carbon nanotube. 10 classes of long carbon nanotubes satisfy the following criteria for the beam-like behavior (Harik, 2001a,b, 2002):

$$\frac{d_{NT}}{L_{NT}} < \frac{1}{10} \quad \text{or} \quad \frac{d_{NT}}{L_{NT}} \ll 1, \tag{5.1}$$

where the diameter, d_{NT}, may vary between 0.2 and 2–3 nm, for SWCNTs, and 35 and 85 nm, for MWCNTs. Consequently, the length, L_{NT}, should be greater than 2 or 20–30 nm, for SWCNTs, which is often the case. Therefore, most of SWCNTs have the structural properties and mechanical behavior of nanoscale hollow beams or fibers. As a result, the analysis of nanoscale buckling, bending, and deflection of SWCNTs and MWCNTs has to be based on the concepts associated with the nanoscale beams, as well as the theory of nanoscale beams.

The aspect ratio, d_{NT}/L_{NT}, separates any nanoscale beam or the beam-like nanostructure (e.g., SWCNTs) from the nanostructured cylinders made of any nanomaterial, including short MWCNTs. The form and the area of the cross-section of any beam are important both for the analysis and the mechanical behavior of nanoscale beams. A nondimensional ratio of their thickness-to-radius, h_{NT}/R_{NT}, can characterize the size of the cross-sectional area of nanoscale beams, especially, when they are hollow inside. The uniqueness of nanoscale beams is associated with several nanoscale phenomena. The thickness of nanoscale beams, and carbon nanotubes, in particular, depends on the effective thickness estimates for the discrete atomic lattices defining their hollow structure. One has to be also aware of the effective thickness paradox associated with the load dependence of the effective thickness estimates for atomic lattices (Harik, 2014). Second nanoscale phenomenon, which makes nanoscale beams unique, is associated with the effect of the curvature of the atomic lattice and the circumferential buckling of its cells on the material and structural properties of nanoscale beams. The significant role of this nanoscale phenomenon is reflected in the degeneration of cylindrical symmetry of SWCNTs, and the formation of distinct classes of thin SWCNT nanostructures. The effective thickness of atomic lattices may also change when the lattice cells buckle out-of-plane to form thick hollow SWCNT nanobeams and SWCNT nanocrystal rods.

Third nanoscale phenomenon, which distinguishes nanoscale beams from their macroscopic counterparts, is the size dependence of their material properties. It has been noted earlier that the size of any nanomaterial, including nanoscale beams or carbon nanotubes, has to be such that their length, L_{NT}, is larger than the homogenization length, L_H, or much larger than the size of the smallest structural element, i.e., a carbon ring of the size: $a = 2.46$ Å. Material properties of nanoscale beams are well-defined and size-invariant, when these requirements for the length, L_{NT}, are satisfied, as in the nanoscale homogenization criterion. Otherwise, the value associated with the material properties may change with slight changes in the length of the atomic lattice structure.

This chapter is devoted to special nanoscale beams, which are represented by the atomic structures of SWCNTs and MWCNTs. Therefore, the following classes of nanoscale beams will be considered (Harik, 2011):

- the high aspect ratio thin SWCNT shells,
- the high aspect ratio thick SWCNT shells,
- the extralong thin and thick SWCNT fibers,
- the high aspect ratio thin MWCNT shells,
- the high aspect ratio thick MWCNT shells,
- the extralong thin and thick MWCNT fibers,
- the high aspect ratio MWCNT beams,
- the extralong MWCNT beams,
- SWCNT nanobeams of small radii, and
- thin SWCNT nanocrystal rods.

Notice that the extralong SWCNT and MWCNT fibers have the length, L_{NT}, which is greater than the Euler buckling length, L_E, associated with the spontaneous buckling caused by environmental forces, i.e., $L_{\mathrm{NT}} > L_E$. The Euler buckling length, L_E, is larger for the thick MWCNT shells, as compared to that of the thick and thin SWCNT shells. It will be noted later that the long MWCNT shells and MWCNT nanobeams and fibers can be toxic, due to their similarity with the asbestos fibers (Harik, 2014).

5.1.1 Length Scales in Nanoscale Deformation and the Cauchy–Born Rule

Length scales in the atomic lattice structure of the beam-like carbon nanotubes have been considered in details in an earlier chapter. The mechanical deformation response of carbon nanotubes depends on the structural characteristics of their atomic lattice structure and the length scales involved. Characteristics of a deformation response of the length, λ, in a particular atomic lattice structure depend on the underlying atomic lattice elements (i.e., a carbon ring and the C–C bond) and the length scales of the size, a, or l_{C-C}, below the length, λ. The wave length, λ, of the deformation response of a nanoscale surface with an atomic structure has to be much larger than the size of the smallest element in the atomic lattice, a, i.e., $\lambda \gg a$, so that the lattice can be considered as a continuum surface at the local scale (Harik et al., 2002). Moreover, a deformed lattice surface can be considered a deformed physical continuum when it satisfies the above nanoscale homogenization criteria (Harik, 2011). It should be noted that the size requirements for nanoscale homogenization (i.e., $\lambda \gg a$) are also important for the uniqueness and the well-definiteness of all physical properties and macroscopic physical parameters associated with atomic surfaces. The length scale of variations in the stress distribution, λ_σ, in the lattice of SWCNTs, however, can correspond to that in the continuum shells at a smaller length scale, i.e., $\lambda_\sigma \gg l_{C-C}$, or $\lambda_\sigma > 10\ l_{C-C}$, where l_{C-C} is the length of the C–C bond.

The relation between the local deformation of an atomic lattice and the overall deformation response of the lattice structure in a nanoscale beam is very important. The Cauchy–Born approximation provides a kinematic hypothesis about the deformation of a material element used in the mathematical formulation of solid mechanics (i.e., the Cauchy stress tensor), which relates the local movement of atoms in a crystal lattice cell to the overall deformation of the bulk lattice of a solid (Born and Huang, 1954). The Cauchy–Born rule equates the averaged continuum strain energy in a crystal lattice cell to the potential energy stored in the deformed atomic bonds. It ensures the equilibrium of atoms for simple lattices. The continuum strain energy density is obtained from the interatomic potential (Arroyo and Belytschko, 2004) by introducing the geodesics and the exponential transform for the Cauchy–Born rule for curved surfaces. The aforementioned nanoscale homogenization criteria require that the Cauchy–Born approximations should be averaged over the smallest volume, or an area of the atomic lattice of the size satisfying the homogenization criteria. Therefore, the classical Cauchy–Born rule (Born and Huang, 1954) and the exponential Cauchy–Born rule (Arroyo and Belytschko, 2004) should be applied with the nanoscale homogenization correction. The magnitude of such nanoscale homogenization correction will depend on the complexity of atomic lattices and their kinematics. The nanoscale homogenization correction for the Cauchy–Born approximation takes into account the multiple atom interactions in the part of the atomic lattice limited by the homogenization size.

5.1.2 Kinematic Inequalities for Atomic Lattices of Carbon Nanotubes

Modeling of different carbon nanotubes should take into account various characteristics of their atomic lattice structures, such as the length scales, chirality, and geometry. Application of the kinematic Cauchy–Born rule for the deformation of atomic lattices associated with different classes of carbon nanotubes can be simplified by taking into account changes in their atomic lattice structure and their kinematic boundary, i.e., the effective lattice thickness. Kinematics of the deforming atomic lattices can be also described by the lattice-sensitive continuum models, which will be discussed later (Harik, 2011). Restrictions on the kinematic motion of atoms in the lattices of carbon nanotubes also emerge from the separation criteria for different classes of SWCNTs.

The kinematic motion of atoms in radial direction during small strains of the lattice of the thin SWCNTs and thin MWCNTs is restricted within the lattice thickness by the following inequalities:

$$\frac{h_{NT}}{R_{NT}} < \frac{1}{10} \quad \text{and} \quad \frac{h_{NT}}{R_{NT}} \ll 1,$$

where the ratio of the thickness, h_{NT}, to the radius, R_{NT}, of a deforming lattice has been discussed earlier in the context of structural properties. These inequalities have been also used in the criteria for the classes of thin shells. Notice that the kinematic motion of the carbon nanotube deflected during an axial buckling describes the global kinematic motion of atoms in the lattice. The aforementioned inequalities restrict the local kinematic motion of atoms within the wall thickness. Moreover, the aforementioned inequalities hold for the radial positions of atoms within the thickness of lattices during any buckling process of thin lattice shells. These kinematic restrictions become clear for the shearing atomic structures of the thick SWCNTs and thick MWCNTs), as they take the form $h_{NT}/R_{NT} > 1/10$, which allows greater kinematic shearing of atomic lattices. The thickness, h_{NT}, of MWCNTs and its shearing may be considerable with the greater number of internal walls.

Kinematic requirements for the motion of atoms in the lattice with respect to the global deformation of the lattice can be also formulated by using the chirality of carbon nanotubes defined by the (n,m) pair. The aforementioned kinematic restrictions for thin lattice shells can provide the following kinematic inequality bounding the kinematic motion of atoms in the thin lattice, Δh_{NT}:

$$\Delta h_{NT} \ll \frac{a}{2\pi}\sqrt{n^2 + m^2 + nm}, \tag{5.2}$$

where the radius, R_{NT}, of the deformed thin shell is connected with the chiral lattice structure as $R_{NT} = (a/2\pi)\sqrt{n^2 + m^2 + nm}$, where a is the size of a carbon ring. The kinematic inequality (5.2) for the motion of atoms can provide an estimate for the displacement of atoms within the lattice of the thin shells of carbon nanotubes in comparison to the length, l_{C-C}, of the C–C bond:

$$\frac{2\pi\Delta h_{NT}}{l_{C-C}\sqrt{3(n^2 + m^2 + nm)}} \ll 1. \tag{5.3}$$

The kinematic bound (5.3) on the kinematic motion of atoms in the atomic lattice of carbon nanotubes explicitly depends on the chirality of their (n,m) lattice. The kinematic bounds (5.2) and (5.3) on the kinematic motion of atoms in the lattice of SWCNTs can be used to verify the application of the Cauchy–Born rule before the nanoscale homogenization correction is determined for the minimum size of their lattice. The minimum size of SWCNT lattice, for which the nanoscale homogenization correction for the Cauchy–Born rule should be applied, is determined by the criteria for nanoscale homogenization discussed earlier.

5.2 Analysis of the Beam-Like Carbon Nanotubes

Carbon nanotubes can be examined under several loading conditions: axial or radial loading (uniform or a point force), cantilever bending, the middle-point bending and the side bending (Fig. 5.2). Distribution of the applied forces over the local area of a carbon nanotube involves a local and a global deformation response and the corresponding length scales. Separation of the length scales associated with the local lattice deformation and the global lattice response is important for the nanoscale analysis, similar to the distribution of forces during macroscopic mechanical testing (i.e., "St. Vincent length"). The local lattice distortions by the applied forces are minimized, when the external forces are applied over several atoms or a circumferential layer of atoms. The nanoscale St. Vincent length for the transfer of the applied forces into the lattice structure of a carbon nanotube may be affected by the chirality of SWCNTs and the structure of MWCNTs.

The motion of atoms in the lattice of carbon nanotubes responding to the applied forces can monitored with the axial and circumferential strains of SWCNT lattices and the Poisson ratio, ν, $\nu = -\varepsilon_r/\varepsilon_x$, where ε_x is the axial strain, $\Delta L_{NT}/L_{NT}$, and ε_r is the radial strain, $\delta r/R_{NT}$, or the circumferential strain, $2\pi\delta r/a_c N_c$, of the lattices depending on how

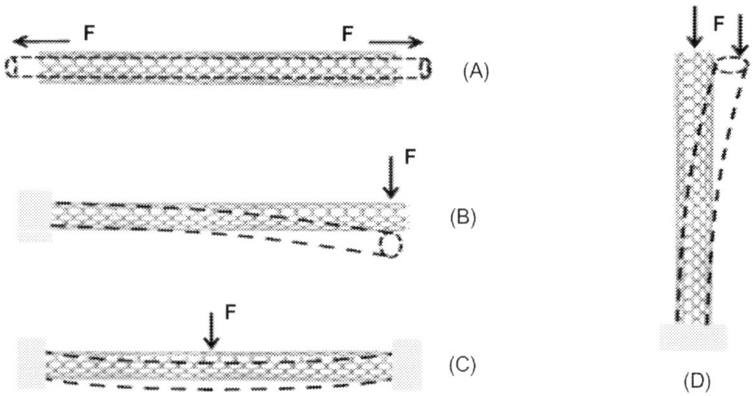

Figure 5.2
Types of deformation of the beam-like lattices of carbon nanotubes: axial tension (A), bending of a cantilever beam (B), the middle point bending of a long beam (C), and the axial buckling of the high aspect ratio lattice beam. *© 2011 by Nanodesigns Consulting (reprinted with a permission).*

the Poisson ratio is defined; N_c is the number of carbon rings in the circumference of SWCNT lattice and a_c is the circumferential size of a carbon ring. The lateral carbon ring deformation, Δa_c, can be determined by

$$\Delta a_c = \frac{2\pi\delta r}{a_c N_c^2},$$

or by $\Delta a = 2\pi\delta r/aN_c^2$ for the zig-zag SWCNT lattices. The kinematic motion of atoms in the atomic lattice can be monitored, along with the deformation of carbon rings and verification of the local evaluation of Poisson's ratio. Poisson's ratio can be also defined as ν_a, i.e., Poisson's ratio based on the ratio of the lateral, Δa_c, to the axial, Δa_l, deformation, $\Delta a_c/\Delta a_l$, of the carbon rings normalized by their longitudinal, a_l, and circumferential, a_c, sizes (the ring size, $a = 2.46$ Å). For the thick SWCNT lattice shells, with the carbon rings well circumscribed into their circumference, the aforementioned formula for Poisson's ratio involving the lateral deformation of carbon rings, and the circumferential strain is appropriate. The above formulae indicate the uniqueness of SWCNT nanobeams due to the somewhat different mechanisms of deformation of carbon rings affecting the value of their Poisson's ratio.

Axial deformation of SWCNT lattices, changes in their radius, $\delta r/R_{\mathrm{NT}}$, and the deformation of carbon rings can be used not only to define and calculate Poisson's ratio of carbon nanotubes, but also to study unique nanoscale phenomena. These changes in the radius can be used to study the onset of structural circumferential carbon ring buckling (SCCRB) in the lattices of carbon nanotubes. For the SWCNT lattices of small radii (around 2.46 Å, for instance), changes in the radius during the axial deformation may either induce or increase the carbon ring corrugation. For example, the circumference of the zig-zag (6,0) SWCNT

Figure 5.3
Atomic lattice structure of the zig-zag (6,0) carbon nanotube with the six carbon rings spanning the hexagon faces without the structural carbon ring buckling, which can be induced by either thermal vibrations or applied forces during the DCCRB. *DCCRB*, dynamic circumferential carbon ring buckling. *© 2011 by Nanodesigns Consulting (reprinted with a permission).*

(Fig. 5.3) includes exactly six carbon rings spanning the hexagon faces without the structural carbon ring corrugation, which may be induced, however, by thermal vibrations or by axial deformation. Circumference of the (6,0) SWCNT lattice can undergo dynamic circumferential carbon ring buckling (DCCRB) during a radial deformation of the lattice under its global axial extension. The SCCRB occurs naturally in the SWCNT lattices, when the normalized circumference equals to 2π:

$$2\pi\frac{R_{\mathrm{NT}}}{a} = \sqrt{n^2 + m^2 + nm} \approx 2\pi,$$

or the normalized radius, R_{NT}/a, is close to unity or right below, $R_{\mathrm{NT}}/a \leq 1$, as in the class of SWCNT nanobeams and SWCNT nanocrystal rods described in Chapter 4, Classification of Carbon Nanotubes: Twenty Classes of Atomic Structures. The DCCRB is a unique nanoscale phenomenon associated with the SWCNT nanobeams and thick SWCNT lattices under axial tension. For thinner SWCNT nanobeams, circumferential carbon rings are not well circumscribed into the circumference, and the carbon rings undergo the structural buckling during their formation.

5.2.1 Atomistic Analysis Methods for Carbon Nanotubes

Atomistic analysis methods for carbon nanotubes are based on the solution of systems of the Schrodinger equations for the wave functions, ψ_i, describing the probability density for the systems of atoms in SWCNT lattice, and the interfacial layer of adjacent materials. Atomistic and Monte Carlo methods (Ohno et al., 1999) can describe deformation of carbon nanotubes and the molecular structure in nanocomposites. However, atomistic methods including ab initio and the tight binding methods are computationally very expensive. Therefore, a system of atoms analyzed by these methods is usually relatively small as compared to molecular dynamics (MD) simulations, atomistic finite element methods, or Monte Carlo methods. Note that MD simulations can be also carried out by using Monte Carlo methods.

In nanoscale mechanics of atomic lattices, atomistic methods are based on the Born–Oppenheimer approximation (Ohno et al., 1999). According to this approximation, the electrons are assumed to be in their steady state associated with their averaged motion around the nuclei. The spatial distributions of electrons change rapidly as compared to the motion of atoms or their nuclei during a mechanical deformation. Interactions between the systems of electrons between two or more atoms can be described by a system of Schrodinger equations:

$$H\psi_i = E\psi_i$$

involving the Hamiltonian operator, H, composed of different contributions to the energy, E, of such interactions (i.e., the kinetic energy, the potential energy of the electron–electron Coulomb interactions, etc.). The form of Schrodinger equations, the mathematical forms of the systems of equations, the form of Hamiltonian operator, and the approximations used for different contributions to the energy of interactions are adjusted for a particular material system (e.g., a particular carbon nanotube).

A systems of N atoms forming a carbon nanotube or graphene lattice has the total internal energy, E_{total}, composed of the sum of the kinetic energies, T_i, of individual atoms and their potential energy, U_i:

$$E_{\text{total}} = \sum_i^N T_i + \sum_i^N U_i$$

where the kinetic energy, T_i, of each atom is $T_i = (m_i V_i V_i)/2$, and U_i is their potential energy represented in the above Hamiltonian operator, H. The law of conservation of energy is applicable at nanoscale to the entire system of atoms. The total energy of a system of N atoms can be used to evaluated the internal stresses resulting from the applied forces due to the gradients of the potential energy. The mechanical stress on a system of carbon atoms can be defined as the change in their internal energy, E, with respect to the strain per unit volume. The small mechanical strain is associated with the increase of the potential energy of reversible elastic deformation under the applied stress. Therefore the stress tensor, σ_{ij}, may have the following form for a linear elastic material:

$$\sigma_{ij} = \frac{1}{V_0}\left(\frac{\partial E}{\partial \epsilon_{ij}}\right)_{S=\text{const}}$$

where V_0 is the volume of a nanoscale material element determined by the nanoscale homogenization criteria, E is the total internal energy of the atomic lattice structure, and ε_{ij} is the strain tensor. The subscript S denotes that the entropy, S, of the strained system is constant during the deformation. During an elastic deformation of a nanostructured material, the internal energy of a nanoscale element can be equal to the strain energy of the deformed material element, the size of which has to satisfy the nanoscale homogenization criteria. Nanomechanics of carbon nanotubes and graphene develops multiscale methods and approaches, which link the atomistic methods, molecular mechanics, and MD simulations with nanoscale analysis of mechanical interactions.

5.2.2 Molecular Dynamics Simulations of Carbon Nanotubes

MD of atomic lattice structure of carbon nanotubes can be described by a system of momentum equations for the acceleration, a_i, of carbon atoms, which are derived from Newton's second law: $\mathbf{F}_i = m_i\mathbf{a}_i$, where $\mathbf{F}_i$ is the force acting on the ith atom of mass, m_i. The displacement of the ith atom at the time t can be described for the time increment Δt by monitoring the changes of its position vector, $\mathbf{r}_i$. Acceleration, $\mathbf{a}_i$, of atoms is determined from Newton's law as force $\mathbf{F}_i$ divided by mass, m_i. The force, $\mathbf{F}_i$, acting on each atom can be evaluated as the gradient of a potential energy function, $V(\mathbf{r})$, which depends on interatomic interactions:

$$\mathbf{a}_i(t) = -\frac{1}{m_i}\nabla_{\mathbf{r}_i} V(\mathbf{r}_1, \ldots, \mathbf{r}_N). \tag{5.4}$$

The acceleration, $\mathbf{a}_i$, of individual atoms can be evaluated as the first time derivative of their velocity, $\mathbf{v}_i$, or the second time derivative of their position vector, $\mathbf{r}_i$. Therefore, the governing momentum equation for the position vector, $\mathbf{r}_i$, of each atom has the following form:

$$\frac{d^2\mathbf{r}_i}{dt^2} = -\frac{1}{m_i}\nabla_{r_i} V(\boldsymbol{r}_{i,} \ldots, \boldsymbol{r}_N), \tag{5.5}$$

where the choice of the potential function, $V(r)$, determines the type of interatomic interactions, as well as the internal stresses in a system of atoms as noted above. Interactions between individual atoms are determined by the potential energy of interatomic interactions described by the atomic potentials. Atomic potentials can be empirical, tight binding, or based on ab initio calculations. The many-body potential, $V(\boldsymbol{r})$, can be presented as a sum of pairwise terms with the energy of pair interactions, $\phi(\boldsymbol{r}_{ij})$, depending on their relative distance, r_{ij}. The bonding interactions between atoms in carbon nanotubes can be described by the atomic potential for the bonding of carbon atoms in hydrocarbons (Brenner, 1990):

$$V = \frac{1}{2}\left[\sum_{i \neq j} \varphi_R(r_{ij}) + \sum_{i \neq j} B_{ij}\varphi_A(r_{ij})\right], \tag{5.6}$$

where the energy of pair interactions, $\phi(\boldsymbol{r}_{ij})$, has attractive, $\phi_A(\boldsymbol{r}_{ij})$, and repulsive, $\phi_R(\boldsymbol{r}_{ij})$, terms. The weight function, B_{ij}, reduces the strength of the ij-bond with the j-atom in the lattice of carbon nanotube, when it has other bonds. The weight function also describes the angular dependence with a cutoff that restricts the pair potential (5.6) to the nearest neighbors. The Tersoff–Brenner potential (5.6) for carbon atoms also includes a switch-type function:

$$G(r) = \begin{cases} 1, & r < r_1 \\ \frac{1}{2}\left(1 + \cos\left[\frac{\pi(r - r_1)}{r_2 - r_1}\right]\right), & r_1 \le r \le r_2, \\ 0, & r > r_2 \end{cases}$$

which leaves the interatomic interactions unchanged for the distances between pairs of atoms smaller than the inner cutoff distance: $r_1 = 1.7$ Å. It is smoothed down to zero at the second cutoff distance: $r_2 = 2$ Å.

Atomic interactions during elastic deformation of carbon nanotubes produce the dynamic stress on individual atoms, which can be evaluated by using the energy contributions associated with the kinetic energy of the nth atom and the potential energy of its interactions with the m-atom. Stress in a nanoscale volume, V_0, of a group of atoms in the atomic lattice of a carbon nanotube is given by

$$\sigma_{ij} = -\frac{1}{V_0}\sum_n \left(m_n v_i^n v_j^n + \sum_m F_i^{nm} r_j^{nm} \right), \tag{5.7}$$

where m_n is the mass of the nth atom, v_n is the velocity of the of nth atom with the i- and j-components, F^{nm} is the force of interactions between the nth atom and the mth atom with the i-components, and $\boldsymbol{r}^{nm}$ is the distance between the nth atom and the mth atom with the j-components. The velocity, v_n, of the nth atom in formula (5.7) is evaluated the change of the position vector, $\boldsymbol{r}^n$, with time. The proper nanoscale averaging of stress components requires that the size of the chosen nanoscale volume, V_0, satisfies the criteria for nanoscale homogenization. Inside this nanoscale volume element, interatomic interactions between individual atoms in the lattice should be described by the atomic potentials.

5.2.3 Solution of the Thickness Paradox for Long Carbon Nanotubes

Carbon nanotubes having high aspect ratios (Fig. 5.1) appear and behave as the beam-like nanostructures. The continuum beam model was used in the analysis of experimental data (Treacy et al., 1996; Wong et al., 1997) for the vibrating nanotubes attached to a tip of the atomic force microscope (AFM). In these experiments, a hollow cylinder model was used for the nanotube geometry, and the corresponding continuum beam model was utilized to predict Young's modulus of a vibrating carbon nanotube. Thickness of carbon nanotubes in the hollow beam models was assumed to be 3.4 Å, as in the most of other studies. These vibration experiments have demonstrated that carbon nanotubes behave similarly to the high aspect ratio beams during bending and deflection (Fig. 5.2). Effective thickness of the atomic lattice of carbon nanotubes is, indeed, the basic parameter describing the size of all SWCNT and MWCNT shells and hollow nanobeams. Spatial interactions between atoms of adjacent lattices and their kinematic motion are connected with the physical thickness of atomic lattices. Effective thickness of SWCNTs is also important for the evaluation of their elastic modulus, which is the critical material parameter affecting their constitutive behavior and their stiffness.

The first experimental results (Iijima, 1991) and theoretical modeling results of carbon nanotube crystals (Ruoff et al., 1993) have demonstrated that the effective thickness of the carbon nanotube lattices is the same as the thickness of graphene sheets in graphite: 3.4 Å. In all of these cases, the graphene sheets and carbon nanotubes have been subjected to radial loading through the van der Waals forces. Moreover, the modeling results for carbon nanotubes with different values of their radius have shown that under transverse loading the effective thickness of carbon nanotubes, h_{NT}, is the same for all nanotube lattices regardless of their diameter, d_{NT}. The radial stiffness of carbon nanotubes with smaller radii seem to be increasing due to either material or structural properties, while their thickness was assumed to be the same.

The ranges of applicability of the continuum beam model for carbon nanotubes have been examined for their constitutive modeling (Harik, 2001a,b, 2002), and to answer the question: Should SWCNTs be treated as nanotubes or nanobeams? The scaling analysis used for the atomic structure of SWCNTs has considered all existing estimates for the effective thickness of carbon nanotubes and presented a new estimate: 0.72 Å, which was adjusted later to 0.71 Å or the half of the bond: 1/2 l_{C-C} (Harik, 2001b, 2011, 2014). The new estimate for the effective thickness of carbon nanotubes was based on an approximation for the thickness of the load-transferring bonds (Harik, 2001b, 2002), that is, the half-length of the covalent bond, 1/2 l_{C-C}, which was close to the estimate of 0.66 Å (Yakobson et al., 1996) and 0.89 Å (Yakobson and Dumitrica, 2003). MD simulations by Yakobson et al. (1996, 2003) and their continuum shell model have shown that the thickness of the (10,10) carbon nanotube under axial loading is about 0.66 Å (adjusted later to 0.89 Å). These estimates for the effective thickness of SWCNTs are almost five times smaller than the earlier predicted thickness of nanotube lattices, 3.4 Å, or the thickness of graphene. This discrepancy in the evaluation of the thickness of carbon nanotubes results in different estimates of Young's modulus and constitutes the so-called effective thickness paradox for carbon nanotubes (Huang et al., 2006). The effective thickness paradox for carbon nanotubes became more explicit with the two independent estimates pointing to much thinner effective thickness of the nanotube lattices.

Resolution of the effective thickness paradox is based on a methodology for the evaluation of contributions from the van der Waals forces and the elastic C—C bonds associated with the two limiting values of the effective thickness of atomic lattices of carbon nanotubes (Harik, 2011), which depend on the complexity of the applied loads (Harik, 2014):

> The effective thickness of the hexagonal atomic lattice of carbon nanotubes may vary from 3.4 Å, which is the effective thickness of the graphene under the transverse loading through van der Waals forces, to 0.66 Å, for axial loading through the load transferring covalent bonds; these variations are load dependent and proportional to the number of carbon rings subjected to a particular loading.

These estimates for the effective thickness of carbon nanotube lattices are applicable to the lattice shells, the atomic structure of which is not affected by the high curvature effects, and the internal van der Waals forces, as in the case of thin SWCNT nanocrystals. The effective thickness of SWCNT nanobeams and nanocrystals is certainly higher than 0.66 Å and closer to the estimate of $l_{C-C}/2$, where the length of the covalent bond is being stretched. However, a more detailed analysis is required beyond this initial approximation.

The value of the effective lattice thickness, h_{NT}, of SWCNTs is affected by the degree of the balance of the elastic C—C bond interactions associated with the approximation: $h_{NT} \approx l_{C-C}/2$, and the van der Waals forces associated with the experimental graphene value of $h_{NT} \approx 3.4$ Å. Therefore the solution to the effective thickness paradox for carbon nanotubes can be formulated in terms of the number of carbon rings subjected to either axial forces or the radial van der Waals forces. In the case of carbon nanotubes subjected to complex loading, the value of the effective thickness, h_{NT}, of SWCNT lattice shells can be estimated by the number of carbon rings, N_{vdW}, subjected to predominately the van der Waals forces (Harik, 2011, 2014):

$$h_{NT} = \frac{1}{2} l_{C-C} + \frac{l_{C-C}}{2\pi R_{NT}} \frac{3aN_{vdW}}{2L_{NT}} \left(h_{vdW} - \frac{1}{2} l_{C-C} \right), \tag{5.8}$$

or by the number of carbon rings, N_{C-C}, mostly involved in the elastic C—C interactions:

$$h_{NT} = h_{vdW} - \frac{l_{C-C}}{2\pi R_{NT}} \frac{3aN_{C-C}}{2L_{NT}} \left(h_{vdW} - \frac{1}{2} l_{C-C} \right). \tag{5.9}$$

The value of the effective thickness, h_{NT}, of SWCNTs in a particular deformation response is such that $l_{C-C}/2 \leq h_{NT} < 3.4$ Å.

5.2.4 Continuum Models for the Beam-Like Carbon Nanotubes

Continuum models have been used to describe the mechanical behavior of long carbon nanotubes. The experimental data (Treacy et al., 1996; Wong et al., 1997) for the vibration of the high aspect ratio SWCNTs demonstrated that a hollow cylinder model for the geometry of nanotubes provides a reasonable estimate for Young's modulus of a vibrating carbon nanotube. These early vibration experiments have indicated that carbon nanotubes behave as the beam-like nanostructures. Govindjee and Sackman (1999) have considered an elastic multisheet model to show the explicit dependence of material properties on the cylindrical system size, when a continuum cross-section assumption is made for a multishell system (e.g., MWCNTs) subjected to bending. The continuum assumption is shown to hold for their cross-sectional area, when more than 201 shells are present in the macromechanical system considered. This was an attempt to examine how the continuum concepts hold for the discrete material systems. However, the ranges of applicability of the continuum concepts can be also examined by considering not only geometric parameters of SWCNTs but also the length scales in the atomic lattice structure of SWCNTs (Harik, 2001a,b, 2002), in order to define a set of restrictions on a series of assumptions that are used in the continuum beam and shell models.

A continuum beam or shell models "may serve as a useful guide, but its relevance for a covalent bonded system of only a few atoms in diameter is far from obvious" (Yakobson et al., 1996). The MD buckling strain predictions of 0.05% indicate hyperelastic rather than elastic behavior of SWCNT lattice shells. To ensure the robustness of data reduction schemes for MD simulations that are based on continuum mechanics, a careful analysis of continuum approximations used in macromechanical models and possible limitations of this approach at the nanoscale level is required. When the classical shell (Yakobson et al., 1996) and beam theories (Treacy et al., 1996; Wong et al., 1997) are used for the atomic lattices, the role of lattice structure, its structural parameters, and the length-scale limitations have to be taken into account.

Applicability of the continuum beam models for carbon nanotubes has been investigated along with the limitations of the continuum mechanics at nanoscale (Harik, 2001a,b, 2002). The axial buckling of carbon nanotubes (Fig. 5.4) was chosen to test the effectiveness of the continuum beam models at nanoscale and the validity of assumptions of the continuum beam theory. Numerical results have indicated that the Euler beam theory (Timoshenko and Gere, 1961) can be used as an approximation for the analysis of small elastic strains of carbon nanotubes with the radius smaller than 2 nm. The aspect ratio of carbon nanotubes with the larger radius has to be exceptionally high in order to apply the continuum beam model as an approximation for the axial buckling of SWCNTs.

The Euler beam equation for small elastic deflections, $w(x)$, of a long continuum beam (Timoshenko and Gere, 1961) can be written as

$$EI\frac{\partial^2 w}{\partial x^2} + Pw = 0, \tag{5.10}$$

where E is the elastic modulus of a nanoscale beam, I is its moment of inertia, and P is the load describing the force per cross-sectional area, A. It should be noted that the deflection

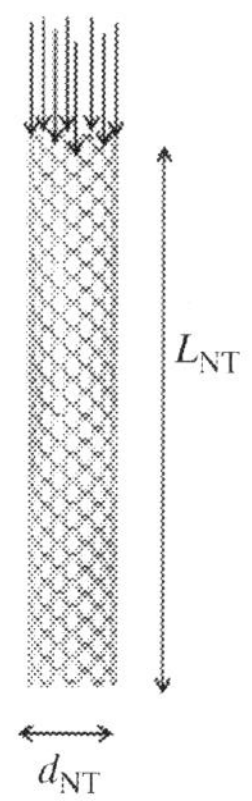

Figure 5.4
Atomic lattice structure of a long carbon nanotube with a small diameter, d_{NT}, and the length, L_{NT}, and the arrows identifying the direction of axial forces applied along the circumference. © *2011 by Nanodesigns Consulting (reprinted with a permission).*

of the nanoscale beam has to be very small, because the second derivative of the deflection of such a beam is associated with the smooth continuous change in the curvature of the deflected beam along its length. The discrete atomic lattice of the beam-like carbon nanotubes can be deflected with smooth change in the curvature of carbon rings along the length of SWCNTs, for very small forces and very high aspect ratios. The ranges of applicability of the Euler beam Eq. (5.10) to carbon nanotubes depend on a long list of assumptions (Harik, 2001a,b, 2002), which are discussed below.

The Euler beam theory and the resulting Eq. (5.10) for long beams are based on a number of assumptions, which provide a rigorous framework for deriving the governing equations for the deformation of elastic beams. In order to examine the possible length-scale limitations of the continuum beam models on nanoscale, all underlying continuum-based assumptions[1] must be scrutinized. The range of validity of different assumptions can be defined by using geometric parameters of the atomic lattice structure of SWCNTs (Harik, 2001a,b, 2002). Assumption (B1) for the atomic lattices of SWCNTs is extended for the hyperelastic deformation. Assumptions (B1) and (B4) allow the use of a one-dimensional

[1] *The continuum beam theory* is based on the following assumptions (Timoshenko and Gere, 1961): (B1) deformation of a beam is linear elastic; (B2) the direction of the applied load remains constant during deformation; (B3) the cross-section of the beam does not vary along its length; (B4) the length, L, is much larger than the radius, R, of the beam; (B5) the axial stiffness of the beam is large compared to the bending stiffness; (B6) all deformations of the column occur in the $X-Y$ plane and all cross-sections of the beam remain planar during deformation; (B7) transverse shear deformation is negligible in the beam (the Timoshenko beam theory allows shear); (B8) strains in the column are small, but the rotations of the cross-section may be finite; (B9) all stresses are negligible as compared to the axial and shear stresses that act on each cross-section of a beam and in the $X-Y$ plane.

theory for capturing the dominant global response, and for the derivation of the elastic beam equation. Therefore the length of SWCNTs and their diameter should be such that $d_{NT}/L_{NT} < 1/10$, or $d_{NT}/L_{NT} \ll 1$. Assumption (B2) restricts the direction of displacement of carbon atoms located near the edges of SWCNTs and the orientation of the axis of SWCNTs. The displacement boundary conditions are more appropriate and easier to implement for the atomic lattices of SWCNTs than the traction boundary conditions. The force-displacement relation and the applicability criteria involve either a change in the length of SWCNTs or the value of the length itself. Note that for the type of loading in the assumptions (B1) and (B2), the work done by the applied load is independent of the path it follows during the reversible elastic deformation.

Note that the existence of continuum cross-section [requirement (B3)] is not required for the analysis of axial buckling of SWCNTs, since they may be considered as a hollow beam. Carbon nanotubes have very high axial stiffness that satisfies the condition (B5). Assumption (B6) implies that the beam cross-sections do not deform in their planes and their products of inertia are negligible. It requires that all cross-sections remain perpendicular to the original image of the beam reference axis during deformation. To control stresses according to the assumptions (B7/B9) is problematic for the atomic lattices of SWCNTs with small radii. In the initial state of relaxed lattice structures, the radial stresses are tensile, while the tangential or circumferential stresses are compressive. These stresses are noticeable for the diameters, such that $d_{NT} < 1$ nm; however, their value is small compared to the elastic modulus of SWCNTs. These internal stresses diminish as the diameter of SWCNTs increases.

In order to use the continuum beam theory, one has to ensure that the geometry of carbon nanotubes allows the definition of continuum media, or equivalent continuum, for the discrete atomic lattice structures (Harik, 2001a,b, 2002). The definition of a continuum media is possible when the length of carbon nanotube satisfies the nanoscale homogenization criteria: $L_{NT} \gg a$, or $L_{NT}/a \gg 1$, where a is the size of a carbon ring in the atomic lattice of SWCNTs. Moreover, the value of the elastic modulus of carbon nanotubes is unique and well-defined when the length, L_{NT}, is long enough to satisfy the above criteria for nanoscale homogenization of the atomic lattices. The linear stability analysis for continuum elastic beams (Timoshenko and Gere, 1961) can be also used to determine the critical buckling strain of long carbon nanotubes (Harik, 2001a,b, 2002). The critical buckling strain, ε_{cr}, for the long beam-like SWCNTs is close to that for a nanoscale beam (Harik, 2001a,b, 2002):

$$\varepsilon_{cr} = (\pi^2/2)\left(d_{NT}/L_{NT}\right)^2 \tag{5.11}$$

or four times less for a carbon nanotube with hinged (or unclamped) ends. Here, a special form of the formula for the critical buckling strain is used in order to emphasize the well-known importance of the aspect ratio, d/L, in the buckling of the beams

(Timoshenko and Gere, 1961). It should be noted that formula (5.11) for the critical buckling strain of carbon nanotubes is valid only for small axial strains, i.e., the change in their length, δL_{NT}, is such that $\delta L_{NT}/L_{NT} \ll 1$. This is an important requirement, which has to be verified during numerical simulations. Predictions of Eq. (5.10) and formula (5.11) for the critical buckling strain have been examined for carbon nanotubes of small radius (Fig. 5.4) and compared with the data from MD simulations (Fig. 5.5).

The analysis of the axial buckling of carbon nanotubes demonstrates that the continuum beam approximation can be used for the atomic lattices of small radii (Fig. 5.5). For carbon nanotubes with larger values of the radius, the aspect ratio, d_{NT}/L_{NT}, or its equivalent, R_{NT}/L_{NT}, is still the main nondimensional length scale parameter governing the nanoscale buckling. However, the critical buckling strain, ε_{cr}, of the high aspect ratio thin SWCNT shells has to take into account the relatively small thickness, h_{NT}, of the thin SWCNT shells with respect to their radius, R_{NT}. Therefore, the critical buckling strain, ε_{cr}, of the thin SWCNT shells is given by the following formula (Harik, 2001b, 2002):

$$\varepsilon_{cr} = 4\pi^2 \frac{R_{NT}^2}{L_{NT}^2}\left[\left(1+\frac{h_{NT}}{2R_{NT}}\right)^2 - \left(1-\frac{h_{NT}}{2R_{NT}}\right)^2\right], \tag{5.12}$$

where h_{NT} is the equivalent thickness of the thin or thick lattice shells of carbon nanotubes. Note that the nanoscale buckling strain, ε_{cr}, depends on the thickness-to-radius ratio, h_{NT}/R_{NT}, of high aspect ratio SWCNTs, but only on the second-order effects. The radius-to-length ratio, R_{NT}/L_{NT}, is the equivalent of the aspect ratio, d_{NT}/L_{NT}, of

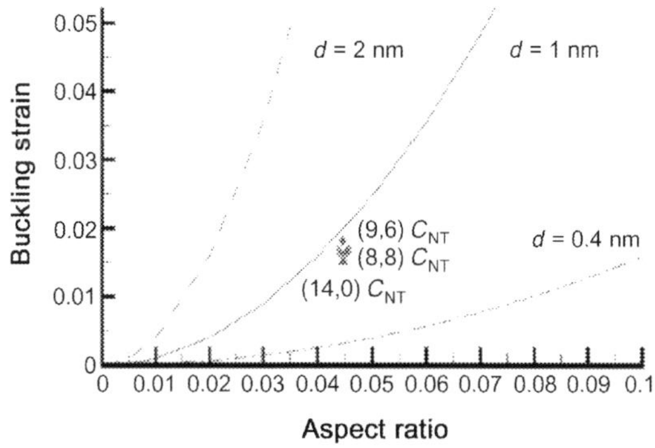

Figure 5.5

Approximation of the representative values of the critical buckling strain, ε_{cr}, of SWCNT *nanobeams* ($d \approx$ 0.4 nm), *SWCNT lattice shells* ($d \approx$ 1 nm), and the high aspect ratio *thin SWCNT shells* ($d \approx$ 2 nm) by the equivalent-continuum beam theory compared with the molecular dynamics simulation data for the (9,6), (8,8), and (14,0) SWCNTs (Chang and Chiang, 2009) for different aspect ratios, d_{NT}/L_{NT}. *SWCNTs*, single wall carbon nanotubes. © *2011 by Nanodesigns Consulting (reprinted with a permission).*

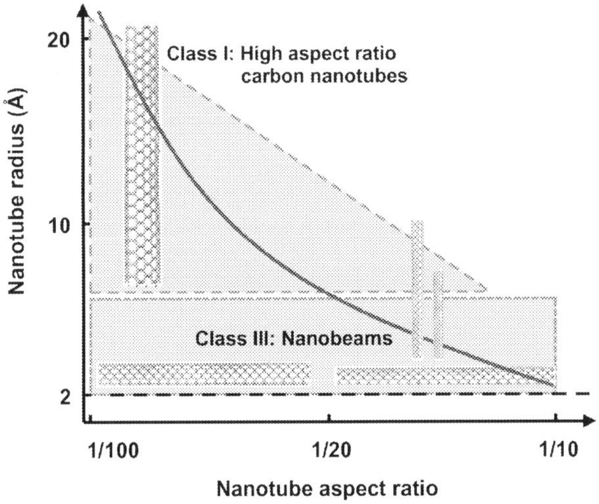

Figure 5.6

A material map for two classes of the high aspect ratio carbon nanotubes: *long beam-like shells* (class I with up to 10 subclasses) and the class of nanobeams and nanocrystals *of* small radii, i.e., class III with up to five subclasses. The thick line ($\sim 1/(10\ d_{NT}/L_{NT})$) represents a boundary for the optimal material structure of the high aspect ratio lattices, for which this material map also serves as the model applicability map for the nanoscale beam models. © *2011 by Nanodesigns Consulting (reproduced with a permission).*

SWCNTs, which is an important structural parameter in many applications such as the carbon-nanotube-based AFM probes. The nondimensional ratios d_{NT}/L_{NT} and h_{NT}/R_{NT} are critical for the analysis of any buckling process for long beam-like SWCNT shells.

Validity of the aforementioned assumptions used in the equivalent-continuum beam models depends on the characteristics of atomic lattice structure of carbon nanotubes. While the maximum strain of 2% is allowed in the linear elastic models, in nanomechanics of SWCNT lattices 5% may be allowed, due to their hyperelastic properties. For the ranges of the atomic lattice parameters specified in the material map (Fig. 5.6) for the high aspect ratio lattices of SWCNT nanobeams, SWCNT nanocrystals and the high aspect ratio SWCNT lattice shells, various equivalent-continuum beam models can be extended to the nanoscale mechanics under appropriate conditions.

5.2.5 Continuum Models with the Lattice Chirality for Carbon Nanotubes

This chapter is focused on the beam-like structures and elongated lattice shells that have one dominant dimension, as opposed to short SWCNT lattice shells and SWCNTs of large diameters and graphene sheets, for instance. The possible length-scale limitations of a linear beam model and the underlying continuum assumptions have to be analyzed within the framework of scaling analysis and hierarchical dimensional analysis of the deformation response of SWCNTs and the influence of their geometric and material parameters (e.g., the bond length: $l_{C-C} = 1.42-1.44$ Å, the carbon ring size, $a = 2.46$ Å, the radius, R_{NT}, and the chirality of the atomic lattices).

Continuum models for carbon nanotubes may include the chirality of the atomic lattices of SWCNTs and the dependence of the bond length and the size of a carbon ring. These are called lattice-sensitive continuum models, in which the size of the lattice shell is connected with the type of chirality of SWCNTs and the number of carbon rings in their circumference. These lattice-sensitive models allow one to demonstrate an explicit dependence of the critical buckling strain, ε_{cr}, on the helicity of the atomic lattices of carbon nanotubes. The equivalent-continuum representation of the atomic lattices of SWCNTs, along with their chirality, yields the following formula for the critical buckling strain, ε_{cr}, of the thin and thick lattice shells of SWCNTs (Harik, 2011, 2014):

$$\varepsilon_{cr} = 4\pi^2\left(\frac{a\sqrt{n^2+nm+m^2}/\pi+h_{NT}}{2L_{NT}}\right)^2 - 4\pi^2\left(\frac{a\sqrt{n^2+nm+m^2}/\pi-h_{NT}}{2L_{NT}}\right)^2, \tag{5.13}$$

where L_{NT} is aN_a or $a\sqrt{3}\,N_a$ for the armchair and the zig-zag carbon nanotubes, respectively; N_a is the number of carbon rings along the length, L_{NT}. Here, the critical buckling strain depends on the chirality of SWCNT lattices, on the number of the carbon rings involved in their circumference and their length, as well as on the size of a carbon ring. Since the effective thickness, h_{NT}, can be estimated as a half of the C–C bond length, 0.71 Å, formula (5.13) for the critical buckling strain of the (n,m) lattice shell can be written as

$$\varepsilon_{cr} = 4\pi^2\left[\left(\frac{2a\sqrt{n^2+nm+m^2}/\pi+l_{C-C}}{4L_{NT}}\right)^2 - \left(\frac{2a\sqrt{n^2+nm+m^2}/\pi-l_{C-C}}{4L_{NT}}\right)^2\right], \tag{5.14}$$

where the C–C bond length is not necessarily 1.42 Å, as it can be elongated. Here, the length, L_{NT}, is equal to aN_a or $a\sqrt{3}\,N_a$ for the armchair and the zig-zag carbon nanotubes, respectively, and N_a is the number of carbon rings along the length, L_{NT}. Under extensive elongation of the C–C bonds, the approximation: $h_{NT} \approx l_{C-C}/2$, has its limitations, which are also affected by the balance between the elastic C–C bond interactions and the van der Waals forces.

Formula (5.14) for the critical buckling strain for the helical (n,m) lattice shells is valid only for the high-aspect-ratio shells satisfying the following criteria:

- the nanoscale homogenization criterion: $L_{NT}/a > 10$; and
- the high aspect ratio criterion: $d_{NT}/L_{NT} < 1/10$.

These scaling criteria also set certain requirements for the MD simulations. The coupling between these two applicability criteria for the geometric parameters of SWCNT lattices has been examined to show the ranges of allowable aspect ratios for different diameters of SWCNTs in the model applicability map (Fig. 5.6), which can also serve as the materials map. The key nondimensional parameters that control the onset of buckling of SWCNTs and the change in buckling modes are identified by the scaling analysis of the SWCNT lattice structure. Applicability of the equivalent-continuum beam models is restricted to the two classes of the high aspect ratio carbon nanotubes specified in the material map shown in Fig. 5.6.

A mechanical law of geometric similitude for the buckling of SWCNTs and its generalization for MWCNT nanostructures applies to all different atomic lattice structures, which will have identical buckling strain when they have the same aspect ratio, d_{NT}/L_{NT}, and the thickness-to-radius ratio, h_{NT}/R_{NT}. The law of geometric similitude and the model applicability map provide a guide for the extension of MD simulations from one SWCNT structure to a broad range of SWCNT geometries specified in the model applicability map (Fig. 5.6). When the law of geometric similitude is violated for a particular case of a SWCNT lattice, other new nondimensional physical parameters have to be taken into account.

5.3 Nanoscale Buckling and the Lattice Waves or Phonons

Nanoscale buckling of atomic lattices of carbon nanotubes involves generation of lattice waves at the bifurcation point at the onset of the axial buckling of SWCNTs and WMCNTs. The equivalent-continuum models, with the lattice-sensitive presentation of the dependence of the critical buckling strain on the chirality of atomic lattices of SWCNTs, can address the problem of deformation of individual carbon rings and generation of the lattice waves or phonons. The nanoscale phenomenon of the buckling of atomic lattice structures can be investigated by the aforementioned equivalent-continuum beam models, the lattice-sensitive formulation, MD simulations, and AFMs with the SWCNT-based tips. The equivalent-continuum beam models provide a useful approximation for the analysis of nanoscale deformation of carbon nanotubes, including their buckling in nanoscale probes. The lattice-sensitive formulation provides a framework for studying the influence of chirality of atomic lattices on their deformation through the discrete changes of radial dimension of SWCNTs. Theoretical modeling of nanoscale buckling of SWCNTs and MWCNTs can be verified experimentally by examining the buckling response of the carbon nanotube-based AFM probes.

5.3.1 Axial Buckling of the Carbon Nanotube AFM Probes

Nanoscale phenomenon of the buckling of atomic lattice structures can be investigated by using experimental setups with the AFMs, numerical experiments with the MD simulations,

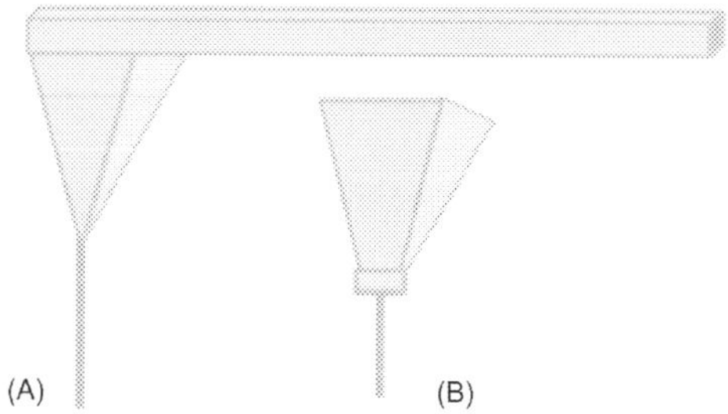

Figure 5.7
The carbon nanotube-based AFM probes for the AFM: (A) the high aspect ratio SWCNT tip on the pyramidal tip and the detachable SWCNT tip (B), which can be manufactured separately for the trapezoidal AFM tip. *AFM*, atomic force microscopy; *SWCNTs*, single wall carbon nanotubes.

and the equivalent-continuum models. The nanoscale buckling of the carbon nanotube-based AFM probes (Fig. 5.7) is important for the characterization of nanostructured surfaces. After the analysis of the axial buckling of SWCNTs, it is evident that their critical buckling strain greatly depends on the value of their aspect ratios and, to a lesser extent, on their diameter, as well as on the effective thickness, as was shown in formulae above. The beam-like buckling modes of the SWCNT-based AFM probes can be described by the Euler buckling problem for the equivalent-continuum beams under appropriate conditions (Harik, 2001a,b, 2002). The main buckling mode is the lateral deflection as in the side bending.

Dependence of the initial beam-like Euler buckling mode of carbon nanotubes on their aspect ratio is shown in Fig. 5.5 for different diameters and aspect ratios, which can be used for AFM probes. The critical buckling force, F_{cr}, is proportional to the axial stiffness of the AFM probe and its aspect ratio, d_{NT}/L_{NT}, as can be shown in the nanoscale buckling problem.[2] The MWCNTs with additional internal lattice shells (Fig. 5.8) have higher axial stiffness and larger external nanotube diameters for thin MWCNTs and thick MWCNTs (e.g., 30–85 Å). The onset of axial buckling in the MWCNTs can be understood to some extent by using the buckling and scaling analysis (Harik, 2001a,b, 2002) similar to the analysis developed for the SWCNT lattice shells. The inner MWCNT lattice shells with higher aspect ratios and smaller cross-sectional area (and higher axial stresses per one carbon ring) undergo the buckling first with the beam-like lateral bending (Fig. 5.9). The outer lattice shells with lower aspect ratios are more prone to the shell-like buckling, or a

[2] The equilibrium equation for the deflection, $w(y)$, of the continuum beam is $EI\left(d^2w/dx^2\right) + Fw(y) = 0$, where E is the Young's modulus, I is the moment of inertia and F is the buckling load or the critical force, F_{cr} (Timoshenko and Gere, 1961). A general solution of the governing equilibrium equation is given by $w(y) = A \sin ky + B \cos ky + Cy + D$, where $k^2 = F/EI$ and A, B, C, D are constants, those values are determined by the boundary conditions. Typical nontrivial solutions are $F_{cr} = n^2\pi^2EI/L^2$, where $n = 1, 2, \ldots$, and $F_{cr} = P_{cr}$ is the critical or the Euler load (for $n = 1$) for a beam with free ends. For pin-joined beams, the critical load is four times lower. For a beam with the ends fixed, it is four times higher.

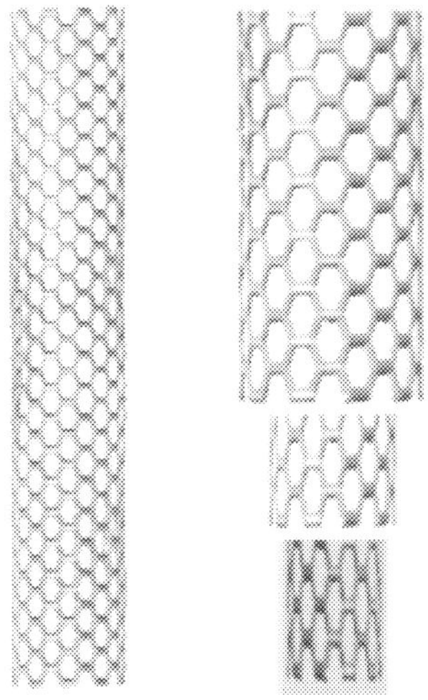

Figure 5.8
Atomic lattice structure of a thin lattice shell of the single wall (10,10) carbon nanotube (left) with the length: $L_{NT} = 52.5$ Å, and the diameter: $d_{NT} = 13.6$ Å, as well as a schematics of the thick lattice structure of a thin multiwall carbon nanotube (right), which can be used for atomic force microscopes as nanoscale probes. © *2011 by Nanodesigns Consulting, reproduced with a permission.*

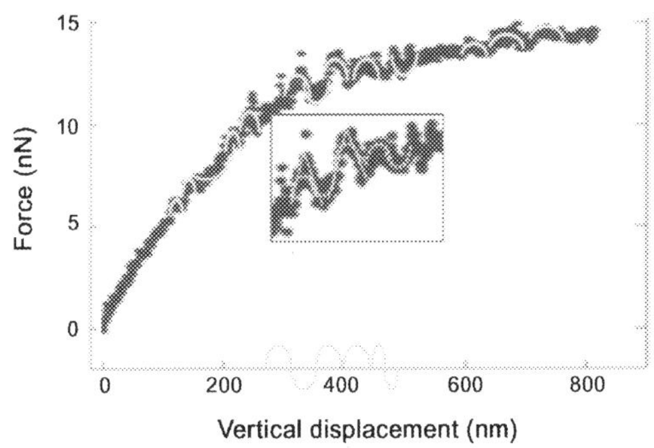

Figure 5.9
Dependence of the AFM force (nN) on the vertical displacement of the AFM MWCNT tip of the radius (35 nm) as shown by Buchoux et al. (2009). The initial buckling force is about 6 and 7 nN for the two inner lattice shells, and about 10 and 11 nN for the other two MWCNT shells. The shell-like buckling with the semisinusoidal force-displacement variations starts at about 13 nN. *AFM*, atomic force microscopy; *MWCNT*, multiwall carbon nanotube.

mixed-mode buckling involving both the beam-like mode (with a smaller value of the deflection amplitude) and the dynamic shell-like buckling modes with the associated lattice phonons (Harik, 2011). The elastically deforming MWCNT lattice should pass through a sequence of the corresponding equilibrium states. This sets a time-scale limit for a continuum model applied to the molecular structures and atomic lattices that have a significant relaxation time, Δt_k, i.e., $\Delta t_k < \Delta T$. This time scale also ensures the dissipation of the generated and scattered lattice phonons involving the corresponding background distributions of the vibration modes.

5.3.2 Nanoscale Deformation of Nanobeams and Phonons

The axial loading of one-dimensional SWCNT lattices of small radii is the simplest example of unidirectional mechanical deformation under a compressive or a tensile force. Tensile deformation of carbon rings in the zig-zag lattice of SWCNTs (Fig. 5.10A) is a simple illustration of the usefulness of the equivalent-continuum lattice-sensitive models for the analysis of nanoscale deformation of atomic lattices. The lattice-sensitive formulae (5.13) and (5.14) can indicate the axial deformation of carbon rings in a particular lattice structure with a specific chirality during axial buckling. The chirality of the SWCNT lattice indicates the orientation of the deforming carbon rings subjected to the external axial forces. Axial deformation of the zig-zag lattice includes the deformation of the arch regions in carbon rings (Fig. 5.10A), as opposed to the stiff middle section of carbon rings bounded by the two vertical C–C bonds ($l_{C-C} = 1.42$ Å), which provide the higher stiffness. This deformation of carbon rings illustrates the type of deformation associated with the propagation of longitudinal lattice waves (or phonons) in the zig-zag lattices of SWCNT

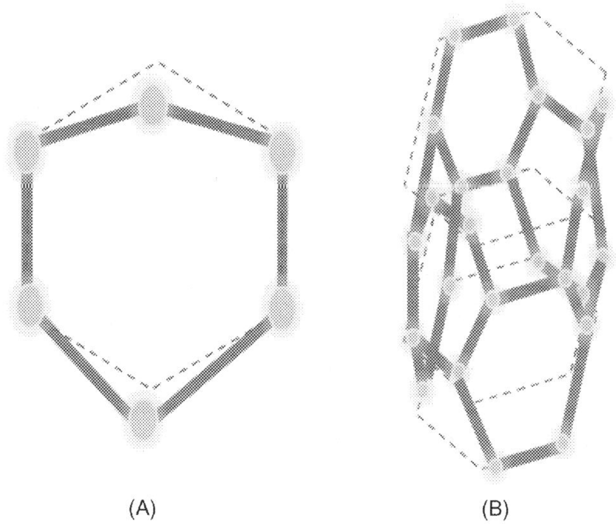

Figure 5.10
Examples of distortions in the atomic lattice of the zig-zag SWCNT lattice (A) under axial tension and changes in the axial buckling of carbon rings in the armchair lattice (B) of the (3,3) *SWCNT nanobeam* with the structural lattice corrugation. *SWCNT*, single wall carbon nanotube.

nanobeams. The two mechanisms in the deformation of individual carbon rings also demonstrate the origins of stiffness and brittleness of the zig-zag SWCNTs.

The deformation of carbon rings in SWCNT nanobeams with the armchair chiralities is illustrated in Fig. 5.10B. Axial forces can change the degree of the structural corrugation of carbon rings in these SWCNT nanobeams. The deformation of carbon rings is more homogeneous across the area of a carbon ring due the existing corrugation of the armchair SWCNTs. Poisson's ratio of the armchair SWCNTs is about 0.19, especially for the atomic lattices of the larger radii, i.e., $R_{NT} > 6$ Å. Poisson's ratio of the armchair nanobeams, which have the carbon ring corrugation of the wave length, $\lambda/2 = a/2$ (Fig. 5.10B), should be somewhat greater due to the corrugation. Poisson's ratio of the zig-zag SWCNT lattices should be a little lower, due to the higher stiffness of the arch and the vertical C–C bonds, which is the stiffer part of the zig-zag lattices. Under the axial compression, the less stiff arch region deforms first axially and then sideways, as can be seen in the Fig. 5.10A, above and below the vertical C–C bonds. The process of axial deformation affects the type of generated phonons under the dynamic conditions in the vertical and arch bonds.

Under the side bending, the lattice of SWCNTs is subjected to the side forces at the ends of a carbon nanotube and an appropriate equilibrium constrain, or a small perturbation in the middle section. The middle section of the bent lattice of SWCNTs involves either one arch atom or the four atoms of the vertical C–C bonds (Fig. 5.10A) in the zig-zag lattices or a bent carbon ring [i.e., the horizontal C–C bond or the middle section of a corrugated carbon ring (Fig. 5.10B), which is less stable] in the armchair lattices. The wave length of the deformation response, $\lambda/2$, of the bent SWCNT lattices depends on the local atomic lattice configuration and its characteristic length scales, i.e., the length of the arch section in a carbon ring, $l_{C-C}/2$, the length of the vertical C–C bonds, l_{C-C}, the size of a carbon ring, a, and the wave length of its corrugation, $a/2$ (Table 5.1). The wave length, $\lambda/2$, of the dynamic deformation response is also related to the wave length of the generated lattice phonons. The changes in the kinetic energy of the atomic lattices are connected to the energy of the generated phonons.

Table 5.1: Length scales of the middle section of the bent SWCNTs

Size of the Bent Carbon Nanotube, L_{NT}	Zig-zag[a] $(n,0)$ Lattice	Armchair (m,m) Lattice
Even number, N_a, of carbon rings, L_{NT}/a Odd number, N_a, of carbon rings, L_{NT}/a	One arch atom, five arch atoms: $\lambda/2 = l_{C-C}/2$ Four C–C bond atoms: $\lambda/2 = 3l_{C-C}/2$ or $\lambda/2 = l_{C-C}$	Two C–C bond atoms: $\lambda/2 = a \cdot 2 + 4$ atoms: $\lambda/2 = a/2$ Two side atoms: $\lambda/2 = a/2$

[a]The deformation wave length, $\lambda/2$, of the zig-zag nanotube involves the two arch sections either on the sides of the two vertical C–C bonds or in the middle at the one arch atom.

This type of generated phonons is clearly affected by the local lattice configuration, as illustrated by the local length scales noted in Table 5.1. It was mentioned earlier that the zig-zag lattices of SWCNTs are prone to the deformation of the less stiff arch regions in carbon rings (Fig. 5.10A) in the case of longitudinal and transverse phonons. The wave length of the generated phonons, $\lambda/2$, associated with these arch deformations in the dynamically bent lattices, is the length of the arch section in the carbon rings, $l_{C-C}/2$. The armchair lattices of SWCNTs are prone to the corrugation-type deformation of carbon rings (Fig. 5.10B) in the transverse and longitudinal phonons. The wave length of the longitudinal and transverse phonons, $\lambda/2$, associated with the local deformation of the dynamically bent or deformed carbon rings, is the half-length of the carbon ring, $a/2$. In the SWCNT lattices of small radii, the dynamic lattice deformation and the corresponding changes in the kinetic energy are also affected by the van der Waals forces.

Under ambient forces and thermal oscillations, the zig-zag lattices of the (3,0) SWCNT nanocrystals (Fig. 5.11) are prone to the axial deformation of the less stiff arch regions of the corrugated carbon rings associated with the longitudinal "arch" phonons, which are different from the in-plane "arch" phonons in the SWCNT lattices of larger radii. The difference is due to the local carbon ring corrugation, i.e., the out-of-plane structural extension of the two arch atoms (Fig. 5.11). The corrugated carbon rings circumscribe the (3,0) SWCNT circumference as a hexagon. The elastic stretching of the C−C bonds is influenced by the thermodynamic requirements for the closed circumference and the intratubular van der Waals forces. The dynamic stability of the (3,0) lattice structure can be affected by the out-of-plane "arch" phonons caused by the ambient forces and thermal vibrations.

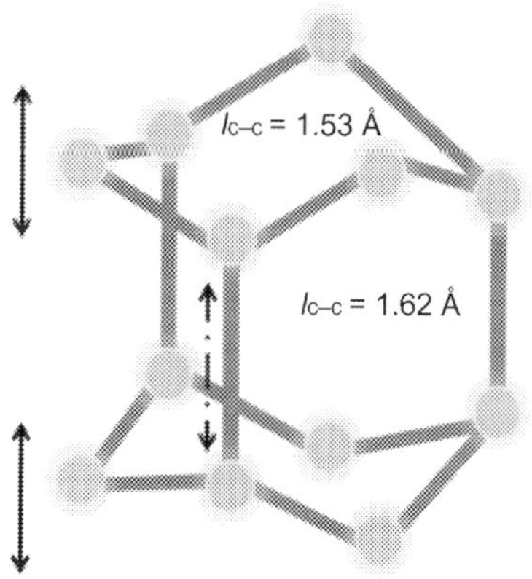

Figure 5.11
Atomic lattice of the (3,0) *SWCNT nanocrystal* and the dominant axial lattice vibrations of the side atoms (solid arrows) and the atoms of the vertical C−C bonds (the dash-point arrows). The bond length is determined by the molecular dynamics simulations (Stojkovic et al., 2001). *SWCNTs*, single wall carbon nanotubes.

5.4 Conclusions

In this chapter, the fundamentals of nanoscale mechanics of the high aspect ratio carbon nanotubes have been presented, along with the analysis of the unique material properties of nanoscale beams. The relation between the length scales in the atomic structure of the beam-like carbon nanotubes and the Cauchy–Born rule is discussed. Kinematic inequalities for the atomic lattices of carbon nanotubes are presented for thin lattice shells of SWCNTs. The need for the nanoscale homogenization correction for the Cauchy–Born approximations has been explained. Uniqueness of nanoscale beams and carbon nanotubes with small and large radii and their distinct material properties are reflected in the characteristics of beam models. Ranges of applicability of the continuum beam theories to the mechanics of carbon nanotubes are discussed, along with the assumptions used in the beam theories and the nanoscale beam models.

Atomistic analysis methods and the equations for MD simulations of carbon nanotubes have been examined, along with the expressions for the internal stresses in the atomic lattice structures. Solution of the thickness paradox for long carbon nanotubes has been presented. Continuum beam models for the beam-like carbon nanotubes and the continuum models with the lattice chirality of carbon nanotubes have been presented for the analysis of nanoscale buckling. Nanoscale buckling is examined for the high aspect ratio carbon nanotubes and the carbon nanotube probes for atomic force microscopy. Nanoscale buckling and the bending of atomic lattices have been studied to analyze the deformation mechanisms of carbon rings in the lattice waves, or phonons. Nanoscale deformation of the zig-zag and the armchair SWCNT nanobeams has been described, along the generation of phonons.

Sample Problems

1. Find a few examples of one-dimensional nanostructures, nanoscale beams or nanoscale wires with some of their lengths scales similar to those of carbon nanotubes.
2. Find examples of the SWCNT lattice structures with the following values of radius: ~2, 3, 4, 5, and 6 Å, and determine their critical buckling strain for different lengths and the aspect ratios.
3. Find examples of the SWCNT lattice structures with the following values of radius: ~6, 8, 10, 15, and 20 Å, and find their critical buckling strain.
4. What is the physical basis associated with the values of the effective thickness, h_{NT}, of 0.71 and 3.4 Å for the lattices of SWCNTs? What is the role of the C–C bond interactions and the van der Waals forces?
5. Determine the values of the thickness-to-radius ratio, h_{NT}/R_{NT}, for the effective thickness, h_{NT}, of 0.71 and 3.4 Å for the SWCNT lattices listed in Problem 3.

6. Determine the values of the thickness to the radius ratio, h_{NT}/R_{NT}, for the effective thickness, h_{NT}, of 0.71 and 3.4 Å for the SWCNT lattices listed in Problem 2.
7. Specify the values of the length of SWCNTs, for which the critical buckling strain of some carbon nanotubes in the Problems 2 and 3 is identical.
8. Find the values of the critical buckling strain of carbon nanotubes in Problem 3 for two aspect ratios below 1/100.
9. What is the definition of the Poisson ratio? What is the value of the Poisson ratio for the carbon nanotubes? Is it the same for all carbon nanotubes?
10. List the arguments why the axial stiffness of the zig-zag carbon nanotubes is higher than the axial stiffness of the armchair lattices of similar radius. When would the value of their length make a difference?
11. Describe the nanoscale phenomenon of "the dynamic carbon ring corrugation."
12. What is the nanoscale homogenization criterion?
13. List the key mechanical assumptions required in the Euler continuum beam theory.
14. Write different versions of the lattice-sensitive formula for the critical buckling strain. Why is it important?
15. Determine the width and the length of the arch regions in the atomic lattice structure of the carbon rings in terms of the carbon ring size, a, and the C–C bond length, l_{C-C}.

References

Arroyo, M., Belytschko, T., 2004. Finite crystal elasticity of carbon nanotubes based the exponential Cauchy–Born rule. Phys. Rev. B 69, 115415-0.

Born, M., Huang, K., 1954. Dynamical Theory of Crystal Lattices. Oxford University Press, Oxford.

Brenner, D.W., 1990. Empirical potential for hydrocarbons for use in simulating the chemical vapor deposition of diamond films. Phys. Rev. B 42, 9458–9471.

Buchoux, J., Aime, J.-P., Boisgard, R., Nguyen, C.V., Buchaillot, L., Marsaudon, S., 2009. Investigation of the carbon nanotube AFM tip contacts: free sliding versus pinned contact. Nanotechnology 20, 475701.

Chang, I.-L., Chiang, B.-C., 2009. Mechanical buckling of single-walled carbon nanotubes: atomistic simulations. J. Appl. Phys. 106, 114313.

Govindjee, S., Sackman, J.L., 1999. On the use of continuum mechanics to estimate the properties of nanotubes. Solid State Commun 110, 227.

Harik, V.M., 2001a. Ranges of applicability for the continuum beam model in the mechanics of carbon nanotubes and nanorods. Solid State Comm. 120 (7–8), 331–335.

Harik, V.M., 2001b. Ranges of applicability for the continuum beam model in the constitutive analysis of carbon nanotubes: nanotubes or nano-beams? NASA/CR-2001-211013. NASA Langley Research Center, Hampton, VA.

Harik, V.M., 2002. Mechanics of carbon nanotubes: applicability of the continuum-beam models. Comput. Mater. Sci. 24 (3), 328–342.

Harik, V. M. et al., 2002. Applicability of the continuum shell theories in the mechanics of carbon nanotubes, Technical Report NASA/CR-2002-211460, ICASE, NASA Langley Research Center, Hampton, Virginia.

Harik, V.M., 2011. Mechanics of Carbon Nanotubes. Nanodesigns Press, Nanodesigns Consulting, Newark, DE.

Harik, V.M. (Ed.), 2014. Trends in Nanoscale Mechanics. Springer, Dordrecht.

Huang, Y., Wu, J., Hwang, K.C., 2006. Thickness of graphene and single-wall carbon nanotubes. Phys. Rev. B. 74, 245413.

Iijima, S., 1991. Helical microtubules of graphitic carbon. Nature 354, 56.
Ohno, K., Esfarjani, K., Kawazoe, Y., 1999. Computational material science: from ab initio to Monte Carlo methods. In: Cardona, M., et al., (Eds.), Solid State Sciences. Springer, Berlin.
Ruoff, R.S., Tersoff, J., Lorents, D.C., Subramoney, S., Chan, B., 1993. Radial deformation of carbon nanotubes by van der Waals' forces. Nature 364, 514–516.
Stojkovic, D., Zhang, P., Crespi, V.H., 2001. Smallest nanotubes: breaking the symmetry of sp^3 bonds in tubular geometries. Phys. Rev. Lett. 87 (12), 125502.
Timoshenko, S.P., Gere, J.M., 1961. Theory of Elastic Stability. McGraw-Hill, New York, NY.
Treacy, M.M.J., Ebbesen, T.W., Gibson, J.M., 1996. Exceptionally high Young's modulus observed for individual carbon nanotubes. Nature 381, 680.
Wong, E.W., Sheehan, P.E., Lieber, C.M., 1997. Nanobeam mechanics: elasticity, strength, and toughness of nanorods and nanotubes. Science 277, 1971.
Yakobson, B.I., Brabec, C.J., Bernholc, J., 1996. Nanomechanics of carbon tubes: instabilities beyond linear response. Phys. Rev. Lett. 76, 2511.
Yakobson, B.I., Dumitrica, T., 2003. Nanomechanics: physics between engineering and chemistry. In: Harik, V. M., Salas, M. (Eds.), Trendsin Nanoscale Mechanics. Kluwer Academic Publishers, The Netherlands, pp. 3–33, Chapter 1.

CHAPTER 6

Mechanics of Carbon Nanotubes: Shell Models, Chiral Formulations and Nanoscale Buckling

Chapter Outline

6.1 Unique Properties of Atomic Lattice Shells of Carbon Nanotubes

Single wall carbon nanotubes (SWCNTs) are hollow cylindrical shell-like nanostructures, which are composed of the periodic hexagonal lattice of carbon rings (Iijima, 1991; Iijima and Ichihashi, 1993). Carbon rings are the smallest structural elements in the atomic lattice structure of SWCNTs. Early studies of the mechanical behavior of SWCNTs have indicated[1] that these nanostructures have great promise as novel materials with superior mechanical properties such as a stiffness of about 1.2—1.5 TPa and a strength approaching 100 GPa, with a density of about 1.3 g/cm^3. However, because of their nanometer size, direct measurement of their properties has been extremely difficult. Moreover, variations in the geometry of SWCNTs and their atomic lattice structures turned out to be so significant that the material properties of different SWCNTs and their mechanical and structural

[1] Foundations of mechanics of carbon nanotubes have been established between 1992 and 2004 as shown by the references presented in the last chapter of this book.

Mechanics of Carbon Nanotubes.
DOI: https://doi.org/10.1016/B978-0-12-811071-3.00006-8

© 2018 Elsevier Inc. All rights reserved.

properties display considerable deviations. Mechanical behavior of carbon nanotubes have been examined by experiments with the atomic force microscopes (AFM) (Treacy et al., 1996; Wong et al., 1997), theoretical modeling (Ruoff et al., 1993), molecular dynamics (MD) simulations (Yakobson et al., 1996), and nanoscale equivalent-continuum modeling (Harik, 2001a,b, 2002; Arroyo and Belytschko, 2004). In order to utilize both measured and predicted properties for the design of nanoscale sensors and material systems (e.g., nanocomposites), a full in-depth understanding of mechanical properties of carbon nanotubes is required.

Early experiments with long carbon nanotubes (Treacy et al., 1996; Wong et al., 1997) demonstrated that they behave as hollow nanoscale beams due their elongated atomic lattice structure. Examples of the long beam-like carbon nanotubes with the high aspect ratios associated with the flexibility and spontaneous Euler buckling due to environmental forces are shown in Fig. 6.1. Theoretical models, on the other hand, have indicated (Ruoff et al., 1993; Yakobson et al., 1996) that SWCNTs deform as cylindrical atomic lattice shells. The length, L_{NT}, of SWCNTs was noted as an important parameter in determining the beam-like response under axial loading. Fig. 6.1 illustrates examples of short sections of SWCNT shell structure as a part of long carbon nanotubes. However, later it was shown that it is not the length, L_{NT}, of SWCNTs, but their aspect ratio, d_{NT}/L_{NT}, which is the key nondimensional parameter, which determines when carbon nanotubes behave as either thin or thick hollow beams (Harik, 2001a,b, 2002). The value of the aspect ratio, d_{NT}/L_{NT}, when

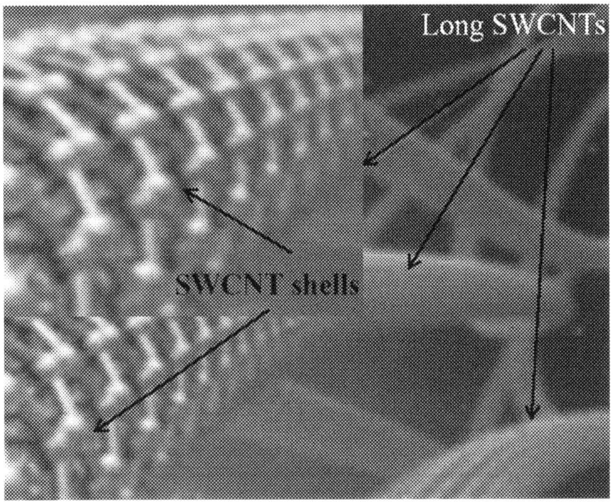

Figure 6.1
Examples of short SWCNT shells and the shell-like sections in the structure of the long beam-like carbon nanotubes with very high aspect ratios associated with the flexibility of SWCNTs and the spontaneous Euler buckling due to environmental forces around and within the loose agglomerate of SWCNTs. *SWCNT*, single wall carbon nanotube.

the transition between the beam-like behavior and the shell-like response occurs in SWCNTs for different atomic lattice structures, is especially useful.

The high aspect ratio cylindrical shells of carbon nanotubes have unique structural properties when they have one or multiple shells. Mechanical behavior of the high aspect ratio SWCNTs was examined in the previous chapter. Classification of carbon nanotubes separates 10 classes of the high aspect ratio nanotubes from the two classes of thin and thick SWCNTs and the two classes of the thin and thick multiwall carbon nanotubes (MWCNTs). All these classes of carbon nanotubes have different material and structural properties. However, there exists a critical value of the aspect ratio, d_{NT}/L_{NT}, of carbon nanotubes, when different atomic lattice structures of SWCNTs and MWCNTs start to behave as either thin or thick lattice shells, excluding, of course, the classes of nanobeams and nanocrystal rods. The critical value of the aspect ratio, $(d_{NT}/L_{NT})_{cr}$, of carbon nanotubes satisfies the following condition (Harik, 2001a,b, 2002; Harik et al., 2002):

$$\left(\frac{d_{NT}}{L_{NT}}\right)_{cr} \geq \frac{1}{10}, \tag{6.1}$$

where the diameter, d_{NT}, may vary between 0.2 and 2–3 nm for SWCNTs, and 35 and 85 nm for MWCNTs. Consequently, the length, L_{NT}, of nanoscale shells should be less than 2 or 20–30 nm for SWCNTs, and less than 350–850 nm for MWCNTs. Therefore most of the carbon nanotubes used in computer industry and nanoelectronics have the structural properties and mechanical behavior of the nanoscale hollow beams or nanoscale shells. As a result, the analysis of nanoscale buckling and bending of SWCNTs and MWCNTs in nanodevices has to be based on the concepts associated with the nanoscale beams and nanoscale shells. The minimum value of the aspect ratio, $(d_{NT}/L_{NT})_{min}$, of carbon nanotubes is associated with the collapse of the longitudinal and transverse length scales in the atomic lattice structure of SWCNTs: $(d_{NT}/L_{NT})_{min} \approx 1$. Naturally, the minimum value of the aspect ratio, $(d_{NT}/L_{NT})_{min}$, satisfies the condition (6.1) for the shell-like behavior of carbon nanotubes. It is clear that the sections of the carbon nanotubes shown in Fig. 6.1, whose lengths are such that $d_{NT}/L_{NT} \approx 1$, display the structure of lattice shells. The range of values of the diameters, d_{NT}, and the lengths, L_{NT}, which satisfy the condition (6.1) for the shell-like behavior is different for SWCNTs and MWCNTs. These ranges of values: $(d_{NT}/L_{NT})_{shell} \leq (d_{NT}/L_{NT})_{cr}$, and the transition lines: $(d_{NT}/L_{NT})_{shell} = (d_{NT}/L_{NT})_{cr}$ can be identified on the material maps discussed in the two previous chapters. Note that the condition (6.1) only approximates the beam-shell transition line on the material maps.

The condition (6.1) for the shell-like behavior is important, but there is another condition which should be satisfied by the atomic lattice of carbon nanotubes. The form and the area of the cross section of carbon nanotubes are also important for determining when their atomic lattice structure will behave as short hollow beams or thin (or thick) lattice shells. It means that the condition (6.1) for the shell-like behavior of carbon nanotubes is the necessary condition, but it is not the sufficient condition. A nondimensional ratio of the thickness-to-radius, h_{NT}/R_{NT}, of carbon nanotubes can characterize the size of their cross-sectional area. The uniqueness of nanoscale lattice shells and nanoscale beams is associated with several nanoscale phenomena.

The thickness of nanoscale shells, graphene, and carbon nanotubes in particular, depends on the effective thickness estimates for their discrete atomic lattice structure. There exists a range of estimates for the effective thickness of the discrete atomic lattices of carbon nanotubes (Harik, 2001b, 2002, 2011, 2014). Existence of such a range for the estimates of such an important parameter constitutes the effective thickness paradox for carbon nanotubes. A solution of this paradox is associated with the evaluation of the load dependence of the effective thickness estimates for atomic lattices (Harik, 2011, 2014). The second nanoscale phenomenon, which makes nanoscale shells unique, is associated with the effect of the curvature of the atomic lattice, the circumferential buckling of carbon rings on the material, and structural properties of nanoscale shells during the breakdown of their cylindrical symmetry. This nanoscale phenomenon is connected not only with the degeneration of cylindrical symmetry of SWCNTs, but also with the formation of distinct classes of thin SWCNT nanostructures. During this transition, the effective thickness of atomic lattices changes, when the carbon rings in the lattice buckle out-of-plane to form thick hollow SWCNT nanobeams and SWCNT nanocrystal rods with the distinct polygonal symmetry.

The third nanoscale phenomenon, which distinguishes nanoscale shells from macroscopic shells, is the size dependence of their material properties. Material properties of different classes of carbon nanotubes and their lattices become unique and independent of their size or the number of carbon rings when their atomic structure satisfies the nanoscale criteria for the unique averaging (or homogenization) of their discrete atomic lattices over an extended surface:$L_{NT} \gg a$, or $L_{NT}/a \gg 1$, where the length of the lattice, L_{NT}, should be considerably greater than the size of a carbon ring: $a = 2.46$ Å. The inequalities in the homogenization criteria mean roughly an order of magnitude difference, i.e., $L_{NT}/a \approx 10$. It has been noted earlier that the size of any nanomaterial, including nanoscale shells and carbon nanotubes, has to be such that their length, L_{NT}, is larger than the homogenization length, L_H. It turns out that the molecular structure of the fullerene C60, i.e., the Buckminster buckyball with $a \approx 2.46$ Å, represents the smallest stable nanostructure that uniquely satisfies the aforementioned homogenization criteria (Harik, 2011). This chapter presents the analysis of atomic lattice structures of carbon nanotubes whose length satisfies the nanoscale homogenization criteria, as well as the condition (6.1) for the shell-like behavior. The classes of carbon nanotubes whose atomic lattice structures behave as nanoscale shells are represented by the thin and thick SWCNTs and the thin and thick MWCNTs.

6.1.1 Deformation of Atomic Lattice Shells and the Cauchy–Born Rule

Atomic lattice shells of carbon nanotubes respond to the applied forces by the deformation of covalent bonds and carbon rings, along with changes in their potential energy. Carbon nanotubes of large diameters (Fig. 6.1) have sections of atomic lattices which deform as lattice shells, even for the high aspect ratios. Therefore the analysis of their deformation response has to take into account the structural characteristics of their atomic lattices and all of the length scales involved. The key length-scale parameters, which describe the atomic structure of SWCNT lattices (Fig. 6.2), include the length of the covalent σ-bond, l_{C-C}, 1.42–1.44 Å, and the size of a carbon ring, $a = 2.46$ Å, the radius, R_{NT}, of SWCNT lattices (~0.2–2 nm, for the single wall nanotubes, and ~35 nm for the multiwall nanotubes), or the diameter, d_{NT}, and their length, L_{NT}, varying between tens of nanometers and hundreds of microns (or even centimeters and longer). The length scale associated with the length of the C–C bond, l_{C-C}, is important for the evaluation of strain of C–C bonds and distribution of stress through elastic interactions. The C–C bonds can be stretched from its minimum size of 1.42 Å in the graphene sheets to longer lengths during elastic deformation. The length scale associated with the size of a carbon ring, a, is essential for nanoscale

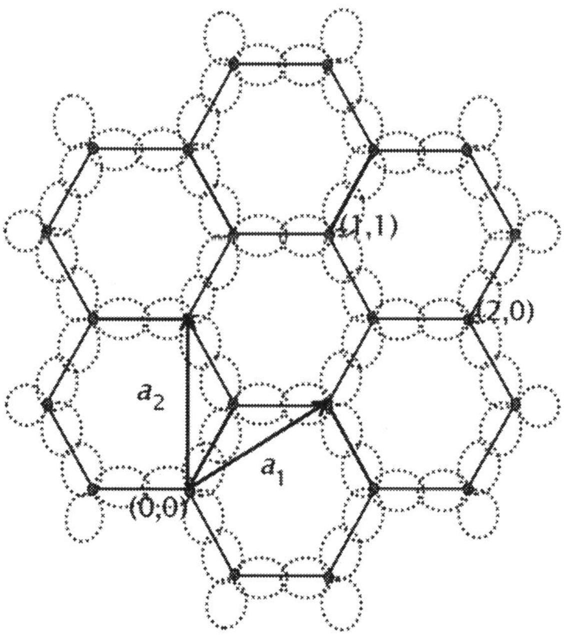

Figure 6.2
Atomic lattice structure of carbon nanotubes with the hexagonal carbon rings of carbon atoms and approximate electron distribution in the C–C bonds. The (n, m) chirality of the lattice is based on the coordinate system with the unit vectors $\boldsymbol{a}_1$ and $\boldsymbol{a}_2$.

homogenization and deformation of large shell-like sections of carbon nanotubes (Fig. 6.1). Analysis of the length scales in the atomic structure of carbon nanotubes with large diameters is important for understanding the deformation of the bundles of the long shell-like SWCNTs (Fig. 6.3). Table 6.1 presents the length scales describing the lattice structure of such SWCNTs.

Length scales of the atomic structure of SWCNTs (Table 6.1) are useful for the characterization of their deformation response of the length, λ, because the elastic deformation of any atomic lattice depends on the underlying lattice elements (i.e., a carbon ring and the C–C bond), and the length scales of the size, a, or l_{C-C}. Sections of SWCNTs shown in Fig. 6.3, whose lengths are such that $d_{NT}/L_{NT} \approx 1$, display smooth deformation response of lattice shells. The wave length, λ, of the flexural deformation response of SWCNTs in Fig. 6.3 is much larger than the size of the smallest element in the atomic

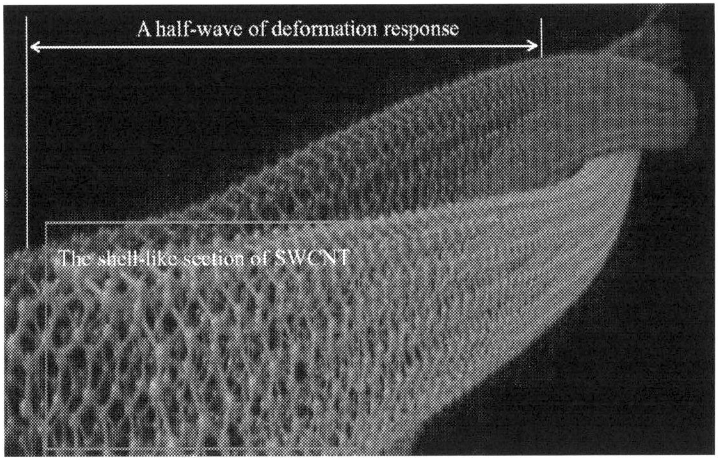

Figure 6.3
Atomic lattice structure of a bundle of long carbon nanotubes with large diameters and very high aspect ratios associated with the flexibility of SWCNTs, their deformation response with a large wave length, λ, and the spontaneous Euler buckling due to small external forces in the bundle (after Ogwen). *SWCNT*, single wall carbon nanotube.

Table 6.1: Length scales in the structure of single wall carbon nanotubes

Parameters	Typical Values	Ranges of Values
C–C bond length, l_{C-C}	1.42 Å	1.41–1.62 Å
Carbon ring size, a	2.46 Å	~2.22–2.70 Å
Effective thickness, h_{NT}	0.66–3.4 Å	Load path dependent
Diameter, d_{NT}	~1 nm	4–50 Å
Length, L_{NT}	~20–100 nm	Vary in AFM/NEMS
Length, L_{NT}	100 nm–1 μm–3 cm	Vary in nanocomposites

lattice, a, i.e., $\lambda \gg a$, so that the lattice can be approximated as an equivalent-continuum surface (Harik et al., 2002). In this case, the atomic lattice satisfies the nanoscale homogenization criteria (Harik, 2001a,b, 2002).

The Cauchy–Born approximation for atomic lattices provides a kinematic rule for the deformation of a lattice element, a carbon ring, and relates the local movement of its atoms to the overall deformation of the lattice (Born and Huang, 1954). The Cauchy–Born rule ensures that elastic deformation passes through the equilibrium states in simple lattices. The Cauchy–Born approximation equates the average strain energy of an equivalent-continuum carbon ring to the potential energy stored in its deformed covalent bonds. The continuum strain energy density can be also obtained from the interatomic potential (Arroyo and Belytschko, 2004) by introducing the exponential transformation for the Cauchy–Born rule on curved surfaces. However, in order to avoid instabilities during complex deformation, the aforementioned nanoscale homogenization criteria require that the Cauchy–Born approximation should be averaged over the smallest area of the atomic lattice of the size satisfying the homogenization criteria. As a result, the Cauchy–Born approximation (Born and Huang, 1954) and the exponential Cauchy–Born rule (Arroyo and Belytschko, 2004) should be averaged over a group of carbon rings satisfying the homogenization criteria presented in the previous section. Alternatively, an appropriate nanoscale homogenization correction should be applied. An averaging correction for the nanoscale homogenization size has to be applied to the Cauchy–Born rules in order to ensure that the transformation of material elements is based on the unique and well-defined material properties. The magnitude of the nanoscale homogenization correction will depend on the complexity of atomic lattices, their homogenization size, and the kinematics of atoms. The nanoscale homogenization correction for the Cauchy–Born approximation should take into account the interatomic interactions in a group of carbon rings limited by the size of homogenization area.

6.1.2 Kinematic Inequalities for the Atomic Lattices of Carbon Nanotubes

Kinematic motion of atoms in the deforming lattices of carbon nanotubes varies between different classes of SWCNTs and MWCNTs, including the classes of thin and thick lattice shells. Restrictions for the kinematic motion of atoms in the lattice with respect to the global deformation of the lattice can be formulated by using the chirality of atomic lattices defined by the pair of integers: (n, m). Kinematic restrictions for the kinematic motion of atoms, Δh_{NT}, in the lattice of the thin lattice shells can described by the following kinematic inequality:

$$\Delta h_{\mathrm{NT}} \ll \frac{a}{2\pi}\sqrt{n^2 + m^2 + nm}, \qquad (6.2)$$

where the radius, R_{NT}, of the deformed thin shell is connected with the chirality of the lattice by the formula: $R_{NT} = (a/2\pi)\sqrt{n^2 + m^2 + nm}$, where a is the size of a carbon ring. The kinematic inequality (6.2) for the motion of atoms can also provide an estimate for the displacement of atoms within the lattice of the thin shells of carbon nanotubes in comparison to the length, l_{C-C}, of the C–C bond:

$$\frac{2\pi\Delta h_{NT}}{l_{C-C}\sqrt{3(n^2 + m^2 + nm)}} \ll 1. \tag{6.3}$$

The kinematic restriction (6.3) on the kinematic motion of atoms in the atomic lattice of carbon nanotubes explicitly depends on the chirality of their (n, m) lattice. The kinematic inequalities (6.2) and (6.3) restrict the local kinematic motion of atoms within the wall thickness, while the kinematic motion of the carbon nanotube deflected during an axial buckling describes the global kinematic motion of atoms in the lattice. The atomic lattice of SWCNTs oscillates before the onset of its axial buckling. Oscillations of the lattice structure of SWCNTs are represented by the lattice waves or phonons (Fig. 6.4). Atomic lattice structure of the armchair (10,10) SWCNT with the radius, $R_{NT} = 6.8$ Å, and the length, $L_{NT} = 52.5$ Å, shows a deformation response represented by the lattice phonons with the wave lengths: $\lambda_{ph} \approx 4a$ and $\lambda_{ph} \approx 5a$ (Fig. 6.4). Restrictions on the kinematic motion of atoms in the lattice of SWCNTs can be used to verify the application of the Cauchy–Born rule before or after the nanoscale homogenization correction is applied. The minimum size of SWCNT lattice, for which the nanoscale homogenization correction for the Cauchy–Born rule should be applied, is determined by the criteria for nanoscale homogenization, which were discussed earlier.

The motion of atoms in the lattice shells deforming under the external forces can be monitored by expressing their axial and circumferential strains in terms of the deformation of individual carbon rings. The axial and circumferential strains of the atomic lattices determine their Poisson ratio, ν, which can be defined as $\nu = -\varepsilon_r/\varepsilon_x$, where ε_x is the axial strain, $\Delta L_{NT}/L_{NT}$, and ε_r is the radial strain, $\delta r/R_{NT}$, or the circumferential strain, $2\pi\delta r/a_cN_c$. Here, N_c is the number of carbon rings in the circumference of SWCNT lattice and a_c

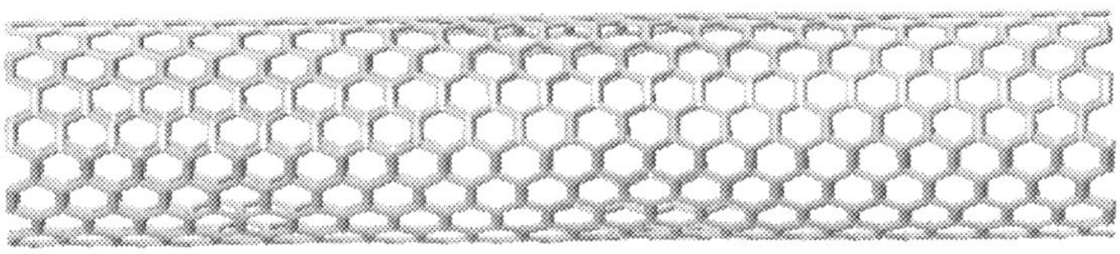

Figure 6.4
Atomic lattice structure of the armchair (10,10) single wall carbon nanotube of the radius, $R_{NT} = 6.8$ Å, and the length, $L_{NT} = 52.5$ Å, shown with the lattice phonons, e.g., along the bottom with the wave lengths: $\lambda_{ph} \approx 4a$ and $\lambda_{ph} \approx 5a$.

is the circumferential size of a carbon ring. The lateral deformation of a carbon ring, Δa_c, can be determined by

$$\Delta a_c = \frac{2\pi\delta r}{a_c N_c^2},$$

or by $\Delta a = 2\pi\delta r / a N_c^2$ for the zigzag SWCNT lattices. The kinematic motion of atoms in the atomic lattice can be monitored along with the deformation of carbon rings and verification of the Poisson ratio. The Poisson ratio can be also defined as ν_a, i.e., the Poisson ratio based on the ratio of the lateral, Δa_c, to the axial, Δa_l, deformation, $\Delta a_c / \Delta a_l$, of the carbon rings normalized by their longitudinal, a_l, and circumferential, a_c, sizes (the ring size, $a = 2.46$ Å).

6.2 Analysis Methods for the Thin and Thick Carbon Nanotube Shells

Analysis methods presented in this chapter apply to the atomic lattice shells belonging to the classes of thin and thick lattice shells. The two classes of the thin SWCNT shells and the thin MWCNT shells are defined by the criteria for the thin shells, which are given by the following inequalities:

$$\frac{h_{\mathrm{NT}}}{R_{\mathrm{NT}}} < \frac{1}{10} \text{ and } \frac{h_{\mathrm{NT}}}{R_{\mathrm{NT}}} \ll 1, \tag{6.4}$$

where the ratio of the effective thickness, h_{NT}, to the radius, R_{NT}, has been discussed earlier in the context of structural properties of SWCNTs. Atomic lattice structure of the armchair (10,10) SWCNT with the radius, $R_{\mathrm{NT}} = 6.8$ Å, represents the smallest thin shell in the class of thin lattice shells of SWCNTs. The deformation response of a thin lattice shell may involve oscillations such as lattice phonons (Fig. 6.4), e.g., phonons of the wave lengths: $\lambda_{\mathrm{ph}} \approx 4a$ and $\lambda_{\mathrm{ph}} \approx 5a$. The criteria for the classes of the thick SWCNTs and thick MWCNTs have the forms: $h_{\mathrm{NT}}/R_{\mathrm{NT}} > 1/10$, or $h_{\mathrm{NT}}/R_{NT} \gg 1$, which have been presented earlier in the chapter on classification of carbon nanotubes (Table 6.2). Atomic lattice structure of the largest thick lattice shell, the armchair (9,9) SWCNT shell, is shown in

Table 6.2: Classes of the thin and thick single wall carbon nanotubes

	Class II-a	Class II-b
Estimates of Thickness, h_{NT}	**Thin Shells[a] (R_{NT})**	**Thick Shells (R_{NT})**
Graphene analog: 3.4 Å (van der Waals spacing)	37 Å–...	2–34 Å
MD shell analog: 0.66 Å (axial buckling)	7.2 Å–...	2–6.6 Å
Half of the C–C bond length, l_{C-C}: 0.72 Å	7.9 Å–...	2–7.2 Å

[a]A transitional range of 10% is assumed.

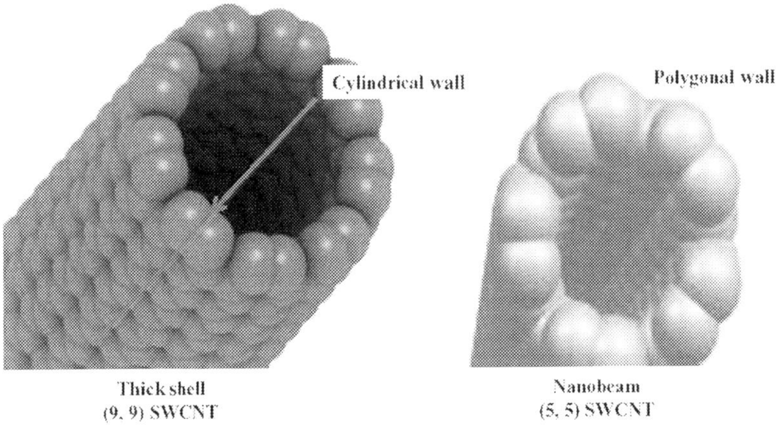

Figure 6.5

Atomic lattice structures of one of the largest thick shells, the (9,9) SWCNT with a cylindrical wall, and one of the largest nanobeams, the (5,5) SWCNT with a polygonal wall and a smaller diameter (enlarged) (after Petrushenko). *SWCNT*, single wall carbon nanotube.

Fig. 6.5 with a cylindrical wall. The cylindrical symmetry of the thick lattice shells degenerates for smaller values of the radius into the polygonal symmetry of hollow nanobeams. Atomic lattice of one of the largest nanobeams, the armchair (5,5) SWCNT, is also shown in Fig. 6.5 with the polygonal wall and a smaller diameter. The ranges of values corresponding to the classes of thin and thick SWCNT shells are presented in Table 6.2 for different estimates of the effective thickness of their atomic lattices.

The radius, R_{NT}, of atomic lattice shells can be expressed in terms of the number of carbon rings along their circumference: $R_{NT} = (a/2\pi)\sqrt{n^2 + m^2 + nm}$, where the pair (n, m) defines the chirality of their atomic lattice. Therefore the condition (6.4) for the thin lattice shells can be written as

$$\frac{h_{NT}}{R_{NT}} = \frac{2\pi h_{NT}}{a\sqrt{n^2 + m^2 + nm}} \ll 1 \quad \text{or} \quad h_{NT} \ll \frac{a}{2\pi}\sqrt{n^2 + m^2 + nm}. \tag{6.5}$$

The value of the effective thickness, h_{NT}, of the thin lattice shells can be excluded from the condition (6.5) by using an approximation: $h_{NT} \approx l_{C-C}/2$ (Harik, 2001a,b, 2002):

$$\frac{\pi}{\sqrt{3(n^2 + m^2 + nm)}} \ll 1. \tag{6.6}$$

Conditions (6.5) and (6.6) define the class of thin lattice shells for carbon nanotubes with different chiralities, which are defined by the integers: n and m. The atomic lattice structure of the armchair (10,10) SWCNT ($R_{NT} = 6.8$ Å) satisfies the conditions (6.5) and (6.6) for the thin lattice shells. It so happens that the (10,10) lattice shell shown in Fig. 6.4 is the smallest carbon nanotube, which fulfill this condition for the class of thin SWCNTs. SWCNT lattices of larger radii (both with the armchair and the zigzag chiralities) also satisfy these conditions. Note that the conditions (6.5) and (6.6) also define the thin lattice shells inside MWCNTs, which is important for the mechanics of the thin and thick MWCNT shells.

6.2.1 Molecular Dynamics Simulations of Carbon Nanotube Shells

MD simulations of carbon nanotubes are described by a system of momentum equations for the acceleration, $\mathbf{a}_i$, of carbon atoms, which are derived from Newton's second law: $\mathbf{F}_i = m_i\mathbf{a}_i$, where $\mathbf{F}_i$ is the force acting on the ith atom of mass, m_i. The initial positions of carbon atoms are defined by the atomic lattice structure of a particular carbon nanotube (Fig. 6.4). The coordinates (x_i, y_i, z_i) of each carbon atom are computed to produce a stable atomic structure of a carbon nanotube with a cylindrical symmetry. The displacement of the ith atom at the time t is described for the time increment Δt by monitoring the changes of its position vector, $\mathbf{r}_i$. The origin of the position vector, $\mathbf{r}_i$, coincides with the origin of the coordinate system (x, y, z). Acceleration, $\mathbf{a}_i$, of each carbon atom is determined from Newton's law as the force, $\mathbf{F}_i$, divided by the mass, m_i. The magnitude of the force, $\mathbf{F}_i$, acting on each atom can be evaluated as the gradient of a potential energy function, V ($\boldsymbol{r}$), which depends on interatomic interactions:

$$\mathbf{a}_i(t) = -(1/m_i)\nabla_{\mathbf{r}_i} V(\mathbf{r}_1, .., \mathbf{r}_N). \tag{6.7}$$

The acceleration, $\mathbf{a}_i$, of individual atoms can be found as the rate of change (or the first time derivative) of their velocity, $\mathbf{v}_i$, or the second time derivative of their position vector, $\mathbf{r}_i$. Therefore the governing momentum equation for the motion of carbon atoms subjected to the forces proportional to the gradient of potential energy, $V(\boldsymbol{r})$, describes the changes in their position vector, $\mathbf{r}_i$, as

$$\frac{d^2\mathbf{r}_i}{dt^2} = -\frac{1}{m_i}\nabla_{r_i} V(\boldsymbol{r}_i, \ldots, \boldsymbol{r}_N), \tag{6.8}$$

where the form of the potential function, $V(r)$, depends on specific contributions to the interatomic interactions, including the internal stresses generated by elastic deformation in a system of atoms by the external forces. Interactions between individual atoms through the covalent bonds are determined by the potential energy of interatomic interactions and the associated elastic deformation of the covalent bonds described by the atomic potentials.

Atomic potentials can be based on empirical considerations and on tight binding or ab initio calculations, as was mentioned earlier. For example, the many-body potential, $V(\boldsymbol{r})$, can be formulated as a sum of pairwise terms with the energy of pair interactions, $\phi(\boldsymbol{r}_{ij})$, which depends on their relative distance r_{ij}. The bonding interactions between atoms in carbon nanotubes can be described by the atomic potential for the bonding of carbon atoms in hydrocarbons (Brenner, 1990):

$$V = \frac{1}{2}\left[\sum_{i \neq j} \varphi_R(r_{ij}) + \sum_{i \neq j} B_{ij}\varphi_A(r_{ij})\right], \tag{6.9}$$

where the energy of pair interactions, $\phi(\boldsymbol{r}_{ij})$, has attractive, $\phi_A(\boldsymbol{r}_{ij})$, and repulsive, $\phi_R(\boldsymbol{r}_{ij})$, terms. The weight function, B_{ij}, reduces the strength of the ij-bond with the j-atom in the lattice of carbon nanotube when it has other bonds. The weight function also describes the angular dependence with a cutoff that restricts the pair potential (6.9) to the nearest neighbors. The Tersoff–Brenner potential (6.9) for carbon atoms also includes a switch-type function:

$$G(r) = \begin{cases} 1, & r < r_1 \\ \frac{1}{2}\left(1 + \cos\left[\frac{\pi(r - r_1)}{r_2 - r_1}\right]\right), & r_1 \le r \le r_2, \\ 0, & r > r_2 \end{cases}$$

which leaves the interatomic interactions unchanged for the distances between pairs of atoms smaller than the inner cutoff distance: $r_1 = 1.7$ Å. It is smoothed down to zero at the second cutoff distance: $r_2 = 2$ Å. MD simulations of a small deformation response by the atomic lattice structure of an armchair SWCNT is shown in Fig. 6.4. Atomic lattice structure of the armchair (10,10) SWCNT with the radius, $R_{NT} = 6.8$ Å, and the length, $L_{NT} = 52.5$ Å, shows a deformation response represented by the lattice phonons with the wave lengths: $\lambda_{ph} \approx 4a$ and $\lambda_{ph} \approx 5a$ (Fig. 6.4).

A system of N carbon atoms in the atomic lattice of a carbon nanotube has the total internal energy, E_{total}, composed of the sum of the kinetic energies, T_i, of individual atoms and their potential energy, U_i:

$$E_{total} = \sum_i^N T_i + \sum_i^N U_i$$

where the kinetic energy, T_i, of each atom is $T_i = (m_i\, V_i V_i)/2$, and U_i is their potential energy. The total energy of a system of N atoms can be used to evaluate the internal stresses resulting from the applied forces and the gradients of the potential energy. For example, the mechanical stress on a system of carbon atoms can be defined as the change in their internal energy, E, with respect to the strain per unit volume. The small mechanical strain is associated with the increase of the potential energy of reversible elastic deformation under the applied stress. Therefore the stress tensor, σ_{ij}, may have the following form for a linear elastic material:

$$\sigma_{ij} = \frac{1}{V_0}\left(\frac{\partial E}{\partial \epsilon_{ij}}\right)_{S=\text{const}}$$

where V_0 is the volume of a nanoscale material element determined by the nanoscale homogenization criteria, E is the total internal energy of the atomic lattice structure, and ε_{ij} is the strain tensor. The subscript S denotes that the entropy, S, of the strained system is constant during the deformation. During an elastic deformation of a nanostructured material, the internal energy of a nanoscale element can be equal to the strain energy of the deformed material element, the size of which has to satisfy the nanoscale homogenization criteria.

Atomic interactions during elastic deformation of carbon nanotubes produce the dynamic stress on individual atoms, which can be evaluated by using the energy contributions associated with the kinetic energy of the nth atom and the potential energy of its interactions with the m-atom. Stress in a nanoscale volume, V_0, of a group of atoms in the atomic lattice of a carbon nanotube is given by:

$$\sigma_{ij} = -\frac{1}{V_0}\sum_n (m_n v_i^n v_j^n + \sum_m F_i^{nm} r_j^{nm}), \tag{6.10}$$

where m_n is the mass of the nth atom, v_n is the velocity of the of nth atom with the i- and j-components, F^{nm} is the force of interactions between the nth atom and the mth atom with the i-components, and $\boldsymbol{r}^{nm}$ is the distance between the nth atom and the mth atom with the j-components. The velocity, v_n, of the nth atom in the formula (6.10) is evaluated the change of the position vector, $\boldsymbol{r}^n$, with time. The proper nanoscale averaging of stress components requires that the size of the chosen nanoscale volume, V_0, satisfies the criteria for nanoscale homogenization. Inside of this nanoscale volume element, interatomic interactions between individual atoms in the lattice should be described by the atomic potentials. The size of the nanoscale volume, V_0, determines the size of the material element in the multiscale modeling of nanostructured materials, which has to be described by the atomic potentials. The aforementioned nanoscale homogenization correction to the Cauchy–Born approximation (in the Cauchy–Born rule) has to be also calculated for the atomic lattice of this volume.

Extensive atomistic and MD simulations of carbon nanotubes remain computationally expensive. As a result, the continuum models that are appropriately tailored for a particular molecular structure and specific loading conditions may be useful for the qualitative analysis of constitutive behavior of carbon nanotube lattice shells. Moreover, approximations based on the equivalent-continuum models may be useful for the data reduction and analysis of computational data in MD simulations. Since mechanical behavior of SWCNTs is likely to depend on their atomic lattice structure, a blend of nanoscale scaling analysis and the continuum models based on atomic potentials, whenever possible, seems appropriate for the development of a methodology for the extension of continuum models to the nanoscale level.

6.2.2 Solution of the Thickness Paradox for Carbon Nanotube Shells

Deformation of carbon nanotubes, the kinematic motion of carbon atoms in the wall of SWCNTs, and the data of MD simulations can be examined and described by using the concept of thickness for the two-dimensional atomic lattices of cylindrical SWCNTs. The thickness of any nanostructure is the basic geometric parameter describing its shape and size. The thickness of atomic lattice shells is also critical for the evaluation of Young's modulus of SWCNTs, which describes their material properties and affects their mechanical behavior. Moreover, the estimates of thickness of SWCNTs can alter their evaluation as either thin or thick shells, and, hence, change the predictions of their structural response.

The first experimental results (Iijima, 1991; Iijima and Ichihashi, 1993) have suggested that the thickness of individual shells in MWCNTs, and, by extension, in SWCNTs, is the same as that in graphite. Theoretical modeling of carbon nanotube crystals has also indicated that the thickness of SWCNT lattices seem to be the same as the thickness of graphene sheets in the graphite, that is, 3.4 Å (Ruoff et al., 1993). It should be noticed, however, that carbon nanotubes in a SWCNT crystal have been subjected to radial loading through van der Waals forces. The modeling results have also shown that under transverse loading the effective thickness of carbon nanotubes, h_{NT}, is the same for all nanotubes regardless of their diameter, d_{NT}. However, MD simulations of Yakobson et al. (1996) and their continuum shell model have shown that the effective thickness of the (10,10) carbon nanotube is about 0.66 Å (Table 6.3).

The shell-based estimate of 0.66 Å for the effective SWCNT thickness is five times smaller than the earlier predicted thickness of nanotube lattices and the thickness of graphene sheets. This discrepancy in the evaluation of the thickness of carbon nanotubes constitutes thickness paradox for SWCNTs (Harik, 2001a,b, 2002, 2011, 2014; Huang et al., 2006). In the analysis of the equivalent-continuum beam model for the modeling of SWCNTs, a new estimate for the effective thickness, i.e., 0.72 Å, was presented (Harik, 2001a,b, 2002) and adjusted later to 0.71 Å or 1/2 l_{C-C} (Harik, 2011, 2014). The effective thickness, h_{NT}, can be estimated by evaluating the transverse dimension of a symmetric electron distribution in a half of the σ-bond transferring the mechanical load i.e., approximately a half of the C–C bond length, 0.71 Å or 0.72 Å, depending on the initial length of the σ-bond (Table 6.3). Therefore the effective thickness, h_{NT}, can be directly linked to the lowest length scale in the carbon ring, l_{C-C}, i.e., the length of the load-transferring covalent bond.

The new estimates for the thickness of SWCNTs have further exposed the existence of the thickness paradox for carbon nanotubes. The effective thickness paradox for carbon nanotubes became more explicit with many independent estimates pointing to much thinner effective

Table 6.3: Values of the effective thickness estimates for SWCNTs

Types of Analysis	Thickness Estimates, h_{NT} (Å)	Authors[a] of the Thickness Estimates
MD	0.66	Yakobson et al. (1996)
Scaling analysis[b]	0.71–0.72	Harik (2001a, 2011)
Tight-binding method	0.74	Zhou et al. (2000)
Local density approx.	0.75	Tu and Ou-Yang (2002)
Continuum shell theory	0.75	Pantano et al. (2004b)
Ab initio computations	0.665	Wang et al. (2005)
Continuum ring theory	0.617	Vodenitcharova and Zhang (2003)
Atomic potential	0.62–0.87	Huang et al. (2006)
Continuum model	0.87	Goupalov (2005)
Ab initio computations	0.89	Kudin et al. (2001)
Continuum hollow cylinder	0.98	Sears and Batra (2004)
SWCNT bundle crystal	3.42	Tersoff and Ruoff (1994)
MD	~3.4	Avouris et al. (1999)

MD, Molecular dynamics.
[a]Zhou et al. (2000); Kudin et al. (2001); Tu and Ou-Yang (2002); Vodenitcharova and Zhang (2003); Pantano et al. (2004a,b); Wang et al. (2005); Goupalov (2005); Huang et al. (2006); Avouris et al. (1999); Sears and Batra (2004); Batra and Gupta (2008).
[b]The effective lattice thickness, h_{NT}, is estimated as a half of the length of the C–C bond, $l_{C-C} = 1.42–1.44$ Å, by an approximation for the distribution of σ-electrons in the load-transferring bonds.

thickness of the nanotube lattices (Table 6.3). It has not been noticed, however, that the higher thickness estimate corresponds to the transverse loading through van der Waals forces, while the lower estimates are often associated with the axial loading (Harik, 2011, 2014). This load dependency of the thickness estimates is an important phenomenon associated not only with carbon nanotubes, but also with other shell-like atomic lattices (e.g., graphene).

Resolution of the effective thickness paradox for carbon nanotubes is based on a methodology for the evaluation of contributions to the change of the potential energy, U, from the van der Waals forces (Harik, 2011, 2014):

$$\sum_{i=1}^{n_{vdW}} \frac{1}{V_{vdW,i}} \left| \left(\frac{\partial^2 U}{\partial \varepsilon^2}\right)_{\varepsilon=0,i} \right|_{vdW}$$

and the elastic interactions through the C–C bonds (Harik, 2011, 2014):

$$\sum_{i=1}^{n_C} \frac{1}{V_{C-C,i}} \left| \left(\frac{\partial^2 U}{\partial \varepsilon^2}\right)_{\varepsilon=0,i} \right|_{C-C}$$

associated with the two limiting values of the effective thickness of the atomic lattices (Harik, 2011), which depend on the complexity of the applied loads at the onset of the initial strain, $\varepsilon = 0$ (Harik, 2014). The value of the effective lattice thickness, h_{NT}, in a particular deformation response of the atomic lattice (e.g., axial buckling with $h_{NT} \approx l_{C-C}/2$, or other types of deformations with h_{NT} such that $l_{C-C}/2 < h_{NT} < 3.4$ Å) is associated with the specific change in the potential energy, U, of the considered atomic lattice. In a physical setting, the balance between the elastic C–C bond interactions and the van der Waals forces in a deformation response of the atomic lattices is associated with a particular loading path. The value of the effective lattice thickness, h_{NT}, of SWCNTs is affected by the degree of the balance between the elastic C–C bond interactions associated with the approximation: $h_{\mathrm{NT}} \approx l_{\mathrm{C-C}}/2$, and the van der Waals forces associated with the experimental graphene value of $h_{\mathrm{NT}} \approx 3.4$ Å. Therefore the solution to the effective thickness paradox for carbon nanotubes can be formulated in terms of the number of carbon rings subjected to either axial forces or the radial van der Waals forces: the effective thickness of the hexagonal atomic lattice of carbon nanotubes may vary from 3.4 Å, which is the effective thickness of the graphene under the transverse loading through van der Waals forces, to 0.66 Å, for the axial loading through the load transferring covalent bonds; these variations are load dependent and proportional to the number of carbon rings subjected to a particular loading.

In the case of carbon nanotubes subjected to complex loading, the value of the effective thickness, h_{NT}, of SWCNT lattice shells can be estimated by the number of carbon rings, N_{vdW}, subjected to predominately the van der Waals forces (Harik, 2011, 2014):

$$h_{\mathrm{NT}} = \frac{1}{2} l_{\mathrm{C-C}} + \frac{l_{\mathrm{C-C}}}{2\pi R_{\mathrm{NT}}} \frac{3aN_{vdW}}{2L_{\mathrm{NT}}} \left(h_{vdW} - \frac{1}{2} l_{\mathrm{C-C}} \right), \tag{6.11}$$

or by the number of carbon rings, $N_{\mathrm{C-C}}$, mostly involved in the elastic C–C interactions:

$$h_{\mathrm{NT}} = h_{vdW} - \frac{l_{\mathrm{C-C}}}{2\pi R_{\mathrm{NT}}} \frac{3aN_{\mathrm{C-C}}}{2L_{\mathrm{NT}}} \left(h_{vdW} - \frac{1}{2} l_{\mathrm{C-C}} \right). \tag{6.12}$$

The value of the effective thickness, h_{NT}, of SWCNTs in a particular deformation response is such that $l_{\mathrm{C-C}}/2 \leq h_{\mathrm{NT}} < 3.4$ Å. These estimates for the effective thickness of carbon nanotube lattices are applicable to the lattice shells, the atomic structure of which is not affected by the high curvature effects, and the internal van der Waals forces, as in the case of thin SWCNT nanocrystals.

6.2.3 Continuum Models for the Lattice Shells of Carbon Nanotubes

Mechanical behavior of carbon nanotubes can be described by the equivalent-continuum shells with certain restrictions imposed on the characteristics of their atomic lattice structure. Assumption of the continuum surface area forces limitations on size of the homogenized lattices, while the concept of shell stiffness requires a resolution of the effective thickness paradox for carbon nanotubes. The global mechanical behavior of atomic lattices can be analyzed by representing their discrete molecular structure as an equivalent-continuum shell, which is done by equating the potential energies of the elastic deformation of the two corresponding material systems, as in the Cauchy–Born approximation. However, the nanoscale homogenization correction is required for the Cauchy–Born rules in order to satisfy the nanoscale homogenization criteria for the uniqueness and the well-definiteness of material properties. Applicability of the continuum models for carbon nanotubes also depends on several basic assumptions of the continuum shell theory (Timoshenko and Gere, 1961; Harik et al., 2002). Requirements for the relative wavelength of structural deformation and the characteristic dimensions of the atomic lattices associated with different classes of carbon nanotubes have to be satisfied as well. Such requirements are important, because continuum models of discrete atomic lattices have certain limitations in their ability to resolve the local details of discontinuous anisotropic material structures (Harik, 2001a,b, 2002).

Continuum models have been used for the analysis of mechanical behavior of carbon nanotubes and the data of experiments with AFM (Treacy et al., 1996) and MD simulations (Yakobson et al., 1996). Experimental results on the atomic structure of SWCNTs (Iijima and Ichihashi, 1993) have pointed out their "single shell" structure, which has to be analyzed and properly taken into account. Early predictions of the applicability of continuum models for carbon nanotubes have been quite optimistic: the (hollow structure of fullerenes) should also manifest itself in dynamic properties of molecules, resembling so much the macroscopic objects known in continuum elasticity as shells (Yakobson et al., 1996). "A less obvious property of CNTs is that the specific features of large nonlinear strain can be understood and predicted in terms of continuum theories." Applicability of the continuum models and their effectiveness for nanoscale modeling of carbon nanotubes depends on many important assumptions, restrictions and nanoscale requirements (Harik, 2001a,b, 2002, 2011, 2014).

Mechanical behavior of "the molecular structure can be put in the framework of continuum elasticity" (Yakobson and Dumitrica, 2003), in which "a curved sheet (of graphene) can be approximated by a uniform shell with only two elastic parameters: flexural rigidity D, and its in-plane stiffness, C." It should be noted that material parameters of two-dimensional cylindrical nanoscale shells are well-defined, when their length and their circumference are large enough to satisfy the nanoscale homogenization criteria (Harik, 2001a,b, 2002). According to the Cauchy–Born approximation, the elastic energy of a continuous shell (Timoshenko and Gere, 1961) representing a carbon nanotube can be described by a surface integral of the quadratic form associated with the potential elastic energy of local deformation (Yakobson et al., 1996):

$$E = \frac{1}{2}\iint \{D[(\kappa_x+\kappa_y)^2 - 2(1-\nu)(\kappa_x\kappa_y - \kappa_{xy}^2)] + \frac{C}{(1-\nu^2)}\left[(\varepsilon_x+\varepsilon_y)^2 - 2(1-\nu)(\varepsilon_x\varepsilon_y - \varepsilon_{xy}^2)\right]\}ds,$$

where κ is the curvature variation, ε is the in-plane strain, and x and y are local coordinates; and the dimensions of the area of surface integration have to satisfy the nanoscale homogenization criteria (Harik, 2001a,b, 2002). To derive the Cauchy–Born approximation, the elastic strain energy of a continuous shell (Timoshenko and Gere, 1961) "can be adapted" to a sheet of graphene in the structure of the armchair (10,10) carbon nanotube (Yakobson et al., 1996). The values of material parameters D and C can be evaluated by using data from ab initio and semiempirical predictions for the strain energy of this carbon nanotube at small strains. Equivalence between the potential energy of the atomic lattice and its elastic strain energy represented by the continuum shell model can be used to obtain the value of the axial and flexural bending stiffness: D and C.

The in-plane stiffness, C, of a continuum shell (Timoshenko and Gere, 1961) can be evaluated for small strains as the second derivative of the total elastic strain energy with respect to the axial strain ($C = 56$ eV/atom $= 340$ J/m^2) of the atomic lattice (Yakobson et al., 1996). The strain energy is also related to the flexural rigidity, D ($D = 1.46$ eV) and the shell radius, R, as $D/(2R^2)$, according to the linear stability analysis (Timoshenko and Gere, 1961). The Poisson ratio ($\nu = 0.15$) is estimated by using the change of the diameter of the strained carbon nanotube during axial extension. This value of Poisson ratio is supported by the experimental data for the elastic constants obtained for the single crystal graphite. The in-plane stiffness, C, of continuum shells depends on their wall thickness, h, and their elastic modulus, E, as $C = Eh$ (Timoshenko and Gere, 1961). The flexural rigidity, D, of continuum shells is determined by $D = Eh^3/12(1 - \nu^2)$ (Timoshenko and Gere, 1961). These two equations and the MD data for the shell parameters: C and D, yield the following estimates for the elastic modulus of the (10,10) carbon nanotube: $E_{NT} = 3.9$ TPa, and its effective thickness: $h_{NT} = 0.089$ nm. Yakobson et al. (1996) have used these estimates and the linear stability analysis for continuum elastic shells (Timoshenko and Gere, 1961) to assess the buckling behavior of the (10,10) carbon nanotube under axial strain. The particular geometry of the (10,10) carbon nanotube is such that the material parameters of its two-dimensional cylindrical nanostructure are well-defined. That is, the length of the (10,10) carbon nanotube used in MD simulations and its circumference are large enough to satisfy the nanoscale homogenization criteria and the criteria for the thin shells (Harik, 2001a,b, 2002). It is understandable that there are atomic lattice shells of carbon nanotubes for which this analysis has significant limitations.

The critical buckling strain, ε_{cr}, for continuum cylindrical shells can be estimated as (Timoshenko and Gere, 1961):

$$\varepsilon_{cr} = \frac{2\sqrt{D/C}}{R} = \frac{1}{\sqrt{3}}(1 - \nu^2)^{-1/2}\frac{h}{R}, \tag{6.13}$$

where the ratio of the shell thickness, h, to its radius, R, plays a critical role (Harik, 2001a,b, 2002), because the thickness-to-radius ratio, h/R, actually determines for what type of carbon nanotubes the estimate (6.13) can apply (Harik, 2011, 2014). It so happens that the radius of the (10,10) carbon nanotube is large enough to satisfy the thin shell criterion for carbon nanotubes (Harik, 2001a,b, 2002). It means that the above model has an error as an approximation for the axial deformation of the armchair (9,9) SWCNTs (Fig. 6.5) and other armchair SWCNTs with smaller radii. It should be noticed that the onset of axial buckling of the (10,10) carbon nanotube has been predicted at the critical strain, $\varepsilon_{cr} \approx 0.05$ (Yakobson et al., 1996), while the typical critical strains, ε_0, for the elastic buckling of continuum shells are below 0.01–0.02 (Timoshenko and Gere, 1961). Hence, the atomic lattices of SWCNTs are inherently hyperelastic in their mechanical behavior, but less so for the small radii.

The linear stability analysis for continuum shells (Timoshenko and Gere, 1961) is very effective for the investigation of different deformation modes of carbon nanotubes with large radii under axial compression. An axial perturbation of a cylindrical shell and its nanoscale counterparts can be evaluated by its Fourier harmonics (Yakobson et al., 1996), with m azimuthal portions and n half waves along the axis of the shell, i.e., sines and cosines of arguments my/R and $n\pi x/L$. If the axial perturbation of a carbon nanotube results in a critical strain, ε_{cr}, the variation of the elastic strain energy vanishes for the most favorable shape disturbance (m, n). Under the critical strain the cylindrical shell of a carbon nanotube becomes unstable, and starts to deform in such a way that results in the lower value of the strain energy by assuming the energetically favorable (m, n)-pattern. The critical strain, ε_{cr}, for a relatively short cylindrical shell occurs for the buckling pattern with $m = 2$ and $n \geq 1$. For a short shell or a corresponding carbon nanotube of the length, $L_{NT} = 6$ nm and diameter $d_{NT} = 1$ nm, the most favorable buckling pattern occurs for $m = 2$ and $n = 2, 3$ (Yakobson et al., 1996). This deformation pattern includes the flattening of the shell in two directions perpendicular to each other and occurs in other atomic shells with the ranges of their parameters such that $(d_{NT}/L_{NT})_{shell} \leq (d_{NT}/L_{NT})_{cr}$, as was noted above (Harik, 2011).

The critical buckling strain, ε_{cr}, for carbon nanotubes can be estimated by the formula (6.13) in the form that emphasizes its dependence on the diameter, d_{NT}. The armchair (10,10) carbon nanotube buckles under the axial critical strain, ε_{cr}, which can be expressed as inversely proportional to the diameter, d_{NT} (Yakobson et al., 1996):

$$\varepsilon_{cr} = \frac{0.1\ nm}{d_{NT}}.$$

This form of the formula (6.13) has to be used with the additional requirement for cylindrical shells about the smallness of the thickness-to-radius ratio, h/R (Harik, 2001a,b, 2002). Otherwise, the diameter of small carbon nanotubes can easily fall outside of the range of applicability of the critical strain formula (6.13) and all its forms. In fact, this formula is poorly applicable to the atomic structure of the class of hollow SWCNT nanobeams, especially when their radius is very small. Moreover, the critical strain formula (6.13) is not applicable to the axial buckling of the class of SWCNT nanocrystal rods.

6.2.4 Limitations of the Continuum Shell Models for Atomic Lattice Shells

Analysis of applicability of the continuum shell theories (Timoshenko and Gere, 1961) for the discrete atomic structure of carbon nanotubes is an important topic in nanoscale mechanics (Harik, 2011, 2014). Criteria for nanoscale homogenization of atomic lattices and the condition for the value of the thickness-to-radius ratio, h_{NT}/R_{NT}, for the thin atomic shells (Harik, 2001a,b, 2002) have to be satisfied by the range of parameters associated with the atomic lattice structure of a particular group of SWCNTs. An averaging correction for the nanoscale homogenization size has to be applied to the Cauchy–Born rule in order to ensure that the transformation of a material element of an atomic lattice is based on the unique and well-defined material properties of carbon nanotubes, as opposed to that of the fragments of graphene. Moreover, one has to consider a long list of assumptions used in the derivation of the continuum thin shell theory (Timoshenko and Gere, 1961; Harik, 2001a, 2002, 2011). There are carbon nanotube shells which can be described by the thin shell theories (Harik, 2001a, 2002, 2011, 2014). However, there are more than 10 classes of carbon nanotubes which cannot be treated as thin shells (Harik, 2001a,b, 2002). In order to ensure the robustness of data reduction schemes in MD simulations, which are based on continuum mechanics (Yakobson et al., 1996), a careful analysis of continuum approximations used in macromechanical models and possible limitations of this approach at the nanoscale level is required.

Extension of the continuum shell models to nanoscale mechanics and carbon nanotubes can be done in a variety of ways. For example, Govindjee and Sackman (1999) have considered an elastic multishell model, which might be applied to MWCNTs, in order to show the dependence of material properties on the number of shells. An assumption about the continuum cross section was shown to hold when more than 201 shells were present in a system of continuum shells. The applicability of different continuum concepts can be tested for different atomic lattices and various discrete material systems, in general. Ru (2000) has proposed an intrinsic bending stiffness for carbon nanotubes in order to decouple the bending shell stiffness from their ill-defined effective thickness, h_{NT}, and to ensure a consistent use of the classical shell theory. The ranges of applicability of the continuum concepts can be also examined by considering not only geometric parameters of SWCNTs, but also the length scales in the atomic lattice structure of SWCNTs (Harik, 2001a,b, 2002).

In order to examine possible length-scale limitations of continuum shell theories (Timoshenko and Gere, 1961) at the nanoscale level, all underlying continuum-based assumptions[2] must be examined, when applied to atomic lattices and taken into account. Assumption (1) for the atomic lattices is extended to the hyperelastic deformation. Assumption (2) allows the use of two-dimensional theory for describing the global response of SWCNTs and for deriving the elastic shell equations. Assumption (3) restricts the direction of displacement of carbon atoms located near the edges of SWCNTs and the orientation of a single nanotube. The displacement boundary conditions are more appropriate and easier to implement for the atomic lattices of SWCNTs than the traction boundary conditions. The force-displacement relation and the applicability criteria involve either a change in the length of SWCNTs or the value of the length itself.

Note that for the type of loading in the assumptions (1) and (3), the work done by the applied load is independent of the path it follows during deformation. The existence of a continuum cross section [assumption (3)] is not generally required for the analysis of buckling of SWCNTs. Similar to other open lattice structures, carbon nanotubes are characterized by their effective thickness. In the basic continuum shell models, the bending stiffness is proportional to the cube of the wall thickness. Thus, adequate representation of the wall thickness of SWCNTs is essential to the success of an equivalent-continuum model. Assumption (4) about the constant radius and cross-sectional area leads to the constant moment of inertia. The requirements for the value of radius and the nature of cross-sectional area along the length of SWCNTs need more investigation. Furthermore, assumptions (3) and (4) are not just shell assumptions, as they are used in the classical beam theory as well (Timoshenko and Gere, 1961). Assumption (5) implies that the shell cross sections do not deform in their planes; instead, they remain perpendicular to the original image of the shell reference surface during axial deformation as required by assumption (6). The Kirchhoff's hypothesis (6) can be linked to the elastic constitutive relation. Assumption (7) require that the displacements of the SWCNT shell, excluding the linear translation, are small compared to the thickness of SWCNTs. Assumption (8) requires that the through-the-thickness normal stresses and the radial components of the applied forces are small (or negligible) as compared to other elastic stresses. To restrict all stresses according to assumption (8) is problematic for SWCNTs of small radii. These

[2] Assumptions of the classical continuum shell theories (Timoshenko and Gere, 1961): (1) the continuum shell deforms elastically; (2) the length, L, and its width or the half-perimeter, πR, are sufficiently larger than the thickness so that a two-dimensional cylindrical model accurately captures the essential deformation response, λ; (3) the direction of the applied load and the dominant forces remains constant during deformation; (4) the radius, R, and the effective cross-section of the shell do not vary along the length; (5) elastic strains and rotations of the shell are small compared to unity, or the gradients of displacements are infinitesimal; (6) the radial material line elements that are straight and perpendicular (or normal) to the effective shell reference surface within the shell remain that way during deformation and are inextensible (i.e., Kirchhoff's hypothesis); (7) displacements of the shell, excluding the linear translation, are small compared to the shell thickness; (8) through-the-thickness normal stresses and the radial components of the applied forces are small (or negligible) as compared to other elastic stresses.

curvature-generated internal stresses set a lower limit on the radius of SWCNTs as far as the applicability of the thin shell model is concerned.

In the continuum shell models for SWCNTs, the equivalent form of the above assumptions has to be satisfied. In the context of mechanics of carbon nanotubes,[3] the range of validity of these different assumptions has to be examined within the framework of the eight requirements presented above by taking into account the molecular structure of SWCNTs, especially below the homogenization length scale, $L_H \sim 10a$ (Harik, 2001a,b, 2002), where most of the nanoscale effects have intrinsically discrete atomistic nature, and the local deformation mechanisms of SWCNT lattices reveal their increasing dependence on the local atomic lattice structure. The nanoscale homogenization approach and the hierarchical scaling analysis of different classes of carbon nanotubes provides a link between the atomic lattice structure of SWCNTs, their mechanical behavior, and the equivalent-continuum shell models.

Assumptions used in the derivation of estimates for the effective thickness of SWCNTs are also important in the analysis of their limitations at the nanoscale. The effective thickness of the atomic lattice of carbon nanotubes, h_{NT}, was estimated to be 0.06 nm for the cylindrical shell model (Yakobson et al., 1996), which was implicitly based on all of the aforementioned assumptions for the thin shell theory (Timoshenko and Gere, 1961). Discrepancy in the estimates of the effective thickness of the atomic lattice of carbon nanotubes, h_{NT}, can significantly change the range of parameters, i.e., values of the radius of SWCNTs, for which the continuum shell theory or, rather, the thin shell theory can applied. Discrepancy in the estimates of the effective thickness of carbon nanotubes results in the inconsistency in the evaluation of the Young's modulus of carbon nanotubes (Table 6.4).

Table 6.4: Young's modulus of single wall carbon nanotubes in early experiments

Methods Used in a Study	Predicted Values (TPa)	Authors of Predictions
MD	5.5[a]	Yakobson et al. (1996)
AFM vibration experiments	1.5–5	Treacy et al. (1996)
AFM bending experiments	1.3	Wong et al. (1997)
AFM tensile experiments	0.32–1.5	Yu et al. (2000)

AFM, atomic force microscopes
MD, Molecular dynamics.
[a]Molecular dynamics and the shell model predictions of Young's modulus are associated with the value of 0.66 Å for the effective thickness, h_{NT}, of carbon nanotubes.

[3] Harik, V. M., "Mechanics of Carbon Nanotubes" © 2001, lecture notes for a short course, which was presented at the Annual Congresses of American Society of Mechanical Engineers (ASME) in 2001 and 2004, and at the ASME Nanosystems Conference (2002) with the ASME Continuing Education Institute and the ASME Nanotechnology Institute.

6.2.5 Continuum Models for the High Aspect Ratio Carbon Nanotube Shells

Carbon nanotubes of large diameters may have the long atomic lattice structures, which have the beam-like (Fig. 6.1) and the shell-like (Fig. 6.3) structural properties. It is understandable that continuum models for such SWCNTs have to satisfy both groups of assumptions for the above continuum shell theories and those for the continuum Euler beam theory discussed in the previous chapter. The complexity of deformation patterns in the atomic lattices of such SWCNTs is demonstrated for the loose bundles in Fig. 6.3. One has to take into account the presence of the shell-like sections in the deformation of the long beam-like SWCNTs (Fig. 6.1), as well as complex deformation response across long deforming sections of the shell-like lattice structure of SWCNTs with large diameters in a bundle (Fig. 6.3). Such a combination of the beam-like and the shell-like effects in the mechanical behavior of SWCNTs is unique and has not been examined in many details before.

In order to capture the beam-like behavior of the long shell-like SWCNTs and their spontaneous Euler buckling shown in Fig. 6.1 and their buckling in a bundle (Fig. 6.3), the importance of their aspect ratio, $d_{\mathrm{NT}}/L_{\mathrm{NT}}$, should be recognized in the process of buckling. The formula for the critical buckling strain in the linear stability analysis (Timoshenko and Gere, 1961) can be written in the form which underscores the importance of the aspect ratio, $d_{\mathrm{NT}}/L_{\mathrm{NT}}$. However, the shell-like behavior of the high aspect ratio SWCNTs with large diameters is governed by another nondimensional parameter, i.e., their thickness-to-radius ratio, $h_{\mathrm{NT}}/R_{\mathrm{NT}}$. Their shell-like deformation response can be different depending on the thin or thick lattice shell structure of a particular SWCNT. For very low thickness-to-radius ratios, $h_{\mathrm{NT}}/R_{\mathrm{NT}}$, SWCNTs with large diameters can deform even under van der Waals forces (Avouris et al., 1999) (Fig. 6.6).

The analysis of the axial buckling of carbon nanotubes (Fig. 6.7) can demonstrate how the beam-like behavior of long SWCNTs occurs with the shell-like deformation patterns of their cross section (Fig. 6.3). It should be noted that for carbon nanotubes with larger values of the radius, the aspect ratio, $d_{\mathrm{NT}}/L_{\mathrm{NT}}$, or its equivalent, $R_{\mathrm{NT}}/L_{\mathrm{NT}}$, is still the main nondimensional length scale parameter governing their nanoscale buckling. However, the critical buckling strain, $\varepsilon_{\mathrm{cr}}$, of the high aspect ratio thin SWCNT shells has to take into account the relatively small thickness, h_{NT}, of the thin SWCNT shells with respect to their radius, radius, R_{NT} (Harik, 2001b, 2002) Therefore the critical buckling strain, $\varepsilon_{\mathrm{cr}}$, of the thin SWCNT shells is given by the following formula (Harik, 2002, 2011, 2014):

$$\varepsilon_{cr} = 4\pi^2 \frac{R_{\mathrm{NT}}^2}{L_{\mathrm{NT}}^2}\left[\left(1+\frac{h_{\mathrm{NT}}}{2R_{\mathrm{NT}}}\right)^2-\left(1-\frac{h_{\mathrm{NT}}}{2R_{\mathrm{NT}}}\right)^2\right], \tag{6.14}$$

where h_{NT} is the equivalent thickness of the thin or thick lattice shells of carbon nanotubes. Note that the nanoscale buckling strain, $\varepsilon_{\mathrm{cr}}$, depends on the thickness-to-radius ratio, $h_{\mathrm{NT}}/R_{\mathrm{NT}}$,

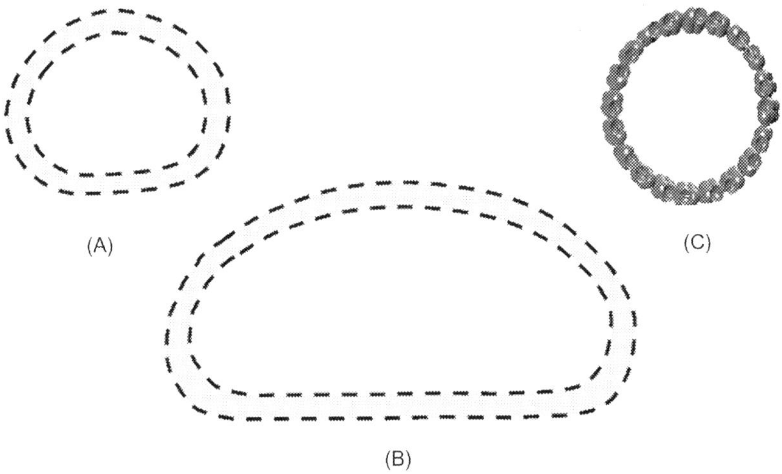

Figure 6.6
Deformed contours of the atomic lattices of SWCNTs with large diameters under the van der Waals forces (after Avouris et al., 1999): (A) the armchair (20,20) SWCNT (B) the (40,40) SWCNT and the armchair (10,10) SWCNT (C). *SWCNT*, single wall carbon nanotube.

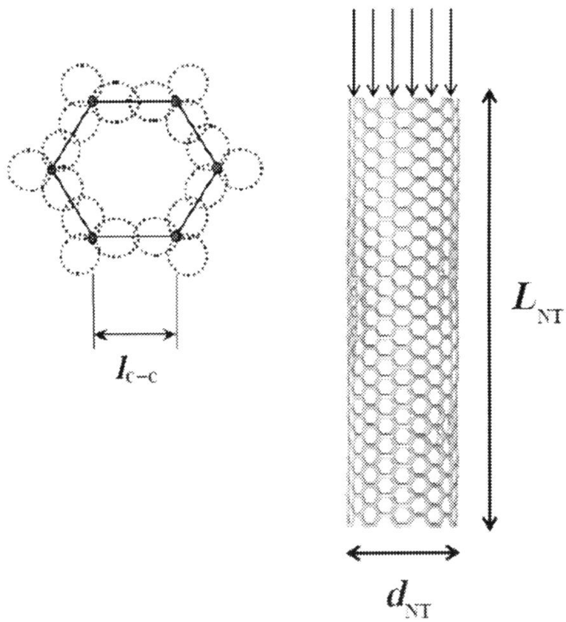

Figure 6.7
Schematic of a carbon ring with the length of the C–C bond, $l_{C-C} = 1.41-1.42$ Å, and the size of 2.46 Å, and the atomic lattice structure of an armchair (10,10) carbon nanotube of diameter, $d_{NT} = 13.6$ Å, and the length, $L_{NT} = 52.5$ Å, which is subjected to the axial loads during nanoscale buckling.

of high aspect ratio SWCNTs, but only in the second-order effects. The nondimensional parameters: d_{NT}/L_{NT} and h_{NT}/R_{NT} are critical for the analysis of any buckling process for the long shell-like SWCNTs. The formula (6.14) for the critical buckling strain of carbon nanotubes is valid only for small axial strains, i.e., for the changes in their length, δL_{NT}, such that $\delta L_{NT}/L_{NT} \ll 1$. While the maximum strain of 2% is allowed in the linear elastic models, in the mechanics of SWCNTs a larger buckling strain may be allowed due to their hyperelastic properties. Note that the length, L_{NT}, equals to aN_a or $a\sqrt{3}\,N_a$, for the armchair and the zigzag carbon nanotubes, respectively, where N_a is the number of carbon rings along the length, L_{NT}. Here, the critical buckling strain can be connected with the number of the carbon rings involved in their length and the size of a carbon ring.

6.2.6 Continuum Models With the Lattice Chirality for Carbon Nanotubes

The continuum shell models, which have been examined in the previous two sections, require the proper definition of continuum media or equivalent continuum for the discrete atomic lattice structures (Harik, 2001a,b, 2002). It has been noted before that the definition of a continuum media is possible when the length of carbon nanotube satisfies the nanoscale homogenization criteria: $L_{NT} \gg a$, or $L_{NT}/a \gg 1$, where a is the size of a carbon ring in the atomic lattice of SWCNTs. It should be noticed that the consequence of the proper definition of a physical continuum is that the value of the elastic modulus of carbon nanotubes is unique and well-defined, because the length, L_{NT}, is long enough to satisfy the above criteria for nanoscale homogenization of the atomic lattices (Fig. 6.8). The equivalent-continuum models can allow the introduction of the lattice-sensitive formalism in order to capture some effects of the chirality of atomic lattices and the deformation of individual carbon rings. For example, first a formula for the critical buckling strain of thin shells can be presented with the nondimensional parameter, the thickness-to-radius ratio, h_{NT}/R_{NT}, in the following form:

$$\varepsilon_{cr} = \frac{1}{\sqrt{3(1-\nu^2)}}\left(\frac{h_{NT}}{R_{NT}}\right),$$

where ν is the Poisson ratio and h_{NT} is the effective thickness of the thin SWCNT shells. Second, the lattice-sensitive formalism can be used for the radius of SWCNTs. As a result, some of the discrepancies in the buckling strain predictions for ε_{cr} between various continuum shell models for the (n, m) lattices of SWCNTs (Fig. 6.7) can be partially avoided by using the following formulae for the nanoscale critical buckling strain, ε_{cr}:

$$\varepsilon_{cr} = \frac{2\pi}{\sqrt{(1-\nu^2)}}\frac{h_{NT}/a}{\sqrt{3(n^2+m^2+nm)}} \tag{6.15}$$

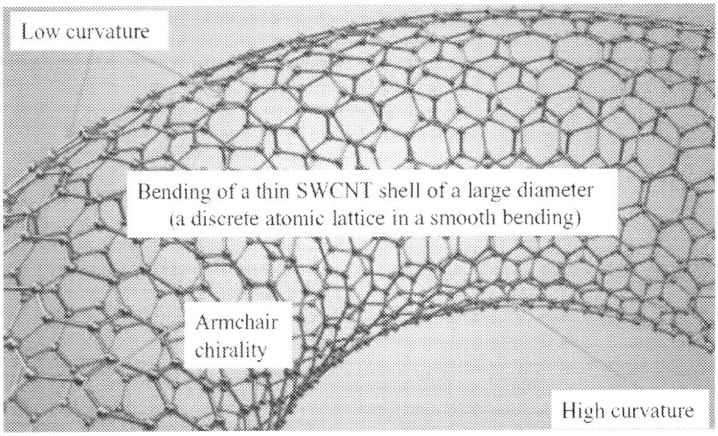

Figure 6.8
Atomic lattice structure of an armchair SWCNT of a large diameter displaying a smooth deformation response of a thin lattice shell during its bending. The high curvature section is showing smooth bending due to the presence of many carbon rings. *SWCNT*, single wall carbon nanotube.

or

$$\varepsilon_{cr} = \frac{2\pi}{\sqrt{(1-\nu^2)}} \frac{h_{\mathrm{NT}}/l_{C-C}}{\sqrt{n^2+m^2+\mathrm{nm}}}, \tag{6.16}$$

where $l_{C-C} = a\sqrt{3}$ and the (n, m) indices represent the chirality of the lattice structure of SWCNTs. This formulation shows that continuum models for carbon nanotubes may include the chirality of the atomic lattices of SWCNTs and the dependence on the bond length and the size of a carbon ring.

In the lattice-sensitive continuum models, the radial size of the deforming lattice shell is connected with the type of chirality of SWCNTs and the number and the local orientation of carbon rings subjected to elastic strains. The atomic lattice structure of thin SWCNT shells of large diameters (Fig. 6.8) displays a smooth deformation response of an equivalent-continuum shell during its bending. The lattice-sensitive models allow one to demonstrate an explicit connection between the critical buckling strain, $\varepsilon_{\mathrm{cr}}$, and the helicity of the atomic lattices of SWCNTs, which defines the complexity of the local deformation. The formula (6.16) for the critical buckling strain, $\varepsilon_{\mathrm{cr}}$, for the helical lattice shells is valid only for SWCNTs with low aspect ratios satisfying the criterion: $d_{\mathrm{NT}}/L_{\mathrm{NT}} > 1/10$. Dependence of the critical buckling strain, $\varepsilon_{\mathrm{cr}}$, of SWCNT lattices on their radius, R_{NT}, for various estimates of the effective thickness, h_{NT}, is shown in Fig. 6.9. MD data (Zhang et al., 2009) for the (10,10) armchair SWCNT with the radius: $R_{\mathrm{NT}} = 6.78$ Å (circle) is in good agreement with the continuum shell model with the effective thickness, $h_{\mathrm{NT}} = 0.72$ Å. The critical strain, $\varepsilon_{\mathrm{cr}}$, for the (17,0) zigzag SWCNT, $R_{\mathrm{NT}} = 6.65$ Å (triangle) is lower, because the MD simulations can account for the earlier arch region buckling. Formulae

(6.15) and (6.16) are applicable for the analysis of MD data for the aforementioned atomic lattices, since the nanoscale homogenization criteria and the scaling condition for the class of thin shells: $h_{NT}/R_{NT} \ll 1$, are fulfilled. The laws of geometric similitude allow one to extend the qualitative applicability of the above results to the buckling deformation of other thin lattice shells (Fig. 6.10). However, the atomic lattice structure of such thin lattice shells

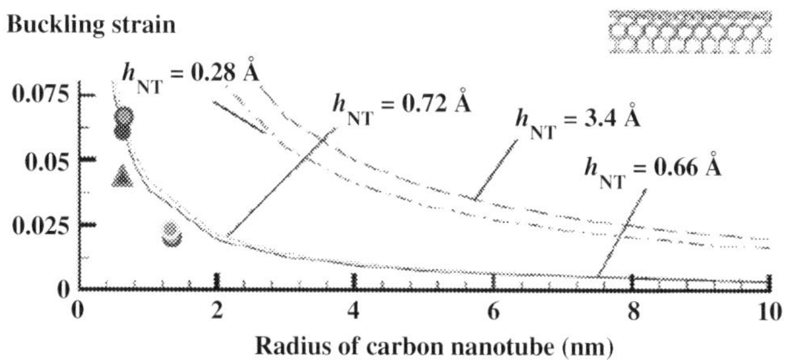

Figure 6.9
Dependence of the critical buckling strain, ε_{cr}, on the radius, R_{NT}, and the effective thickness, h_{NT}, (Harik, 2011) with the molecular dynamics data shown for the (10,10) and (20,20) armchair lattices, $R_{NT} = 6.78$ Å and 13.6 Å (circles), and the (17,0) zigzag lattice, $R_{NT} = 6.65$ Å (triangle). *© 2011 by Nanodesigns Consulting (reproduced with permission).*

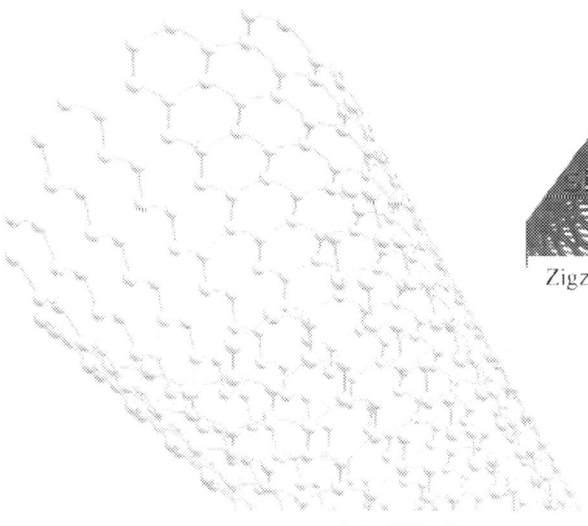

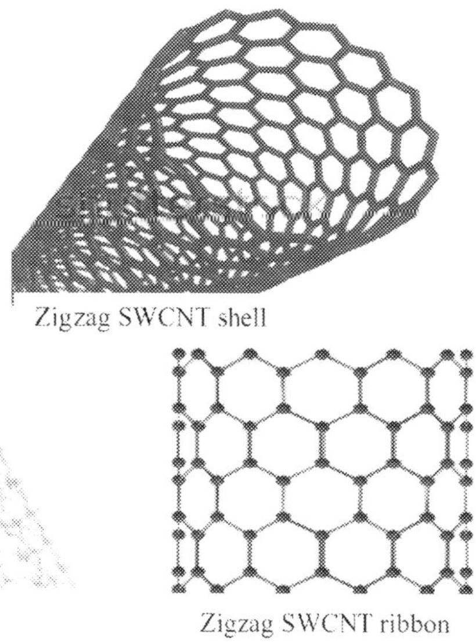

Figure 6.10
Atomic lattice structure of the thin lattice shells of SWCNTs with the armchair and the zigzag chiralities. The zigzag SWCNT ribbon has the structure of the thin lattice shells, but not their material and structural properties, due to its small length and the nanoscale size-dependence of material properties (after Petrushenko). *SWCNT*, single wall carbon nanotube.

has to satisfy all of the aforementioned requirements including the nanoscale homogenization criteria.

6.3 Conclusions

In this chapter, the fundamentals of mechanics of carbon nanotube shells have been presented, along with examples of their atomic lattice shells under axial and transverse loads. The Cauchy–Born rule, its nanoscale homogenization correction, and kinematic inequalities for the atomic lattices of thin SWCNT shells have been discussed. Equations for MD simulations of carbon nanotubes have been reviewed, along with the expressions for stresses in a system of carbon atoms. A solution of the thickness paradox for carbon nanotube shells has been presented. Ranges of applicability of the continuum shell theories and the resulting equivalent-continuum models in the mechanics of carbon nanotubes are discussed, along with the assumptions used. New lattice-based criteria for the applicability of the thin shell models to different classes of SWCNT lattices are presented without the explicit use of the thickness and the radius. A new lattice-sensitive form of a formula for the critical strain of SWCNTs is presented, along with the results of nanoscale buckling of carbon nanotubes for different estimates of their effective thickness. Results of the continuum shell model for carbon nanotubes have been compared with the available data of MD simulations.

Sample Problems

1. Find a few examples of two-dimensional nanostructures: nanoscale shells or graphene sheets or nanoribbons, with some of their dimensions similar to those of carbon nanotubes.
2. Find examples of the atomic lattice structures with the following values of radius: ~ 10, 20, 30, 500, and 800 Å, and determine their critical buckling strain for different lengths and the aspect ratios.
3. Find examples of the SWCNT lattice structures with the following values of radius: ~ 10, 20, 30, and 40 Å, and find their critical buckling strain.
4. What is the physical basis associated with the values of the effective thickness, h_{NT}, of 0.71 and 3.4 Å for the lattices of SWCNTs? What is the role of the C–C bond interactions and the van der Waals forces?
5. Determine the values of the thickness-to- radius ratio, h_{NT}/R_{NT}, for the effective thickness, h_{NT}, of 0.71 and 3.4 Å for the SWCNT lattices listed in Problem 3.
6. Determine the values of the thickness-to-radius ratio, h_{NT}/R_{NT}, for the effective thickness, h_{NT}, of 0.71 and 3.4 Å for the SWCNT lattices listed in Problem 2.
7. Specify the values of the length of SWCNTs for which the critical buckling strain of some carbon nanotubes in the Problems 2 and 3 is identical.

8. Find the values of the critical buckling strain of carbon nanotubes in Problem 3 for two aspect ratios below 1/100.
9. What is the nanoscale homogenization criterion?
10. List the key assumptions required in the continuum shell theory.
11. Write different versions of the lattice-sensitive formula for the critical buckling strain of carbon nanotube shells.
12. Calculate the thickness-to-radius ratios for the armchair (5,5), (8,8), (10,10), and (20,20) SWCNTs and the zigzag (5,0), (10,0), (14,0), (17,0), and (20,0) SWCNTs.
13. Specify the scaling condition for the class of thin lattice shells in different forms.
14. What is the minimum length of carbon nanotube suitable for the nanoscale averaging? What are the two forms of the nanoscale homogenization criteria and its two forms?

References

Arroyo, M., Belytschko, T., 2004. Finite crystal elasticity of carbon nanotubes based the exponential Cauchy–Born rule. Phys. Rev. B 69, 115415-0.

Avouris, P.H., Hertel, T., Martel, R., Schmidt, T., Shea, H.R., Walkup, R.E., 1999. Appl. Surf. Sci. 141, 20.

Batra, R.C., Gupta, S.S., 2008. J. Appl. Mech. 75, 061010.

Born, M., Huang, K., 1954. Dynamical Theory of Crystal Lattices. Oxford University Press, Oxford.

Brenner, D.W., 1990. Empirical potential for hydrocarbons for use in simulating the chemical vapor deposition of diamond films. Phys. Rev. B 42, 9458–9471.

Goupalov, S.V., 2005. Phys. Rev. B 71, 085420.

Govindjee, S., Sackman, J.L., 1999. On the use of continuum mechanics to estimate the properties of nanotubes. Solid State Commun. 110, 227.

Harik, V.M., 2001a. Ranges of applicability for the continuum beam model in the mechanics of carbon nanotubes and nanorods. Solid State Commun. 120 (7-8), 331–335.

Harik V.M., Ranges of applicability for the continuum beam model in the constitutive analysis of carbon nanotubes: nanotubes or nano-beams? NASA/CR-2001-211013, 2001b, NASA Langley Research Center, Hampton, VA.

Harik, V.M., 2002. Mechanics of carbon nanotubes: applicability of the continuum-beam models. Comput. Mater. Sci. 24 (3), 328–342.

Harik, V. M. et al., 2002. *Applicability of the continuum shell theories in the mechanics of carbon nanotubes*, Technical Report NASA/CR-2002-211460, ICASE, NASA Langley Research Center, Hampton, Virginia.

Harik, V.M., 2011. Mechanics of Carbon Nanotubes. Nanodesigns Press, Newark, DE.

Harik, V.M. (Ed.), 2014. Trends in Nanoscale Mechanics. Springer, Dordrecht.

Huang, Y., Wu, J., Hwang, K.C., 2006. Thickness of graphene and single-wall carbon nanotubes. Phys. Rev. B 74, 245413.

Iijima, S., 1991. Helical microtubules of graphitic carbon. Nature 354, 56.

Iijima, S., Ichihashi, T., 1993. Single-shell carbon nanotubes of low diameters. Nature 363, 603.

Kudin, K.N., Scuseria, C.E., Yakobson, B.I., 2001. Phys. Rev. B 64, 235406.

Pantano, A., Boyce, M.C., Parks, D.M., 2004a. Phys. Rev. Lett. 91, 145504.

Pantano, A., Parks, D.M., Boyce, M.C., 2004b. J. Mech. Phys. Solids 52, 789–821.

Ru, C.Q., 2000. Effective bending stiffness of carbon nanotubes. Phys. Rev. B 62 (15), 9973.

Ruoff, R.S., Tersoff, J., Lorents, D.C., Subramoney, S., Chan, B., 1993. Radial deformation of carbon nanotubes by van der Waals' forces. Nature 364, 514–516.

Sears, A., Batra, R.C., 2004. Phys. Rev. B. 69, 235406.

Tersoff, J., Ruoff, R.S., 1994. Structural properties of a carbon nanotube crystal. Phys. Review Letters 73, 676.

Timoshenko, S.P., Gere, J.M., 1961. Theory of Elastic Stability. McGraw-Hill, New York, NY.

Treacy, M.M.J., Ebbesen, T.W., Gibson, J.M., 1996. Exceptionally high Young's modulus observed for individual carbon nanotubes. Nature 381, 680.

Tu, Z.C., Ou-Yang, Z.C., 2002. Phys. Rev. B 65, 233407-4.

Vodenitcharova, T., Zhang, L.C., 2003. Phys. Rev. B 68, 165401.

Wang, L., Zheng, Q., Liu, J.Z., Jiang, Q., 2005. Phys. Rev. Lett. 95, 105501.

Wong, E.W., Sheehan, P.E., Lieber, C.M., 1997. Nanobeam mechanics: elasticity, strength, and toughness of nanorods and nanotubes. Science 277, 1971.

Yakobson, B.I., Brabec, C.J., Bernholc, J., 1996. Nanomechanics of carbon tubes: instabilities beyond linear response. Phys. Rev. Lett. 76, 2511.

Yakobson, B.I., Dumitrica, T., 2003. Nanomechanics: physics between engineering and chemistry. In: Harik, V. M., Salas, M. (Eds.), Trends in Nanoscale Mechanics. Kluwer Academic Publishers, The Netherlands, pp. 3–33. Chapter 1.

Yu, M.F., Lourie, O., Dyer, M.J., Moloni, K., Kelly, T.F., Ruoff, R.S., 2000. Strength and breaking mechanism of multi walled carbon nanotubes under tensile load. Science 287 (5453), 637–640.

Zhang, X.H., Santoro, G.E., Tartaglino, U., Tosatti, E., 2009. Phys. Rev. Lett. 102 (12), 125502.

Zhou, X., Zhou, J.J., Ou-Yang, Z.C., 2000. Phys. Rev. B 62, 13692.

CHAPTER 7

Mechanics of Multiwall Carbon Nanotubes: Unique Structures, Beam and Shell Models, Sliding and Friction

Chapter Outline

7.1 Unique Properties of Multiwall Carbon Nanotubes

Multiwall carbon nanotubes (MWCNTs) are hollow cylindrical nanostructures which are made of the multiple lattice shells composed of the hexagonal carbon rings (Iijima, 1991). Similarly to the single-wall carbon nanotubes (SWCNTs), carbon rings are the smallest structural elements in the atomic lattice structure of MWCNTs. Therefore, many considerations about the length scales in the atomic structure of lattice shells apply for MWCNTs (Fig. 7.1). MWCNTs have many unique characteristics, including their distinctive material properties and their multishell structure. Mechanical behavior of MWCNTs, their strength and the breaking mechanisms under tensile loading have been examined by experiments with the atomic force microscopes (Yu et al., 2000). These

Mechanics of Carbon Nanotubes.
DOI: https://doi.org/10.1016/B978-0-12-811071-3.00007-X

© 2018 Elsevier Inc. All rights reserved.

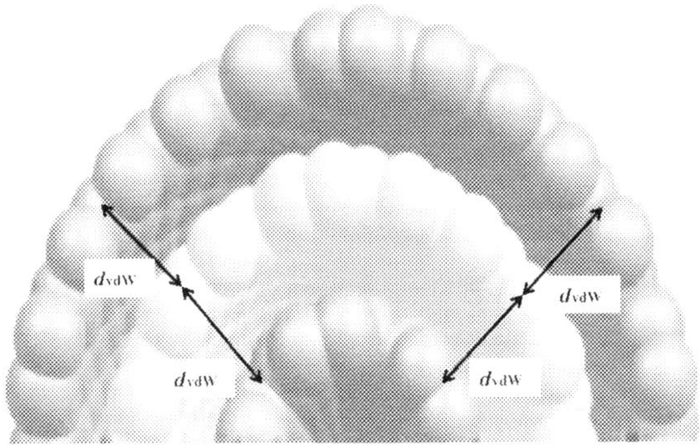

Figure 7.1
One of the simplest atomic lattice structures of MWCNTs with the chiral external wall having the cylindrical symmetry and the armchair single-wall carbon nanotube (a polygonal nanobeam) as its smallest internal wall (both separated by the van der Waals distance, d_{vdW}) (after Petrushenko). *MWCNTs*, multiwall carbon nanotubes.

experiments have shown that the value of Young's modulus of MWCNTs may vary from 270 to 950 GPa, which is somewhat lower than that of SWCNTs.

During experiments with MWCNTs and SWCNTs, Yu et al. (2000) have "pointed out that concepts such as Young's modulus and elastic constants belong to the framework of continuum elasticity; an estimate of these material parameters for nanotubes implies the continuum assumption. Because each individual SWCNT involves only a single layer of rolled graphene sheet, the thickness [h] will not make any sense until it is given based on the continuum assumption." The unique nanoscale characteristics, which distinguish nanoscale shells in MWCNTs from the macroscopic shells, are connected with the discrete atomic structure of nanostructures, as well as with the size dependence of their material properties. In order to use the convenient concepts of Young's modulus and other material constants, it is necessary to introduce the nanoscale homogenization criteria allowing proper averaging of material properties with or without the definition of a physical continuum. However, for nanostructured materials, in general, and for MWCNTs, in particular, the value of their Young's modulus will depend on the size of their atomic lattice structure. In order to resolve all of the above problems with the discrete nature of atomic lattices, the size dependence of their material properties and the desire to use the concepts of continuum elasticity, the size of MWCNTs has to satisfy the nanoscale homogenization criteria.

It is important to note that the material properties of atomic lattices of MWCNTs become unique and independent of their size or the number of carbon rings along their length when the size of their atomic structure satisfies the nanoscale criteria for the unique averaging (or homogenization) of their discrete atomic lattices over an extended surface, i.e.,: $L_{NT} \gg a$, or $L_{NT}/a \gg 1$, where the length of the lattice, L_{NT}, should be considerably greater than the size of a carbon ring: $a = 2.46$ Å. The inequalities in these homogenization criteria mean roughly an order of magnitude difference, i.e., $L_{NT}/a \approx 10$, as for SWCNTs. For MWCNTs, however, it can be noted that their length, L_{NT}, has to be larger than the homogenization length, L_H, while the difference between their external and internal radius can be larger than the so-called radial homogenization size, R_H. MWCNTs with the thickness below the value of radial homogenization, R_H, depend on their individual atomic structure, while those with larger thickness become the unique class of MWCNT beams with the well-defined material properties independent of the number of internal lattice shells or walls. This is the class of the homogenized MWCNT beams.

The uniqueness of MWCNTs is also associated with different combinations of chiralities of the internal lattice shells (Fig. 7.1). When the thickness of MWCNTs is larger than the value of radial homogenization, R_H, then a particular combination of chiralities of the internal lattice shells plays a less important role. However, different types of combinations of chiralities of the internal lattice shells are associated with distinct constitutive behavior of MWCNTs. For example, MWCNTs that have better commensurability of the internal lattice shells have higher Young's moduli. In contrast, MWCNTs with poor commensurability of the internal lattice walls have lower values of Young's modulus. The constitutive properties of MWCNTs are much more versatile than those of SWCNTs. Nanoscale homogenization provides a useful framework for the analysis of the constitutive properties of MWCNTs.

Classification of MWCNTs into different classes, which includes the classes of thin and thick MWCNT shells and MWCNT beams, provides a basic framework for further analysis of their structural and material properties. It has been noted that the definition of the thin and the thick lattice shells depends on the thickness-to-radius ratio, h_{NT}/R_{NT}. In the case of MWCNTs, the definition of their thickness is intimately connected with the number of internal lattice shells in their structure. Moreover, the estimates for the effective thickness for their internal walls with their discrete atomic lattices are not as important as in the case of SWCNTs. However, under axial loading, the estimates of the effective thickness of internal lattice walls are very important for the analysis of the constitutive behavior of MWCNTs, especially, if there is poor commensurability between the internal lattice walls.

The atomic lattice structure of MWCNTs can be quite unique, when some of their internal walls become affected by their high curvature and the circumferential buckling of carbon rings, as in the case of SWCNTs with small diameters. These MWCNTs usually have small diameters or very large thickness, which can be above the radial homogenization size. Such MWCNTs will have the external lattice walls with cylindrical symmetry and the internal lattice walls with the polygonal symmetry of hollow nanobeams. These MWCNTs are, in fact, composite nanostructures that are made of the external cylindrical lattice shells and the internal polygonal lattice shells or nanobeams (Fig. 7.1). The external and internal lattice shells of such MWCNTs have somewhat different material and structural properties. However, strictly speaking, all MWCNTs are composite nanostructures, because of their multiwall structure and the different chiralities of their internal walls.

7.1.1 Length Scales in the Structure of Multiwall Carbon Nanotubes

Atomic lattice shells in MWCNTs have large diameters and the flexibility of thin lattice shells loosely supported by van der Waals forces on one or both sides (Fig. 7.1). Therefore, the key length-scale parameters, which describe the atomic structure of MWCNTs, have to include the distance between the internal walls in MWCNTs, as well as their chiralities and the commensurability of the adjacent atomic lattices (Table 7.1). The distance between the internal walls in MWCNTs, 3.3–3.7 Å (Fig. 7.1), is comparable with the length of the covalent C–C bonds, $l_{C-C} \approx 1.42$ Å, and the size of a carbon ring, $a = 2.46$ Å. It should be noted that the equilibrium distance between the adjacent graphene lattices is about 3.4 Å, and the small deviations caused by radial deformation or the internal buckling of shells in MWCNTs are associated with the increased van der Waals interactions (Fig. 7.2). The length scale associated with the length of the C–C bond, l_{C-C}, is important for the evaluation of strain of C–C bonds and the distribution of stress through elastic interactions. The C–C bonds can be stretched from their minimum size of 1.42 Å in the graphene sheets to longer lengths during elastic deformation (Table 7.1). The length scale associated with the size of a carbon ring, a, is essential for nanoscale homogenization and deformation of large shell-like sections of carbon nanotubes.

The number of internal walls in MWCNTs is very important, since it determines their thickness and their structural and material properties. The larger the number of internal walls in MWCNTs, the thicker and stiffer they are, due to higher Young's modulus. Thickness of multiple shells in MWCNTs affects their mechanical behavior and structural response. It is plausible that the effective thickness of single shells in MWCNTs may

Table 7.1: Length scales in the structure of multiwall carbon nanotubes

Parameters	Typical Values	Ranges of Values
C–C bond length, l_{C-C}	1.42 Å	1.41–1.62 Å
Carbon ring size, a	2.46 Å	~2.22–2.70 Å
Van der Waals distance, d_{vdW}	3.4 Å	3.4 Å
Van der Waals layer, h_{vdW}	1.7 Å	1.7 Å
Layer of electrons, h_e	0.91 Å	0.91 Å
Thickness of electron layers, $2h_e$	1.82 Å	1.82 Å
Effective thickness, h_{NT}	0.66–3.4 Å (single wall)	6.8–34 Å (multiple walls)
Diameter, d_{NT}	~6.0 nm	35–85 Å
Length, L_{NT}	~60–100 s nm	Vary in AFM/NEMS
Length, L_{NT}	100 nm–1 μm–5 cm	Longer in nanocomposites

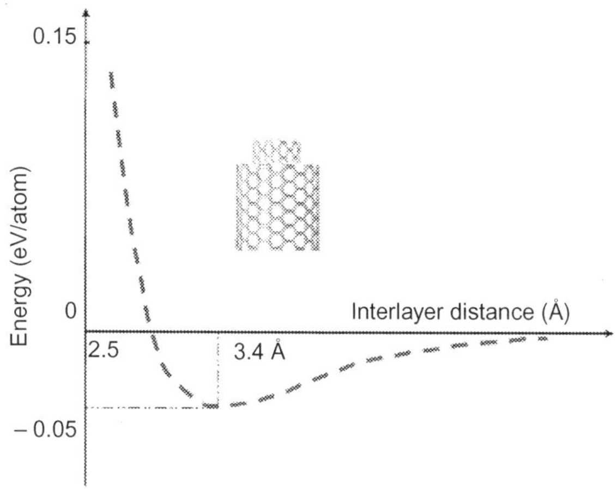

Figure 7.2
Dependence of the energy of van der Waals interactions on the distance between the adjacent atomic lattices of internal walls in the multi wall carbon nanotubes (with the graphene-based equilibrium distance of 3.4 Å).

depend on the type of loading, since the estimates for the thickness of SWCNTs are load-dependent. However, the solution of the effective thickness paradox is more complicated for MWCNTs. Under axial loads, one has to consider the effective thickness of a single shell in MWCNTs with large separations between the internal walls (Table 7.1). Moreover, one has to recognize that single lattice shells in MWCNTs are, in fact, transversely semi-isotropic, having a high in-plane stiffness of the hexagonal lattice and two much more compliant layers of the out-of-plane π-electrons with the thickness of about 1.7 Å, which is equal to the van der Waals radius. The compliance of the van der Waals layers in MWCNTs depends on their interwall separation and changes from one pair of adjacent lattices to another. Under transverse loading, the thickness of MWCNTs is simply the sum of the thicknesses of individual walls.

The thickness of MWCNTs and the number of internal walls determine whether the structure of MWCNTs belongs to the class of thin lattice shells or the class of thick lattice shells. In order to evaluate the mechanical response of a single lattice shell in a MWCNT, its radius, R_{NT}, has to be compared with the thickness of a graphene sheet, 3.4 Å, as demonstrated in Table 7.2. Similarly, the range of radii for a double wall carbon nanotube, a triple wall carbon nanotube and other MWCNTs can be evaluated when their structure will have mechanical and structural properties of the thin or thick lattice shells. According to the ranges specified in Table 7.2, many MWCNTs will likely behave as thin lattice shells or, rather, as the high-aspect-ratio thin MWCNT shells or the long thin MWCNT shells. The diameter of most MWCNTs varies between 35 and 85 nm. MWCNTs may have a variety of multishell structures which belong to different classes of thin and thick MWCNT

Table 7.2: Classes of the thin and thick multiwall carbon nanotubes

	Class II-m	Class II-mb
Estimates of Thickness, h_{NT}	Thin Shells[a] (R_{NT})	Thick Shells (R_{NT})
Single graphene shell, 3.4 Å (with two van der Waals layers)	3.7 nm–...	3.4–34 Å
Single lattice shell, 1.82 Å (with two layers of π-electrons)	2 nm–...	3.4–18 Å
Double-wall carbon nanotubes (with four van der Waals layers)	7.5 nm–...	3.4–68 Å
Triple-wall carbon nanotubes (with six van der Waals layers)	11.2 nm–...	3.4–102 Å
MWCNTs (four walls) (with eight van der Waals layers)	15.0 nm–...	3.4–136 Å
MWCNTs (five walls) (with 10 van der Waals layers)	18.7 nm–...	3.4–170 Å

MWCNTs, multiwall carbon nanotubes.
[a]A transitional range of 10% is assumed.

shells, the high-aspect-ratio MWCNTs, and diverse MWCNT beams specified in the classification of carbon nanotubes.

The relation between the length scales in the atomic structure of MWCNTs changes along their radial direction. The external chiral wall of MWCNT shown in Fig. 7.1 has sections with the length such that $d_{NT}/L_{NT} \approx 1$, which are associated with the smooth shell-like deformation response of its atomic lattice. Note that the wave length, $\lambda/2$, of its structural bending response along the half of its cylindrical circumference, is much larger than the size of the smallest element in its atomic lattice, a, i.e., $\lambda \gg a$. Consequently, the atomic lattice of the aforementioned external wall is very smooth and preserves its cylindrical symmetry, as opposed to its smallest internal wall of a hollow nanobeam with the polygonal symmetry (Fig. 7.1). The cylindrical symmetry of walls in MWCNTs is associated with a shell-like behavior, while the polygonal symmetry of a lattice shell is connected with the beam-like properties. Length scales in the structure of the smallest internal wall in MWCNTs (Fig. 7.1) are such that their diameter, d_{NT}, is much smaller than their length, L_{NT}.

Understanding of different length scales in the atomic lattice structure of MWCNTs is very important for proper modeling and analysis of their material properties. For example, Fig. 7.3 shows the atomic lattice structure of one zigzag carbon nanotube inside of another zigzag carbon nanotube, which is similar to the structure of a double-wall carbon nanotube. However, the much larger van der Waals spacing between the adjacent lattice shells and an erroneous misrepresentation of a continuum rod inside of the smaller carbon nanotube underscore the importance of understanding the length scales in the atomic structure of MWCNTs. The continuum representation shown for a rod without any indication, however approximate, of its atomic structure is simply impossible next to a specific atomic lattice. This is an example of nanotechnology illiteracy among students and the practicing scientists and engineers.

In order to understand the difference between an example of nanotechnology illiteracy about MWCNTs and a simple error associated with the misunderstanding of the atomic

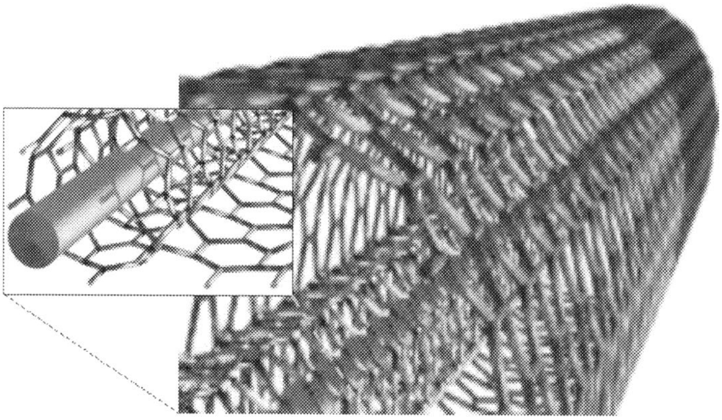

Figure 7.3
An example of the atomic lattice structure of one zigzag carbon nanotube inside of another zigzag carbon nanotube, which is similar to the structure of a double-wall carbon nanotube. The much larger van der Waals spacing between the lattice shells and an erroneous misrepresentation of a continuum rod inside of the smaller carbon nanotube underscore the importance of understanding of the length scales in the atomic structure of MWCNTs. *MWCNTs*, multiwall carbon nanotubes.

structure of MWCNTs, one should consider another example. Fig. 7.4 illustrates a typical multiwall structure of MWCNTs with the armchair lattice shells nested one inside another at a considerably smaller distance between the adjacent shells. The atomic lattice structure of such a triple-wall carbon nanotube is impossible with the armchair lattice shells shown, because their radii are too large. The van der Waals spacing between the adjacent walls in the triple-wall carbon nanotube shown is considerably smaller than normal. In the presented example, the van der Waals spacing is erroneously comparable with the length of the covalent bonds, as shown in the inset of Fig. 7.4. Consequently, the presented atomic lattice structure of this MWCNT is simply impossible with the suggested chirality and the radii shown.

An example of the atomic lattice structure of a MWCNT with four armchair lattice shells nested closely one inside another with the average van der Waals spacing is shown in Fig. 7.5. One can note that the lattice structure of this MWCNT includes the average van der Waals spacing between four armchair lattice walls with very different diameters, unlike those in the model of a MWCNT shown in Fig. 7.4. The example of the lattice structure of a MWCNT presented in Fig. 7.5 is interesting, because it includes the (3,3) armchair nanobeam, which belongs to a special class of nanobeams with their polygonal symmetry. Furthermore, this structure of a MWCNT also consists of the (6,6) wall and the (11,11) wall, which are the thick lattice shell and the thin lattice wall, respectively, belonging to two different classes of carbon nanotubes. Therefore, the presented structure of a MWCNT consists of three lattice walls that belong to three different classes of carbon nanotube lattices.

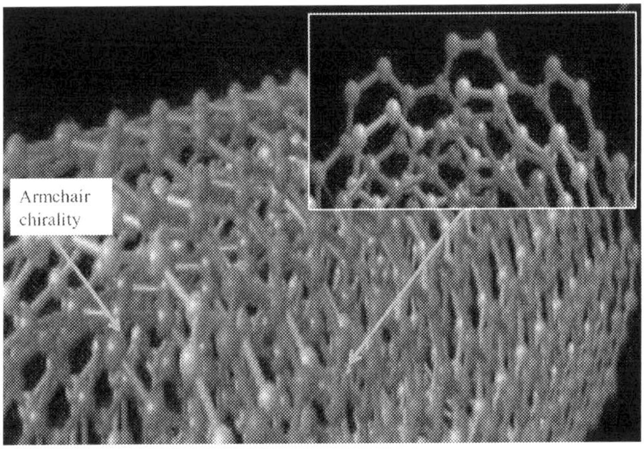

Figure 7.4

Atomic lattice structure of a triple-wall carbon nanotube with the armchair lattice shells nested closely one inside another with the van der Waals spacing, which is considerably smaller than normal and erroneously comparable with the length of the covalent bonds (see the inset) (after Ogwen).

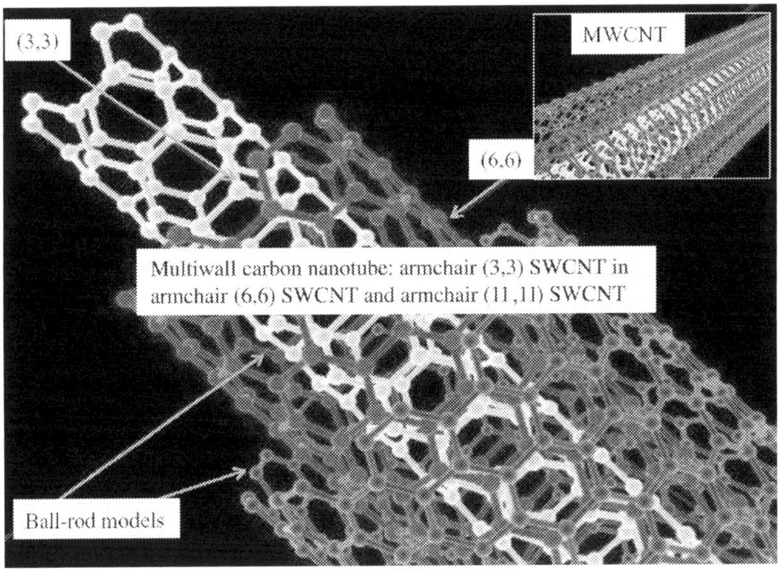

Figure 7.5

Atomic lattice structure of a multiwall carbon nanotube with the armchair lattice shells nested closely one inside another with the average van der Waals spacing [the (3,3) armchair nanobeam has the polygonal symmetry, while the (6,6) wall is the thick lattice shell and the (11,11) wall is the thin lattice wall with the cylindrical symmetry] (after Gollonet).

7.1.2 Kinematic Motion of Atoms in Multiwall Carbon Nanotubes

Kinematic motion of carbon atoms in the deforming lattices of MWCNTs varies between the external and internal walls, large and small van der Waals separations, different chiralities, commensurable and noncommensurable lattices, and distinct classes of MWCNTs. In each case, the local atomic structure involves different length scales with unique interatomic interactions and specific constraints, as well as distinct combinations of forces and unique deformation mechanisms, which are available for the energetically favorable deformation paths. For example, deformation mechanisms are different for the classes of thin and thick lattice shells, which also may involve large or small numbers of internal walls. Understanding of different deformation mechanisms and complex kinematic motion of carbon atoms in MWCNTs depends on the analysis of length scales and kinematic restrictions in their multiwall structure. Restrictions for the kinematic motion of carbon atoms in the chiral lattices with the global deformation of each lattice can be formulated by using the chirality defined by the pair of integers: (n, m). In thin shells of MWCNTs with fewer walls and dense MWCNTs with many walls, separation between adjacent walls is similar to that in graphite. Therefore, the magnitude of van der Waals forces between the adjacent walls is also similar (Fig. 7.2).

Diameters of MWCNTs are usually very large as compared to those of SWCNTs. Therefore, the kinematic inequalities, which have been presented earlier for SWCNTs, for the restriction of the kinematic motion of atoms, Δh_{NT}, in the thin lattice shells, are easily satisfied for MWCNTs. In fact, they become trivial for the large values of radii. Kinematic motion of carbon atoms in each shell satisfies the Cauchy–Born approximation (Born and Huang, 1954), which equates the average strain energy of an equivalent-continuum carbon ring to the potential energy stored in its deformed covalent bonds. It should be noted that in MWCNTs with large diameters and a significant number of internal walls, the Cauchy–Born approximation is updated more often in the areas with larger deformation gradients. It means that during bending of a MWCNT, the kinematic Cauchy–Born rule is applied more often in radial direction for internal lattice shells and less frequently to the neighboring carbon rings along their length. For internal walls with larger curvature, the exponential Cauchy–Born rule (Arroyo and Belytschko, 2004) should be applied on their curved surfaces. The continuum strain energy density of a carbon ring can be obtained from the interatomic potential.

In MWCNTs, the smallest internal shells have higher aspect ratios and smaller values of the critical buckling strain. Therefore, the kinematic motion of atoms in these lattices is significantly affected by the global deformation of the smallest internal walls. The length of internal walls in MWCNTs is large enough in order to satisfy not only nanoscale homogenization criteria, but also the homogenization requirements for the wave length of the deformation response. Consequently, the minimum size of internal and external lattice shells, for which the nanoscale homogenization correction for the Cauchy–Born rule should be applied, is much smaller than the length of any shell in MWCNTs, and the criteria for nanoscale homogenization are satisfied for different sections along their length. The Cauchy–Born rule ensures that elastic deformation of MWCNTs passes through the equilibrium states. Therefore, in order to avoid instabilities during complex bending of MWCNTs, the nanoscale homogenization criteria require that the Cauchy–Born approximation should be averaged over the smallest volume (not area) of the adjacent atomic lattices of the size satisfying the homogenization criteria. As a result, the Cauchy–Born approximation (Born and Huang, 1954) and the exponential Cauchy–Born rule (Arroyo and Belytschko, 2004) should be averaged over a group of internal walls and their carbon rings satisfying the homogenization criteria in the radial and longitudinal directions. An averaging correction for the nanoscale homogenization size has to be applied to the Cauchy–Born rules in order to ensure that the deformation of the material elements made of multiple shells is based on the unique and well-defined material properties of a particular MWCNT. The magnitude of the nanoscale homogenization correction will depend on the number of internal walls in MWCNTs and the complexity of their atomic lattices (e.g., chirality and defects). The nanoscale homogenization correction for the Cauchy–Born approximation should take into account the interatomic interactions, which take place between adjacent internal walls and in the group of carbon rings limited by the size of homogenization area in each shell.

7.2 Deformation of Multiwall Carbon Nanotubes

Deformation of the multishell structure of MWCNTs is complex due to its structural complexity and numerous deformation mechanisms, which are available for the accommodation of the applied forces. While the length, L_{NT}, of MWCNTs is important for understanding the process of their bending, their aspect ratio, d_{NT}/L_{NT}, is the key nondimensional parameter, which forces the internal walls to buckle first. The initial buckling of the smaller internal walls is characteristic behavior for MWCNTs, especially, those considered as thick shells. The aspect ratio of long MWCNTs determines their critical buckling strain, as will be shown later. The high aspect ratio MWCNTs behave as beams under the axial loads regardless of the number of internal walls they include in their structure. However, there are values of the aspect ratio, d_{NT}/L_{NT}, of MWCNTs, which are associated with a unique form of nanoscale buckling that involves the simultaneous beam-like and the shell-like buckling modes.

Nanoscale buckling of MWCNTs also depends on their thickness. It has been noted before that the definition of the thin and the thick lattice shells depends on their thickness-to-radius ratio, h_{NT}/R_{NT}. In the case of MWCNTs, their thickness depends on the number of internal lattice shells in their structure. The thickness of MWCNTs also depends on the van der

Waals spacing between the internal walls in their structure. Larger values of van der Waals spacing near the external walls of MWCNTs can affect the value of their Poisson's ratio, because the radial contraction is easier and the attractive van der Waals forces help such contraction. An estimate of a possible increase of Poisson's ratio is presented below.

7.2.1 Simultaneous Beam- and Shell-Like Buckling Modes

The value of the aspect ratio, d_{NT}/L_{NT}, is important in determining the geometry of MWCNTs, for which the transition between the beam-like behavior and the shell-like response occurs in the internal and external walls. This transition is especially interesting in MWCNTs of large diameters with many internal walls. The process of axial buckling of such MWCNTs involves the beam-like buckling modes of the smallest internal walls, for which the aspect ratio, d_{NT}/L_{NT}, is smaller than 1/10, with the immediate subsequent shell-like buckling of the external walls, for which the aspect ratio, d_{NT}/L_{NT}, is greater than 1/10 (the mathematical ratio is considered here, as opposed to the engineering one). This is unique nanoscale buckling phenomenon occurring only in MWCNTs.

The simultaneous mixture of the beam-like and the shell-like buckling modes may occur more easily during the axial buckling of thick MWCNTs or short MWCNT beams. It means that there exists a critical value of the aspect ratio, d_{NT}/L_{NT}, of MWCNTs, when the external lattice walls start to behave as either thin or thick lattice shells, while the smaller internal walls still behave as the high aspect ratio beam-like shells. The critical value of such aspect ratio, $(d_{NT}/L_{NT})_{cr}$, of the external walls in MWCNTs satisfies the following condition (Harik, 2001a,b, 2002a):

$$\left(\frac{d_{NT}}{L_{NT}}\right)_{cr} \geq \frac{1}{10}, \tag{7.1}$$

where the diameter, d_{NT}, may vary between 10 and 85 nm or more. Consequently, the length, L_{NT}, of MWCNTs should be less than 100 or 850 nm. The minimum value of the aspect ratio, $(d_{NT}/L_{NT})_{min}$, of MWCNTs, is associated with the collapse of the longitudinal and transverse length scales in the atomic lattice structure of MWCNTs: $(d_{NT}/L_{NT})_{min} \approx 1$. Naturally, the minimum value of the aspect ratio, $(d_{NT}/L_{NT})_{min}$, satisfies the condition (7.1) for the shell-like behavior of lattice walls. The range of values of the diameters, d_{NT}, and the lengths, L_{NT}, which satisfy the condition (7.1) for the shell-like behavior is different for the thick MWCNTs and short MWCNT beams. The ranges of values of the diameters, d_{NT}, and the lengths, L_{NT}, such that $(d_{NT}/L_{NT})_{shell} \geq (d_{NT}/L_{NT})_{cr}$, and the corresponding transition lines: $(d_{NT}/L_{NT})_{shell} = (d_{NT}/L_{NT})_{cr}$ can be identified. Note that the condition (7.1) only approximates the beam-shell transition line for the buckling modes of MWCNTs.

A transition to the shell-like buckling modes occurs easier in the external walls of MWCNTs, whose radii have larger values. It is possible to specify the range of geometric parameters for the external walls of MWCNTs when their atomic lattice structure will be prone to the shell-like buckling modes and the onset of the beam-shell transition during their buckling. In fact, the external walls of MWCNTs, which satisfy the criteria for the thin lattice shells:

$$\frac{h_{NT}}{R_{NT}} < \frac{1}{10}, \quad \text{and} \quad \frac{h_{NT}}{R_{NT}} \ll 1, \tag{7.2}$$

have the effective thickness, h_{NT}, and large enough radius, R_{NT}, which lower the critical value of the aspect ratio of MWCNTs for the onset of the beam-shell transition in their buckling. Evaluation of the effective thickness, h_{NT}, of internal walls in MWCNTs will be discussed below.

Simultaneous buckling of walls in MWCNTs with the beam- and shell-like buckling modes may be even more complex with the smaller internal walls buckling as beams, the larger external walls buckling as thin shells, and the intermediate internal walls buckling as thick lattice shells. The lattice structure of a MWCNT shown in Fig. 7.5 would have the simultaneous beam- and shell-like buckling modes, because it has the thin external lattice wall, the thick internal wall, and the smallest nanobeam of a polygonal shape at the center. It is possible to determine the range of geometric parameters for the internal walls in MWCNTs, for which this more complex buckling may take place. The internal walls in MWCNTs will behave as thick lattice shells when their radius will be such that $h_{NT}/R_{NT} > 1/10$, which has been presented earlier in the chapter on classification of carbon nanotubes. In fact, the atomic structure of the triple wall carbon nanotube shown in Fig. 7.1 satisfies the criteria for the onset of axial buckling with the mixture of the beam-like buckling of the smallest internal wall and the shell-like buckling of its external wall.

It is possible to describe the atomic lattice structure of the external walls in MWCNTs, which will be susceptible to the onset of the aforementioned beam-shell transition in their buckling. The radius, R_{NT}, of any of the atomic lattice shells in MWCNTs can be expressed in terms of the number of carbon rings along their circumference, i.e., $R_{NT} = (a/2\pi)\sqrt{n^2 + m^2 + nm}$. Here, the pair (n, m) defines the chirality of the atomic lattice. Therefore, the condition (7.2) for the thin lattice shells can be written as:

$$\frac{h_{NT}}{R_{NT}} = \frac{2\pi h_{NT}}{a\sqrt{n^2 + m^2 + nm}} \ll 1 \quad \text{or} \quad h_{NT} \ll \frac{a}{2\pi}\sqrt{n^2 + m^2 + nm}. \tag{7.3}$$

The condition (7.3) allows to state that the chirality of the external wall of the triple wall carbon nanotube shown in Fig. 7.1 is such that the external lattice shell will behave as a thin shell. Note that the lattice sensitive form of the condition (7.3) allows one to predict a variety of chiralities for the potential external wall in MWCNTs, which can experience the onset of the beam-shell transition in its axial buckling. While evaluation of the effective thickness, h_{NT}, of internal walls in MWCNTs will be discussed below, the conditions (7.3) provide inequalities for the upper bounds on any such estimates.

7.2.2 Poisson's Ratio of Multiwall Carbon Nanotubes

The structure of MWCNTs, the number and the chirality of their walls, the van der Waals spacing between the walls, and their diameter influence their deformation under axial and complex loads. While the compressive axial loads may result in complex buckling behavior, the tensile axial loads may also lead to a variety of unique deformation and failure patterns (Yu et al., 2000). The complex structure of MWCNTs may affect the axial and circumferential strains of their atomic lattices. Atomic lattice structure of MWCNTs with larger van der Waals spacing between the external and internal walls will have higher values of Poisson's ratio. Poisson's ratio is defined as the ratio of the radial contraction over the longitudinal or axial extension. That, is Poisson's ratio, ν, of MWCNTs is calculated as the ratio: $\nu = -(\Delta R_{NT}\, L_{NT}/\Delta L_{NT}\, R_{NT})$ of the change of radius, ΔR_{NT}, due to the axial strain, and the change in the length, ΔL_{NT}. The increase in the value of Poisson's ratio, $\Delta\nu$, of MWCNTs is proportional to the increase of the van der Waals spacing, d_{vdW}, with respect to its average value of 3.4 Å:

$$\Delta\nu \propto \alpha N \sum_{n} \left(d_{vdW}/3.4 - 1\right), \tag{7.4}$$

where α is the proportionality coefficient, N is the number of walls, and the summation is over every van der Waals spacing in MWCNTs. The value of the proportionality coefficient, α, is higher for thicker MWCNTs with greater number of internal walls, larger values of their radii, and larger van der Waals spacing between the walls. In order to illustrate the mechanism for the increase in Poisson's ratio, Fig. 7.6 presents three double-wall carbon nanotubes with large and moderate diameters of their external walls and significant, average and above average van der Walls spacing between the adjacent walls. Poisson's ratio of the double-wall carbon nanotube with the largest van der Waals spacing will be higher due to the ability of its external wall to radially contract with help of the attractive van der Waals forces without the induced repulsion from the internal wall (see Fig. 7.2). In MWCNTs with more internal walls and similar van der Waals spacing: $(d_{vdW}/3.4 - 1) > 0$, a correction to Poisson's ratio estimated by formula (7.4) will be greater. Formula (7.4) provides not only an estimate for the increase of Poisson's ratio in MWCNTs with larger van der Waals spacing; it describes the structure of a distinct class of MWCNTs with higher porosity and higher Poisson's ratios. MWCNTs belonging to the class of porous MWCNTs have distinct material properties associated with higher values of Poisson's ratio and larger van der Waals spacing.

The class of MWCNTs with higher porosity is composed of the external walls of large diameters and the shell-like behavior associated with thin lattice shells (Fig. 7.6C). It should be noted, however, that very thin lattice shells with the thickness-to-radius ratio, h_{NT}/R_{NT}, such that

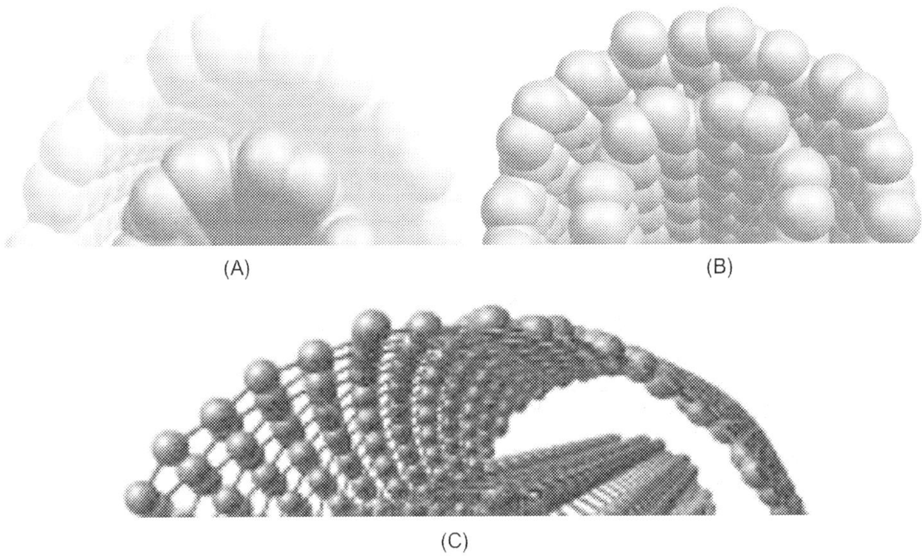

Figure 7.6
Atomic lattice structure of three double-wall carbon nanotubes with large (C) and moderate diameters (A and B) of their external walls and significant (C), slightly above average (B) and average (B) van der Walls spacing between the adjacent walls (after Petrushenko).

$$\frac{h_{\mathrm{NT}}}{R_{\mathrm{NT}}} < \frac{1}{20} \tag{7.5}$$

belong to the class of SWCNT membranes, a subclass of the thin SWCNTs, which deform under van der Waals forces and can collapse into SWCNT ribbons. Condition (7.5) defines a class of lattice membranes which have unique structural properties. SWCNT membranes do not have the cylindrical symmetry of the thin SWCNTs. It means that SWCNT membranes can collapse in the middle forming SWCNT nanoribbons with two layers of graphene separated by van der Waals spacing and two rounded edges of the bent graphene with higher thickness. MWCNTs, which are made of atomic lattice shells satisfying the condition (7.5) for the lattice membranes, form another distinct class of MWCNT membranes. MWCNTs belonging to the class of MWCNT membranes have deformed external walls, which have lost their cylindrical symmetry due to radial flattening. While evaluation of the effective thickness, h_{NT}, of external walls in MWCNT membranes will be discussed below, the condition (7.5) provides an inequality for the upper bound on any such estimates.

MWCNTs with deformed external walls respond differently to the axial strain. Their Poisson's ratio can be still estimated by the radial contraction, as noted above. However, their internal walls may be separated by higher van der Waals spacing associated with some local porosity. Therefore, a correction to Poisson's ratio estimated by formula (7.4) may be necessary. The axial deformation of the external walls in MWCNT membranes can be also analyzed by examining the lateral strain of carbon rings, which can be estimated as $2\pi\Delta R_{\mathrm{NT}}/a_c N_c$, where N_c is the number of carbon rings in the circumference and a_c is the circumferential size of a carbon ring. That is, the lateral deformation of a carbon ring, Δa_c, can be determined by

$$\Delta a_c = \frac{2\pi\Delta R_{\mathrm{NT}}}{a_c N_c},$$

or by $\Delta a = 2\pi R_{\mathrm{NT}}/aN_c$ for the zigzag SWCNT lattices and the size of a carbon ring: $a = 2.46$ Å. Poisson's ratio of the lattice membranes can be also defined as ν_a, i.e., Poisson's ratio based on the ratio of the lateral, Δa_c, to the axial, Δa_l, deformation, $-\Delta a_c/\Delta a_l$, of the carbon rings.

7.3 Analysis Methods for Multiwall Carbon Nanotubes

Analysis of mechanical behavior of MWCNTs can be carried out by using experimental characterization (Yu et al., 2000) and theoretical modeling based on a variety of methods from ab initio and molecular dynamics (MD) simulations to the scaling analysis and equivalent-continuum models. MD data for MWCNTs require multiscale approaches for the analysis of numerical results about material properties and mechanical behavior of atomic lattice structures. Nanoscale homogenization criteria provide the minimum size of the section of MWCNTs, within which the mechanical behavior of carbon rings should be analyzed by atomic potentials. Scaling analysis provides separation criteria for different classes of MWCNT structure and distinct properties of the atomic lattices of their internal and external walls. Equivalent-continuum models can provide information about the expected deformation response (e.g., buckling modes), estimates for Young's modulus of MWCNTs, and their Poisson's ratio. Prediction of material properties of carbon nanotubes by MD simulations depends on the estimate of the effective thickness of their lattice shells, as in the case of Young's modulus. It is important to reduce potential discrepancies in the estimates of the thickness of lattice walls in MWCNTs and the thickness paradox, which result in deviations in the prediction of Young's modulus and other material properties. Evaluation of any material and structural properties of MWCNTs should be carried out over an area or a volume of the atomic lattices that satisfies the nanoscale homogenization criteria. Below the homogenization size or length, any evaluation should be explicitly based on atomic potentials. Any Cauchy–Born approximation should be applied with the nanoscale homogenization correction associated with the well-defined averaging over the minimum size of a volume of a MWCNT.

7.3.1 Atomic Potentials for Multiwall Carbon Nanotubes

Atomic potentials describe the interatomic interactions between carbon atoms in the covalent bonds and in the adjacent lattice walls in MWCNTs. Gradients of atomic potentials determine the force acting on each carbon atom and the acceleration of atoms according to Newton's second law: $\boldsymbol{F} = m\,\boldsymbol{a}$, where the force $\boldsymbol{F}$ is the gradient, ∇U, of the potential energy function, U, which has both attractive, $\phi_A(\boldsymbol{r})$, and repulsive, $\phi_R(\boldsymbol{r})$, terms. Interaction between carbon atoms in the covalent bonds of the internal walls of MWCNTs is described by the so-called Tersoff–Brenner potential (Brenner, 1990), which was described earlier. The van der Waals interactions between carbon atoms in carbon rings and in adjacent lattice walls are described by the Lennard-Jones potential:

$$U = 4\varepsilon_{\mathrm{LJ}}\left(\frac{\sigma_{\mathrm{LJ}}^{12}}{r^{12}} - \frac{\sigma_{\mathrm{LJ}}^{6}}{r^{6}}\right), \tag{7.6}$$

where U is the potential energy between a pair of atoms, r is the distance between a pair of atoms, $\varepsilon_{\mathrm{LJ}}$ is the potential well depth, and σ_{LJ} is the size of van der Waals spacing. The rate of changes of the potential energy, U, with respect to infinitesimal strain, ε, associated with the van der Waals forces (Harik, 2011, 2014):

$$\sum_{i=1}^{n_{\mathrm{vdW}}} \frac{1}{V_{\mathrm{vdW},i}} \left| \left(\frac{\partial^2 U}{\partial \varepsilon^2}\right)_{\varepsilon=0,i} \right|_{\mathrm{vdW}} \tag{7.7}$$

and the elastic interactions through the C–C bonds (Harik, 2011, 2014):

$$\sum_{i=1}^{n_C} \frac{1}{V_{\mathrm{C-C},i}} \left| \left(\frac{\partial^2 U}{\partial \varepsilon^2}\right)_{\varepsilon=0,i} \right|_{\mathrm{C-C}} \tag{7.8}$$

can be used to together to determine the stiffness of the atomic lattices associated with the joined van der Waals interactions and the elastic interactions through the load-transferring C–C bonds. Here, the nanoscale volumes: V_{vdW} and $V_{\mathrm{C-C}}$, which are associated with the van der Waals and elastic interactions, satisfy the criteria for nanoscale homogenization. Under transverse loading, the atomic lattice walls in MWCNTs are subjected to the dominating van der Waals interactions, whose stiffness is described by the formula (7.7). Under axial loads, the atomic lattice walls in MWCNTs experience changes in the potential energy, which are mostly associated with the elastic interactions and changes in the covalent bonds. However, under the axial loads, van der Waals forces do have a constraining boundary effect on internal walls of MWCNTs, which increases the stiffness of lattice walls. The elastic modulus of such joint interactions is estimated by the formulae (7.7) and (7.8). During significant bending of MWCNTs, their walls are also subjected to the joint van der Waals and elastic interactions.

Following similar considerations, the mechanical stress, σ_{ij}, on a system of carbon atoms in MWCNTs subjected to the axial forces and the transverse van der Waals forces can be evaluated by the change in the total internal energy, E, of the atoms dominated by the changes in the potential energy associated with elastic interactions and van der Waals interactions with respect to the strain per unit volume:

$$\sigma_{ij} = \frac{1}{V_{C-C}} \left(\frac{\partial U_{C-C}}{\partial \epsilon_{ij}} \right)_{S=\text{const}} + \frac{1}{V_{vdW}} \left(\frac{\partial U_{vdW}}{\partial \epsilon_{ij}} \right)_{S=\text{const}}, \tag{7.9}$$

where V_{C-C} and V_{vdW} represent the unit volume of a nanoscale material element determined by the nanoscale homogenization criteria for the elastic and van der Waals interactions, respectively; E is the total internal energy of the atomic lattice structure, and ε_{ij} is the strain tensor. The subscript S denotes that the entropy, S, of the strained system is constant during the deformation. During an elastic deformation of a nanostructured material, the internal energy of a nanoscale element can be equal to the strain energy of the deformed material element, the size of which has to satisfy the nanoscale homogenization criteria.

7.3.2 Solution of the Thickness Paradox for Multiwall Carbon Nanotubes

Deformation of MWCNTs and the kinematic motion of carbon atoms with the internal walls of MWCNTs and the displacement of individual walls can be described by using the concept of effective thickness for the atomic lattice shells of MWCNTs. The thickness of atomic lattice shells is also critical for the evaluation of Young's modulus of MWCNTs, which describes their material properties and affects their mechanical behavior. Evaluation of material properties over the volume of MWCNTs, which is smaller than the size specified by the nanoscale homogenization criteria, should be based explicitly on the aforementioned atomic potentials. Moreover, the estimates of thickness of individual walls in MWCNTs determine their structural properties as either thin or thick shells, and, hence, predict their deformation response above the homogenization length. Estimates for the effective thickness, h_{NT}, of external walls in MWCNT membranes provide the range of geometric parameters, which define the class of MWCNT membranes.

The first experimental results (Iijima, 1991) have suggested that the thickness of individual shells in MWCNTs is the same as that in graphite: 3.4 Å. Internal walls in MWCNTs are subjected to van der Waals forces on one or both sides. Assuming that van der Waals spacing between adjacent walls is 3.4 Å, the effective thickness of individual walls in MWCNTs can be estimated as being 3.4 Å for the case of transverse loading. This estimate also applies to the collapsed external walls of MWCNT membranes. Under axial loads, the effective thickness of individual lattice shells in MWCNTs may be lower than 3.4 Å. Under axial forces, the transversely semi-isotropic material properties of the internal walls of MWCNTs have to be taken into account, with their thin hexagonal lattice of the stiff elastic C–C bonds

($h_{NT} < 3.4$ Å), and the thicker out-of-plane layers of electrons and van der Waals forces (Table 7.1). The constraining effect of the van der Waals forces can be accounted for by the application of proper boundary conditions, as opposed to the out-of-plane extension of the lattice. This approach allows one to evaluate the in-plane and out-of-plane stiffness of the internal walls in MWCNTs differently. When van der Waals spacing is larger than 3.4 Å between internal walls, the estimate of the effective thickness of such lattice walls may be also different due to the influence of attractive van der Waals forces. Such influence of van der Waals forces may be also important for the class of porous MWCNTs under axial loading.

7.3.3 Continuum Models for Multiwall Carbon Nanotubes

Mechanical behavior of MWCNTs with many internal walls can be described by the equivalent-continuum beam models with certain restrictions imposed on the characteristics of their atomic lattice structure within their cross-sectional area. Govindjee and Sackman (1999) have considered an elastic multishell model, which might be applied to MWCNTs, in order to show the dependence of material properties on the number of shells. An assumption about the continuum cross section was shown to hold when more than 201 shells were present in a system of continuum shells. Some MWCNTs have many shells in their structure, therefore, a partial continuum cross section can be used as an approximation, in some cases, outside of a hollow shell at the center.

Assumption of the continuum surface area forces limitations on size of the homogenized lattices, while the concept of shell stiffness requires a resolution of the effective thickness paradox for MWCNTs and proper evaluation of the influence of van der Waals forces. The global mechanical behavior of atomic lattices can be analyzed by representing their discrete molecular structure as an equivalent-continuum transversely isotropic shell or a layered shell by equating the potential energies of the elastic deformation and van der Waals interactions of the corresponding material systems, as in the Cauchy–Born approximation. However, the nanoscale homogenization correction is required for the Cauchy–Born rules in order to enforce the uniqueness and the well-definiteness of the local material properties by the nanoscale homogenization criteria and to ensure that the local kinematic motion of carbon atoms precisely reflects the unique and the well-defined material properties of the homogenized MWCNTs. In MWCNTs, the transversely isotropic material properties of the internal walls have to be taken into account with their in-plane hexagonal lattice of the elastic C–C bonds and the two out-of-plane layers of electrons and van der Waals forces. The transversely isotropic material model of the internal walls of MWCNTs, or their representation as a layered shell, also improves the kinematic motion of carbon atoms. This model recognizes that the in-plane and out-of-plane stiffness of the internal walls in MWCNTs is different. Vibrations of the internal walls in MWCNTs and their interactions involving phonons are affected by their complex transversely semi-isotropic material structure.

Applicability of the continuum models for the internal walls of MWCNTs, which are subjected to the van der Waals forces on one of both sides, also depends on several basic assumptions of the continuum shell theory (Timoshenko and Gere, 1961). These assumptions have to be satisfied by the atomic lattice shells (Harik et al., 2002b) with proper nanoscale adjustments. Requirements for the relative wavelength of structural deformation and the characteristic dimensions of the atomic lattices associated with different classes of carbon nanotubes have to be satisfied as well. Such requirements are important because continuum models of discrete atomic lattices have certain limitations in their ability to resolve the local details of discontinuous material structures (Harik, 2001a,b, 2002a).

In order to capture the beam-like behavior of the long shell-like internal walls in MWCNTs, the importance of their aspect ratio, d_{NT}/L_{NT}, should be recognized in the process of buckling. The formula for the critical buckling strain in the linear stability analysis (Timoshenko and Gere, 1961) can be written in the form that underscores the importance of the aspect ratio, d_{NT}/L_{NT}, and the constraining effect of the van der Waals forces. The effect of the van der Waals forces can be separated from the shell-like behavior of the internal walls in MWCNTs by using another nondimensional parameter, i.e., the thickness-to-radius ratio, h_{NT}/R_{NT}. Their shell-like deformation response can be different depending on the effective thickness (i.e., thin or thick lattice shell). In the limiting case of the very low thickness-to-radius ratios, h_{NT}/R_{NT}, the internal walls in MWCNTs with large diameters can deform even under van der Waals forces, as in the class of the MWCNT membranes.

The continuum representation of the internal walls in MWCNTs requires the proper definition of a continuum shell, or the equivalent-continuum for the discrete atomic lattice structures (Harik, 2001a,b, 2002a). It has been noted before that the definition of a continuum shell is possible when the length of MWCNTs satisfies the nanoscale homogenization criteria: $L_{NT} \gg a$, or $L_{NT}/a \gg 1$, where a is the size of a carbon ring. It should be noticed that the consequence of the proper definition of a physical continuum is that the value of the elastic modulus of individual walls in MWCNTs is unique and well-defined, because the length, L_{NT}, is long enough to satisfy the above criteria for nanoscale homogenization of the atomic lattices.

7.3.4 Timoshenko Beam Model for Multiwall Carbon Nanotube

Mechanical response of the multishell lattice structure of MWCNTs involves shear deformation in addition to bending, tension, or compression. Therefore, theoretical models of MWCNTs have to include the ability to describe shearing deformation of elongated structures. Continuum models, which are capable of describing the internal shear in the beam-like structures, are represented by the so-called Timoshenko beam theory (Timoshenko and Gere, 1961). The Timoshenko beam model is more complex than the earlier discussed Euler beam model, which can be used for the high aspect ratio MWCNTs of low diameters. Thick MWCNTs of large diameters include large number of internal walls, which can be sheared during deformation. Therefore, the Timoshenko beam model has to be used, in order to describe the possible shear deformation of such MWCNTs. In this model, the deflection, w, of the thick beam-like structure of MWCNTs can be approximated by the Timoshenko beam formula (Timoshenko and Gere, 1961):

$$w = w_{\text{bending}} + w_{\text{shear}} = \frac{FL_{\text{NT}}^3}{192E_{\text{NT}}^b I} + f_s \frac{FL_{\text{NT}}}{4G_{\text{NT}}A_{\text{NT}}}, \tag{7.10}$$

where F is the applied bending force, I is the moment of inertia, $I = 2\pi h_{\text{NT}}R^3$, G_{NT} is the shear modulus, and A_{NT} is the cross-sectional area of a hollow cylindrical structure where $A_{\text{NT}} = 2\pi R_{\text{NT}}h_{\text{NT}}$, f_s is the geometric factor (10/9, for a cylinder). The bending modulus, E_{NT}^b, can be approximated by the Young's modulus, E_{NT}, when the intertubular shearing is small with respect to the carbon ring size, a.

The shear modulus, G_{NT}, and Young's modulus, E_{NT}, of MWCNTs and their effective thickness, h_{NT}, can be estimated by investigating the frequencies of their vibration (Timoshenko and Gere, 1961):

$$\omega_i = \frac{\beta_i^2}{L_{\text{NT}}^2 R_{\text{NT}}} \sqrt{\frac{EI}{\rho\pi}} \left[\left(1 + \frac{h_{\text{NT}}}{2R_{\text{NT}}}\right)^2 - \left(1 - \frac{h_{\text{NT}}}{2R_{\text{NT}}}\right)^2 \right]^{-1/2} \tag{7.11}$$

and the lattice phonons in MWCNTs, e.g., transverse vibrations in a free–free hollow cylinder (Timoshenko and Gere, 1961):

$$\omega_{t,n} = \frac{\pi n}{L_{\text{NT}}} \sqrt{\frac{E_{\text{NT}}}{\rho_{\text{NT}}}} \tag{7.12}$$

and circumferential chiral shearing or twisting:

$$\omega_{c,n} = \frac{\pi n}{L_{\text{NT}}} \sqrt{\frac{G_{\text{NT}}}{\rho_{\text{NT}}}}, \tag{7.13}$$

where $\omega_{t,n}$ is the frequency of the nth-harmonic of the transverse phonons, ρ_{NT} is the density of about 1.3 g/cm^3, $\omega_{c,n}$ is the frequency of the nth-harmonic of the chiral lattice twisting phonons, and $n = 1, 2, \ldots$. Note that for a linear elastic isotropic material: $\omega_{t,n}/\omega_{c,n} = \sqrt{2(1+\nu)}$ and $\omega_{t,n}/\omega_{c,n} = \sqrt{E/G}$. For the linear elastic isotropic material, the shear modulus, G, is $G = E/2(1+\nu)$, where ν is Poisson's ratio. The vibration frequency of a clamped-free beam (Timoshenko and Gere, 1961) is

$$\omega_{L,n} = \beta_n^2 \sqrt{\frac{EI}{\kappa L^4}}, \tag{7.14}$$

where β_n^2 is the nth harmonics coefficient and κ is the bulk modulus.

The Timoshenko beam theory[1] can be used for the analysis of the MWCNT lateral deflection under the axial stress and lateral pressure (Timoshenko and Gere, 1961; Shames and Dym, 1985; Zhang et al., 2006; Ru, 2000, 2001), with the shear correction factors, K_i, for the ith lattice shell given by

$$K_i = 6(1+\nu)\left[(7+6\nu)+(20+12\nu)\left(1-\frac{1}{2}\frac{h_{\mathrm{NT}}}{R_{\mathrm{NT}}}\right)^2\right]^{-1} \tag{7.15}$$

Note that the ratio of the effective bending modulus, E^b_{NT}, in formula (7.10) and Young's modulus is such that

$$\frac{E^b_{\mathrm{NT}}}{E_{\mathrm{NT}}} \propto \left[\left(1+\frac{1}{2}\frac{h_{\mathrm{NT}}}{R_{\mathrm{NT}}}\right)^4 - \left(1-\frac{1}{2}\frac{h_{\mathrm{NT}}}{R_{\mathrm{NT}}}\right)^4\right].$$

According to the Timoshenko beam model for MWCNTs with the hinged end conditions, the critical axial stress, σ_x is given by

$$\sigma_x = \frac{EI}{A}\frac{\pi^2 n^2}{L^2_{\mathrm{NT}}}\left(1+\frac{T_S}{KG}\frac{EI}{A}\frac{\pi^2 n^2}{L^2_{\mathrm{NT}}}\right)^{-1} + \frac{k}{A}\left(\frac{\pi^2 n^2}{L^2_{\mathrm{NT}}}\right)^{-1}, \tag{7.16}$$

where T_s is the transverse shear parameter: 1, for the Timoshenko shear approximation and 0, for the Euler beam model with no shear; k is the spring constant for the outer shell such that the lateral pressure, p, is described by $p = -k(\boldsymbol{w}\cdot\boldsymbol{n})$, with the normal vector, $\boldsymbol{n}$, and the projected lateral deflection of the beam, $\boldsymbol{w} = (w_x, w_y){:}(\boldsymbol{w}\cdot\boldsymbol{n}) = w_r$, i.e., the normal radial deflection. The Timoshenko shell model is the two-dimensional generalization of the Timoshenko beam theory accounting for the transverse shear approximation.

7.3.5 Interfacial Sliding in Multiwall Carbon Nanotubes

Deformation of MWCNTs may involve interfacial shearing of internal walls and sliding of their atomic lattices (Fig. 7.7). Interfacial sliding of the inner lattice shells in MWCNTs can be described by the Bingham-type model (Harik, 2011, 2014) for the viscoplastic shearing of the interface, which relates the shear stress, τ, to the strain rate, $\dot{\gamma}$, by the Newton's friction law: ($\tau = \mu\dot{\gamma}$) after the critical shear stress, τ_0, is reached:

[1] Equilibrium equations (Timoshenko and Gere, 1961) for the lateral beam deflection, w, are: $(p - \sigma_x A + KGA)(d^2w/dx^2) - KGA\,d\phi/dx = 0$ and $EI(d^2\phi/dx^2) - KGA\phi + KGA\,dw/dx = 0$, where p is the lateral pressure, σ_x is the axial stress, A is the cross-sectional area, K is the shear correction factor that accounts for the constant shear stress state and the parabolic variation of the actual shear stress through the depth of the cross-section, G is the shear modulus, φ is the deflection slope, E is the Young's modulus, and I is the moment of inertia.

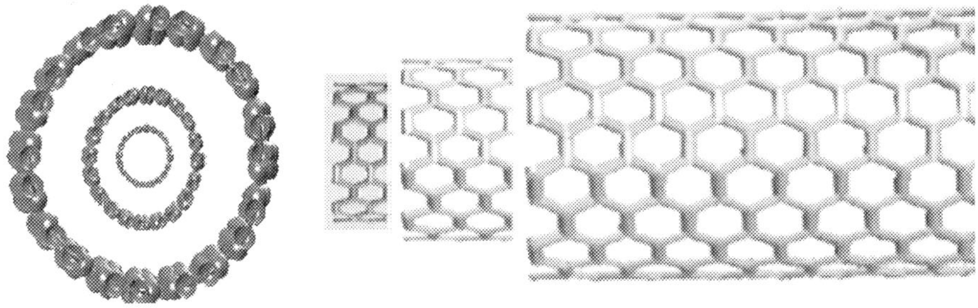

Figure 7.7
Atomic lattice structure of the multiwall carbon nanotube with three lattice walls separated by the van der Walls spacing.

$$\tau = \tau_0 + \mu_m \dot{\gamma}. \tag{7.17}$$

Here, μ and μ_m are the viscosities of a fluid and a viscous material, respectively. The Newton's friction law is defined for the adjacent layers of a continuum fluid subjected to shearing. At the nanometer level, an analog of Newton's friction law can be developed which relates the applied forces and the sliding velocity of a carbon nanotube (Frankland and Harik, 2003), in order to determine an effective viscosity of the nanoscale interface. The stresses and strain rates can be presented in terms of the average forces and the velocity gradients derived from the MD simulations.

In order to model the sliding between an inner lattice wall and the outer lattice shell in MWCNTs on a length-scale level larger than the carbon rings, multiple carbon ring lengths have to be considered for sufficiently meaningful averaging of the sliding of discrete lattice shells. The relative lattice sliding involves two surfaces. For the longitudinal direction, the length, L_{NT}, should be sufficiently long, that is, $L_{\mathrm{NT}}/a > 10$ (Harik, 2001a,b, 2002a), where a is the width of a carbon ring. The number 10 implies an order of magnitude difference. Likewise, for the perimeter of the lattice interface in MWCNTs, the radius of a lattice shell should satisfy an inequality: $2\pi(R_{\mathrm{NT}} + \overline{h}_{\mathrm{vdW}}/2)/a > 10$, where R_{NT} is the radius of an internal wall and $\overline{h}_{\mathrm{vdW}}$ is an average surface separation caused by the van der Waals forces.

At the length scale longer than that of the carbon rings, the nanoscale friction between two atomic lattices can be described as follows. The average shear stress, $\langle \tau_{rz} \rangle$, for axial sliding interactions can be defined as $\langle \tau_{rz} \rangle \approx \langle f \rangle_{\mathrm{pull}}/A_{\mathrm{ss}}$, where $\langle f \rangle_{\mathrm{pull}}$ is the average applied force and A_{ss} is the common interfacial area of steady sliding between the inner and the outer lattices. The average strain rate, $\langle \dot{\gamma} \rangle$, can be estimated as $\langle \dot{\gamma} \rangle \approx \left(\partial \langle V_z \rangle \right) / \partial r$, where $\langle V_z \rangle$ is the average velocity in the axial z-direction and r is the radial distance. Then, the nanoscale analog of Newton's law for steady interfacial sliding of two atomic lattices is given by

$$\langle \tau_{rz} \rangle = \mu_{\mathrm{eff}} \frac{\partial \langle V_z \rangle}{\partial r}, \tag{7.18}$$

where μ_{eff} is an effective viscosity for the interfacial sliding between two atomic lattices in MWCNTs. As a result, the interfacial friction model for the entire sliding process can be developed for the total shear stress, $\langle\tau_{rz}\rangle_{\text{pull}}$, which, in addition to the sliding component $\langle\tau_{rz}\rangle$, includes an initial threshold pull-out stress, τ_0:

$$\left\langle\tau_{rz}\right\rangle_{\text{pull}} = \tau_0 + \mu_{\text{eff}}\frac{\partial\left\langle V_z\right\rangle}{\partial r}. \tag{7.19}$$

After approximating the change in velocity across the interfacial separation, h_{vdW}, as the average sliding velocity, $\langle V_z\rangle$, Eq. (7.19) can be rewritten as

$$\left\langle\tau_{rz}\right\rangle_{\text{pull}} = \tau_0 + \mu_{\text{eff}}\frac{\left\langle V_z\right\rangle}{h_{\text{vdW}}}, \tag{7.20}$$

where the strain rate is $\frac{\partial\langle V_z\rangle}{\partial r} \approx \frac{\langle V_z\rangle}{h_{\text{vdW}}}$. The average surface separation, $\overline{h}_{\text{vdW}}$, is caused by the van der Waals forces, which can be described by the Lennard-Jones potential. In order to complete the description of the interfacial friction model, the effective viscosity, μ_{eff}, should be described. The sliding velocity, $\langle V_{\text{NT}}\rangle = \langle V_z\rangle$, is, on average, linearly related to the average applied force, $\langle f\rangle$ (Frankland and Harik, 2003), and the slope, χ_{eff}, can be determined from the force-velocity dependence:

$$\left\langle f\right\rangle = \chi_{\text{eff}}\left\langle V_{\text{NT}}\right\rangle, \tag{7.21}$$

which is a nanoscale analog of the friction law in Eq. (7.18).[2] The applied force, $\langle f\rangle$, in Eq. (7.21) is related to the shear stress by the force balance: $\left\langle f\right\rangle = \left\langle\tau_{rz}\right\rangle A_{\text{ss}}$. Therefore, the viscosity coefficient, χ_{eff}, can be related to the effective viscosity μ_{eff} by

$$\mu_{\text{eff}} = \chi_{\text{eff}}\frac{h_{\text{vdW}}}{R_{\text{NT}}}\left[2\pi\left(1+\frac{h_{\text{NT}}}{2R_{\text{NT}}}\right)\Delta L_{\text{NT}}\right]^{-1}. \tag{7.22}$$

Then, the interfacial friction model for the entire pull-out process is

$$\left\langle f\right\rangle_{\text{pull}} = f_0 + \chi_{\text{eff}}\left\langle V_{\text{NT}}\right\rangle, \tag{7.23}$$

or

$$\left\langle f\right\rangle_{\text{pull}} = f_0 + 2\pi\mu_{\text{eff}}\left[\frac{h_{\text{vdW}}}{R_{\text{NT}}}\right]^{-1}\left(1+\frac{h_{\text{NT}}}{2R_{\text{NT}}}\right)\Delta L_{\text{NT}}\left\langle V_{\text{NT}}\right\rangle, \tag{7.24}$$

where f_0 is the critical force. The magnitude of the critical force f_0 is on order of pico-Newtons and the value of coefficient χ_{eff} is on order of $(pN\ ps)$/Å, and the average velocity, $\langle V_z\rangle$, is on order of Å/ps or 100 s of m/s at the nanoscale level.

[2] This research on the nanoscale friction modeling was supported in part by the Princeton University based NASA URETI Institute for the Bio-inspired Nanostructured Multifunctional Materials (see Princeton University based web site: http://bimat.org, award No. NCC-1-02037).

7.3.6 Phonons in Multiwall Carbon Nanotubes

Atomic lattice structure of MWCNTs can experience oscillations of the covalent C–C bonds and carbon rings in the lattice waves, which are called phonons. Generation of phonons may occur at the onset of dynamic buckling instability in the atomic lattice of MWCNTs before the aforementioned beam- and shell-like buckling. The stable buckling modes at the nonlinear bifurcation points, however, should not be confused with the transient lattice waves known as phonons. The bifurcation buckling modes are unstable shell configurations that the shell passes dynamically through on its way to the stable buckling configuration. The background lattice phonons oscillate within and around these global deformation modes as they scatter and exchange their energy between different frequency modes.

7.4 Conclusions

In this chapter, unique properties of MWCNTs are examined, along with the analysis of two distinct classes of porous MWCNTs and MWCNT membranes. The separation criteria for these two classes have been discussed. A formula for the correction to Poisson's ratio in porous MWCNTs, which have larger van der Waals spacing between their walls, is presented. A unique nanoscale buckling phenomenon occurring in short MWCNTs, which is associated with the simultaneous beam- and shell-like buckling modes, has been examined. Atomic potentials for MWCNTs have been reviewed, along with the expressions for the stresses. A solution of the thickness paradox for MWCNTs has been discussed, with a review of different approaches for the equivalent-continuum models. The Timoshenko beam model for MWCNTs has been presented, along with the formulae for the analysis of their bending and vibration. Interfacial sliding and friction of the internal walls in MWCNTs has been investigated. A nanoscale analog of Newton's friction law has been presented, along with an interfacial friction model for MWCNTs.

Sample Problems

1. Find and list a few examples of the atomic lattice structure of MWCNTs and describe their structure.
2. Describe the length scales in the atomic lattice structure of MWCNTs in problem 1.
3. Make a PowerPoint viewgraph with an illustration of the atomic lattice structure of MWCNTs.
4. Make a PowerPoint viewgraph with an illustration of the length scales in the atomic lattice structure of MWCNTs.
5. Find examples of MWCNTs which can represent different classes of carbon nanotubes.
6. Describe the atomic structure of MWCNTs which represent different classes of carbon nanotubes.

7. Describe the transversely isotropic material model for the internal walls in MWCNTs.
8. Make a PowerPoint viewgraph with an illustration of the atomic lattice structure of MWCNTs which can represent different classes of carbon nanotubes.
9. What is the physical basis associated with the values of the effective thickness, h_{NT}, of 3.4 Å for the internal walls of MWCNTs?
10. What is the nanoscale homogenization criterion for MWCNTs?
11. Describe the nanoscale analog of Newton's friction law for atomic lattices.
12. Describe the class of porous MWCNTs.

References

Arroyo, M., Belytschko, T., 2004. Finite crystal elasticity of carbon nanotubes based the exponential Cauchy–Born rule. Phys. Rev. B 69, 115415-0.

Born, M., Huang, K., 1954. Dynamical Theory of Crystal Lattices. Oxford University Press, Oxford.

Brenner, D.W., 1990. Empirical potential for hydrocarbons for use in simulating the chemical vapor deposition of diamond films. Phys. Rev. B 42, 9458–9471.

Frankland, S.J.V., Harik, V.M., 2003. Carbon nanotube pull-out from a polymer matrix. Surf. Sci. Lett. 525, L103.

Govindjee, S., Sackman, J.L., 1999. On the use of continuum mechanics to estimate the properties of nanotubes. Solid State Commun. 110, 227.

Harik, V.M., 2001a. Ranges of applicability for the continuum beam model in the mechanics of carbon nanotubes and nanorods. Solid State Commun. 120 (7–8), 331–335.

Harik, V.M., 2001b. Ranges of applicability for the continuum beam model in the constitutive analysis of carbon nanotubes: nanotubes or nano-beams? NASA/CR-2001-211013. NASA Langley Research Center, Hampton, VA.

Harik, V.M., 2002a. Mechanics of carbon nanotubes: applicability of the continuum-beam models. Comput. Mater. Sci. 24 (3), 328–342.

Harik, V.M., et al., 2002b. Applicability of the continuum shell theories in the mechanics of carbon nanotubes. Technical Report NASA/CR-2002-211460. ICASE, NASA Langley Research Center, Hampton, VA.

Harik, V.M., 2011. Mechanics of Carbon Nanotubes. Nanodesigns Press, Newark, DE.

Harik, V.M. (Ed.), 2014. Trends in Nanoscale Mechanics. Springer, Dordrecht.

Iijima, S., 1991. Helical microtubules of graphitic carbon. Nature 354, 56.

Ru, C.Q., 2000. Effective bending stiffness of carbon nanotubes. Phys. Rev. B. 62 (15), 9973–9976.

Ru, C.Q., 2001. Degraded axial buckling strain of multi-walled carbon nanotubes due to interlayer slips. J. Appl. Phys. 89 (6), 3426–3433.

Shames, I.H., Dym, C.L., 1985. Energy and Finite Element Methods in Structural Mechanics. Hemisphere, Washington, DC.

Timoshenko, S.P., Gere, J.M., 1961. Theory of Elastic Stability. McGraw-Hill, New York, NY.

Yu, M.F., Lourie, O., Dyer, M.J., Moloni, K., Kelly, T.F., Ruoff, R.S., 2000. Strength and breaking mechanism of multi walled carbon nanotubes under tensile load. Science 287 (5453), 637–640.

Zhang, Y.Y., Tan, V.B.C., Wang, C.M., 2006. Effect of chirality on buckling behavior of single-walled carbon nanotubes. J. Appl. Phys. 100 (7), 074304.

Further Reading

Timoshenko, S.P., 1921. On the correction for shear of the differential equation for transverse vibration of prismatic bars. Philos. Mag. 41, 744–746.

CHAPTER 8

Carbon Nanotubes in Nanocomposites

Chapter Outline

8.1 Nanocomposites With Different Carbon Nanotubes

Carbon nanotubes can be used as reinforcement for the polymer, metal, or ceramic matrix in nanocomposite materials (Schadler et al., 1998; Thostenson and Chou, 2002; Park et al., 2002). Carbon nanotubes have higher elastic modulus values, strength, and failure strain than that of carbon fibers, which are used for reinforcing polymer matrix composite materials. Therefore the carbon nanotube-reinforced polymer matrix nanocomposites should have superior material properties with respect to the carbon fiber-reinforced composites. Nanocomposites may include both single-wall carbon nanotubes (SWCNTs) and multiwall carbon nanotubes (MWCNTs). Uniform dispersion of carbon nanotubes in polymer matrix is critical for manufacturing nanocomposites with the well-distributed reinforcement, i.e., SWCNTs or MWCNTs. Dispersion of SWCNTs can be improved by in situ polymerization under sonication (Park et al., 2002). Alignment of SWCNTs can be done by the flow-induced alignment in the extrusion flows, oriented SWCNT mats, SWCNT-impregnated polymer films, the in-field processing flows, and other methods. Better alignment of carbon nanotubes in the matrix material results in better material properties of nanocomposites. Dispersion of carbon nanotubes, effectiveness of their reinforcement in the matrix, and their influence on the average material properties of nanocomposites depend on their size and geometry. Long high aspect ratio SWCNTs and MWCNTs significantly improve the

Mechanics of Carbon Nanotubes.
DOI: https://doi.org/10.1016/B978-0-12-811071-3.00008-1

© 2018 Elsevier Inc. All rights reserved.

stiffness of the matrix material in the direction of their orientation. Nanocomposites with an ideal material structure may be reinforced by the very long carbon nanotubes with extremely high aspect ratios, and by the relatively shot carbon nanotubes, both SWCNTs and MWCNTs (Fig. 8.1).

Carbon nanotubes with the higher aspect ratios (Fig. 8.1A) have better load-carrying capability due to their better ability to assist the transfer of the external loads into the matrix reinforcement. However, carbon nanotubes with very high aspect ratios may be toxic (e.g., long MWCNTs). Therefore, it is important to find out what type of carbon nanotubes are safer to use, how, and in what form, shape, and size. Chapter 5, Mechanics of the High Aspect Ratio Carbon Nanotubes: Beam Models, Nanoscale Buckling, Nanobeams, and Phonons examined several parametric criteria for the potential toxicity of different classes of carbon nanotubes depending on their geometry. It is possible to minimize the potential risks by optimizing the structure of nanocomposites and their use in a particular product or structure. Carbon nanotubes with moderate aspect ratios (Fig. 8.1B) can also provide enough reinforcement for polymer matrices, while their size can be optimized to reduce the potential health risks.[1] The potential exposure to carbon nanotubes can be also minimized by controlling the size of a potential fracture, the extent of the fracture surfaces, and the pullout of carbon nanotubes. The size of the potential fracture surfaces can be controlled by the so-called fail-safe designs of composite materials, in which the most likely failure patterns are controlled in order to ensure the safe outcomes. The extent of the pullout of the reinforcing carbon nanotubes can be controlled by the optimal design of the material structure of nanocomposites and their interface (or interphase) regions (Fig. 8.2).

Interfacial adhesion between polymer matrix and carbon nanotubes is important for the effective stress or load transfer and for improving material properties of nanocomposites (Frankland and Harik, 2003; Harik, 2011). Interfacial shear strength of the polymer matrix-carbon nanotube interface can characterize the adhesion strength of the interface involving the nonbonded interactions and what is called π-bonding (Fig. 8.2). Interfacial adhesion can be based on covalent bonding by the σ-electrons and the noncovalent bonding through the bonding interaction of π-electrons. Noncovalent bonding is also associated with the van der Walls interactions between adjacent atomic surfaces and on the registry between their

[1] Carbon nanotube-based nanocomposites require large numbers of SWCNTs or MWCNTs to be used for their reinforcement. MWCNTs are known to be toxic, while some SWCNTs may also have some toxic effects, depending on the dose and exposure (Harik, 2014). Therefore the use of carbon nanotubes in composite materials has to be carefully controlled and optimized to minimize the potential health risks. Nanocomposites and carbon fiber-reinforced composites can be modified by carbon nanotubes only for applications, where the crack or damage size will be such that the cumulative dose of released carbon nanotube fibers during a failure of a material component will be well below the critical dose. Moreover, these carbon nanotubes (SWCNTs or MWCNTs) should be smaller than the critical length (Harik, 2014). Later, one of the chapters provides detailed information about the critical length of carbon nanotubes, which increases the potential toxic effects associated with SWCNTs and MWCNTs.

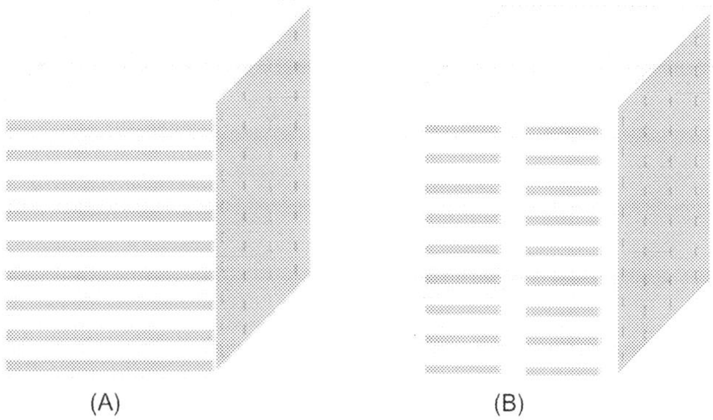

Figure 8.1
Material structure of the carbon nanotube-reinforced polymer matrix nanocomposite material with the long continuous carbon nanotubes (A) and the short carbon nanotubes (B).

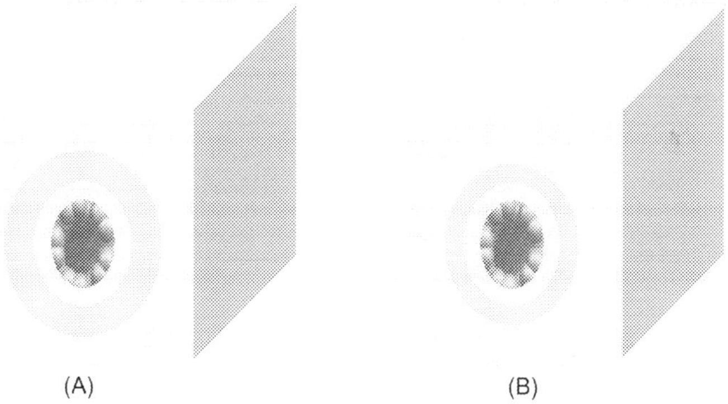

Figure 8.2
Interphase regions around the (5,5) carbon nanotube with van der Waals spacing (white) in the carbon nanotube-reinforced polymer matrix nanocomposite material: (A) a thicker interphase in a polymer matrix with a larger molecular weight and (B) a thinner interphase in a polymer matrix with a low molecular weight.

atomic lattices. The discrete surface of carbon nanotubes is covered by the local distributions of π-electrons that interact with the surrounding molecular structure of the matrix interface. Carbon nanotubes can be functionalized by different molecular groups in order to form the covalent bonding between polymer chains in the matrix and the surface atoms of SWCNTs or MWCNTs. The strength of interfacial adhesion associated with the covalent bonding depends on the density of covalent bonds along the surface of carbon nanotubes. The higher the number of covalent bonds per unit area, the higher the strength of nanoscale interfaces.

The classification of carbon nanotubes identifies 20 main classes of SWCNTs and MWCNTs, as shown in Chapter 4, Classification of Carbon Nanotubes: Twenty Classes of Atomic Structures. Each class of carbon nanotubes is characterized by a particular type of geometry associated with the specific ranges of geometric parameters, which can be used to assess the load-carrying capability and the efficiency of the load transfer of different carbon nanotubes. The nondimensional ratio defined as the aspect ratio, d_{NT}/L_{NT}, is helpful in the evaluation of the load-transfer capability of carbon nanotubes. Another nondimensional ratio defined as the thickness-to-radius ratio, h_{NT}/R_{NT}, is important in the evaluation of the size of interfacial region in the carbon nanotube-reinforced polymer matrix nanocomposites.

8.1.1 Length Scales in the Structure of Nanocomposites

Carbon nanotube-reinforced nanocomposites have different material structure, depending on the type of carbon nanotubes used for their reinforcement and the matrix material. Material properties of nanocomposites are affected by the local microstructure and nanoscale variations in the distribution of polymer chains, their molecular weight, and their entanglement. Dispersion of different classes of carbon nanotubes in polymer matrices and their local agglomeration also contributes to variability in material properties in nanocomposites. The size and form of such agglomerations of carbon nanotubes depends on their geometry and the size of polymer entanglements, all of which affect local variations of material properties in nanocomposites.

The typical length scales characterizing the lattice structure of different carbon nanotubes as compared to the molecular structure of a crystalline polyethylene (PE) polymer and the polymer matrix composites reinforced by the carbon fibers are shown in Table 8.1. The length scales associated with the atomic structure of carbon nanotubes are presented in the first column, since they have been discussed in earlier chapters. It should be noted that MWCNTs may have larger diameters, e.g., 30–100 nm, and much larger lengths, e.g., 5 cm–1 m and more, especially in the form of bundles (e.g., 5 m or more). Polymer chains also have important structural elements and the associated length scales (Table 8.1), which are important in the multiscale modeling of nanocomposites. Carbon nanotube-based nanocomposites or the nanotube-modified polymers have unique structural length scales

Table 8.1: Typical length scales of the material structure in composites

Carbon Nanotubes	Polymers	Polymer Composites
Carbon ring: 2.46 Å	(CH_2) zigzag chain: ~3 Å	Fiber coating: 1–3 μm
Diameter: ~1–2 nm	Crystalline unit: ~7 Å	Fiber/matrix interphase: 0.05–3 μm
Diameter: ~30–85 nm		Carbon fiber diameter: 20 μm
Length: 10 nm–5 μm	Length of chains: ~200 nm	Fiber length: 1–10 m
Length: 20 nm–5 cm	Length of chains: ~600 nm	

shown in the first two columns in Table 8.1, which are different from the hierarchical length scales associated with the microstructure of typical composites shown in the third column in Table 8.1. Therefore, the multiscale modeling of nanocomposites with carbon nanotubes and the fiber-reinforced composite materials is different.

The analysis of nanocomposites requires representative volume elements (RVEs) with discrete atomic structures, as well as other RVEs for the nanoscale and microscale material structures. Analysis of carbon fiber composites requires a microscopic RVE including a fiber, its coating, interface/interphase region, the RVE at the ply level describing the size and volume fraction of fibers in the matrix, and a macroscopic RVE showing the orientation of fibers in the plies and the stacking sequence in the composite. As a result, an RVE in the carbon nanotube-based nanocomposites can be described by the following typical length scales:

1. the average size of an RVE associated with the dispersion of carbon nanotubes and the distribution of their bundles and agglomerates;
2. the average size of the agglomerates (% of dispersion);
3. the typical length of carbon nanotubes;
4. the average size or the distribution of coiling;
5. the average size of the bundles;
6. the average size of the interphase region surrounding the agglomerated bundles;
7. the size of a RVE associated with a 95% dispersion of carbon nanotubes;
8. the size and the distribution of nanotube diameters;
9. the average size of the interphase region surrounding the carbon nanotubes;
10. the average size of the molecular chains, their coiling and their gyration length;
11. the size and the density of interfacial bonding;
12. the size of the van der Waals spacing around the interfacial bonds and away from the areas of interfacial bonding;
13. the molecular length of interfacial bonds and their units;
14. the size of a carbon ring and the size of aromatic rings in the molecular structure of the polymer;
15. the initial estimate(s) of the effective lattice thickness depending on the predominant loading conditions;
16. the length of the C–C bonds in a carbon ring and the length of covalent bonds in the interfacial bonding.

Material properties of the nanotube-based nanocomposites also depend on the structure–property relations and dependencies, which will be discussed below.

Material properties of the carbon nanotube-reinforced nanocomposites also depend on the structure–property relations and the multiscalar dependencies, which are different across the hierarchical length scales. In order to link a particular macroscopic material property with the nanoscale material characteristics or structural elements, an appropriate multiscale "tree" diagram is needed (Table 8.2). Dependence of Young's modulus, E, on the material characteristics at different length scales is illustrated for the polymer nanocomposites based on SWCNTs and MWCNTs, as well as on traditional carbon fibers. The elastic modulus of the carbon nanotube-based nanocomposites depends on the volume or weight fraction of carbon nanotubes, their length and orientation, and other characteristics listed in Table 8.2 that are associated with different length scales. Carbon fiber-reinforced composite materials also depend on their microstructure, as well as on nanoscale effects, which have not been fully understood (see the diagram below).

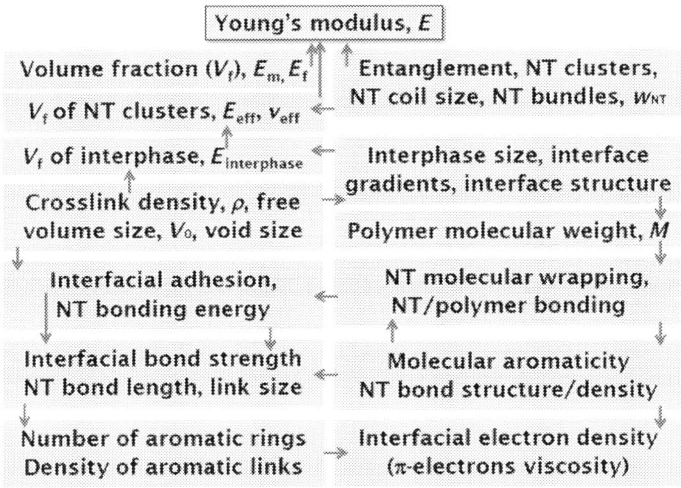

Selection of the nanoscale RVEs for nanocomposite materials should take into account the nanoscale phenomenon of the size-dependence of material properties, which has been discussed in earlier chapters. The nanoscale RVEs may have averaged material properties that may be somewhat different from the microscopic and macroscopic material properties.

Table 8.2: Multiscale structure–property dependencies for Young's modulus

Length Scales	Nanocomposites	Polymer Composites
Macroscopic	Entanglement, SWCNT coil size	Volume fraction, E_m
Microscopic	Volume fraction of SWCNTs	Carbon fiber (size, E)
	SWCNT bundles, clustering	
Submicron	SWCNT vs MWCNT effects	Interphase/interface
Nanoscale	SWCNT/polymer interphase	Crosslink density
	Polymer molecular weight	Interface structure
Atomic	SWCNT/polymer bonding	Polymer structure
	Molecular aromaticity	Molecular weight, etc.

SWCNT, single-wall carbon nanotube; *MWCNT*, multiwall carbon nanotube.

The nanoscale size-dependence of material properties means that small material elements or RVEs may or may not have the macroscopic material properties. Moreover, the polymer interfacial region (Fig. 8.2) around each carbon nanotubes in nanocomposites may or may not have the well-defined material properties. In fact, if material properties of the polymer interfacial region are uniquely defined, then this region can be called the interphase (i.e., the separate distinct phase). The volume of RVEs and their material constituents have to be large enough. The nanoscale phenomenon of size-dependence of material properties is connected not only to the discrete atomic lattices of carbon nanotubes and graphene but also to the size of polymer networks and agglomerations of polymer chains.

8.1.2 Multiscale Homogenization of Nanocomposites

Nanoscale homogenization criteria have been introduced for the atomic lattice structures in order to derive the unique and well-defined material properties of carbon nanotubes. Variability in the lattice structure of carbon nanotubes requires minor adjustments in the application of the nanoscale homogenization for the minimum volume of an RVE or an area of the atomic lattice. The nanoscale correction for the Cauchy–Born approximation is also connected with the adjustments in the kinematic motion of carbon atoms in the atomic lattices of carbon nanotubes, according to the proper averaging of material properties over the minimum nanoscale volume. Nanoscale homogenization of the carbon nanotube-reinforced nanocomposites has to involve different length scales associated with their multiscale material structure.

In the nanoscale homogenization of material properties of nanocomposites, the size of the smallest volume element or RVE is important. The smaller the volume of the RVE representing the periodic multiscale material structure in any nanocomposite material, the smaller size of a nanoparticle of this nanocomposite material may have the unique well-defined material properties. A rigorous representation of this requirement is found in the nanoscale homogenization criteria (Harik, 2001a,b, 2011, 2014), which have to be formulated for the key length scale of nanocomposite materials. The size of the smallest volume element also determines the lower size limit of the unique well-defined nanocomposite. This connection between the material structure of nanocomposites and their material properties is reflected in the form of the nanoscale homogenization criteria, which can mathematically define the lower size limit for the nanocomposites with the well-defined material properties having unique size-invariant values (Harik, 2001a,b, 2002). A criterion for the unique averaging (or nanoscale homogenization) of any material property over a volume element, V_0, of any nanomaterial requires that the smallest dimension, L, of the volume element is noticeably greater than the size of the smallest volume element of the atomic lattice, a, in the direction of the smallest dimension (Harik, 2001a):

$$L \gg a \text{ or } \frac{L}{a} \gg 1. \tag{8.1}$$

These inequalities require that the difference in sizes is roughly an order of magnitude, i.e., $L/a \approx 10$ or more, i.e., $L > 10a$. Nanoscale homogenization criteria (8.1) are applicable for the averaging of any physical property associated with any atomic lattice. Therefore, definition of a nanocomposite, as a nanomaterial, has to include the familiar upper limit of 100 nm for the smallest dimension of its material structure, as well as the lower limit for the dimension, which defines the size of a particular nanocomposite associated with the well-defined material properties with the unique size-invariant values that don't change with the size of the volume element. The lower limit for the dimensions of a nanocomposite, as a nanomaterial, is defined by the nanoscale homogenization criteria (8.1).

Nanoscale homogenization for the carbon nanotube-reinforced nanocomposite has to take into account the size of the carbon nanotube used, which would affect the size of a RVE (Fig. 8.2). Therefore, the nanoscale homogenization criteria (8.1) have to reflect the radial size of a carbon nanotube: $R_{\mathrm{RVE}} > (10a + R_{\mathrm{NT}})$, and the standard requirement for the longitudinal size: $L_{\mathrm{RVE}} > 10a$. However, the existence of the interphase region in nanocomposites, which is caused by the surface interaction of the polymer chains with carbon nanotubes, requires that the radial size of the RVE satisfy the following inequality: $R_{\mathrm{RVE}} > (10a + R_{\mathrm{NT}} + R_{\mathrm{int,o}})$, where $R_{\mathrm{int,o}}$ is the outer radius of the interphase. It should be noted that the material properties of the interphase region are not unique and ill-defined, if its outer radius, $R_{\mathrm{int,o}}$, does not satisfy the nanoscale homogenization criteria (8.1) in the following form: $(R_{\mathrm{int},o} - R_{\mathrm{int},i}) \gg a_{\mathrm{m}}$, or, as a nondimensional ratio: $(R_{\mathrm{int},o} - R_{\mathrm{int},i})/a_{\mathrm{m}} \gg 1$, where a_{m} is the smallest element in the molecular structure of the polymer in the interphase region. As a result, the radial size of the reinforcing SWCNTs and MWCNTs is taken into account for the nanoscale homogenization of nanocomposites, along with the size of the interfacial region, both of which would depend on the range of geometric parameters associated with the particular class of carbon nanotubes.

Material properties of the carbon nanotube-reinforced nanocomposites have to be unique and well-defined on all levels of their multiscale material structure. The volume fraction analysis and the rules of mixture easily satisfy the nanoscale homogenization criteria for the minimum volume of the RVE, unless the volume or the weight fraction of carbon nanotubes is high. The volume fraction analysis has to reflect the nanoscale structure of nanocomposites, including the existence of the interphase region and van der Waals spacing in the absence of the covalent bonding between carbon nanotubes and the matrix material. In a nanocomposite with coiling or clustering of SWCNTs or clustering of MWCNTs, the processing-induced variations in the density of distributed nanotubes, it is also important to average material properties over the RVE associated with a particular element in the material structure of nanocomposites. Averaging of material properties over the next length scale in the multiscale structure of nanocomposites is also important for ensuring that the material parameters of nanocomposites are unique and well-defined in each particular case associated with the material design, processing conditions, etc.

8.2 Analysis Methods for Nanocomposites With Carbon Nanotubes

Analysis of deformation response and the stress distribution in the carbon nanotube-reinforced nanocomposites requires understanding of the length scales in their material structure, which is reflected in the model of the RVE. Analysis methods and the approximations used have to be able to resolve the essential details in the atomic structure of nanocomposites and the degree of nonlinearity in their constitutive behavior and a particular mechanical response. The constitutive behavior of nanostructured materials has not been examined in many details, since such analysis requires large number of atoms to be considered in a material system, as well as the development of proper multiscale approaches for the nanoscale averaging of material properties. Determination of all required material parameters in the complex nonlinear constitutive models requires detailed analysis of the nanoscale effects contributing to various material nonlinearities of the atomic structure of nanocomposites. A careful analysis of the time scales involved in the deformation of nanostructured nanocomposites is also required, since the molecular motion is characterized by much higher velocities that result in much higher strain rates than those found in macroscopic experiments.

Molecular dynamics (MD) simulations of the stress—strain response of nanocomposites can be carried out for specific nanoscale RVEs (Frankland et al., 2003). Computational experiments with MD simulations provide valuable insights into the role of the built-in atomistic assumptions and the effectiveness of parametrization of the intermolecular forces and atomic potentials. Results of MD simulations can be examined by using one of the nanoscale averaging approaches based on a specific RVE. In MD simulations, the nanoscale interactions between pairs of atoms and multiatom configurations are described on the femto- and the picosecond timescales, while the relative deformation of different parts in a nanocomposite may evolve on the nanosecond timescale. These timescales are also associated with the generation of the lattice waves or phonons in the atomic structure of nanocomposites (Harik, 2011). MD simulation results (Frankland et al., 2003) are often associated with very high strain rates ($\sim 10^{10}$), as opposed to those found in macroscopic experiments ($\sim 10^{-2}$).

8.2.1 Molecular Dynamics Simulations of Nanocomposites

Atomic interactions in the carbon nanotube-reinforced nanocomposites depend on the atomic composition and the molecular structure of the polymer matrix, as well as on the atomic lattice structure of nanotubes and their size. Interfacial interactions in nanocomposites with the noncovalent bonding of carbon nanotubes can be approximated by the van der Waals forces represented by the Lennard-Jones potential. More detailed analysis of the interfacial interaction mechanisms, including nanoscale friction of atomic lattices and generation of phonons, will be discussed in the next chapter on interfaces in nanocomposites. Interactions between the carbon atoms in carbon nanotubes and the polymer matrix are described by the Tersoff–Brenner potential (Brenner, 1990).

The periodic atomic and molecular structure of the carbon nanotube-reinforced nanocomposites is composed of thousands and millions of atoms. Computational materials methods allow one to represent the periodic molecular structure by a RVE or a periodic unit cell (Fig. 8.3). Atomic lattice structure of the crystalline PE matrix is represented in the unit cell by the $-CH_2-$ units. In this example, the PE matrix has eight chains of 1095 and 1420 $-CH_2-$ units for the unit cells with the short and long carbon nanotubes (Frankland et al., 2003). The unit cell representing the long SWCNT reinforcement includes an infinite (10,10) SWCNT. The unit cell for the short SWCNT nanocomposite contains a 6-nm long capped (10,10) carbon nanotube, which is approximately half the length of the simulation cell. The overall dimensions for each unit-cell in the MD simulation are $\sim 5 \times 5 \times 10$ nm.

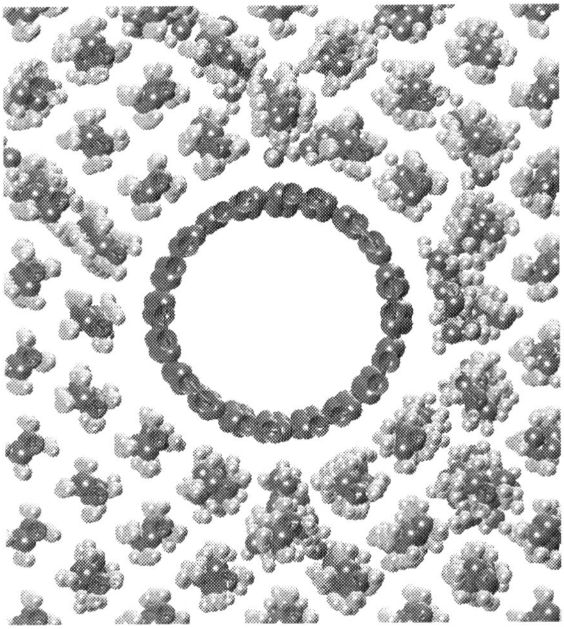

Figure 8.3

Molecular lattice structure of the crystalline part of the PE polymer with the $(CH_2)_n$ zigzag chains surrounding the armchair (10,10) carbon nanotube along the interface with the van der Waals distance of 3.4 Å (after Frankland and Harik, 2003). *PE*, polyethylene.

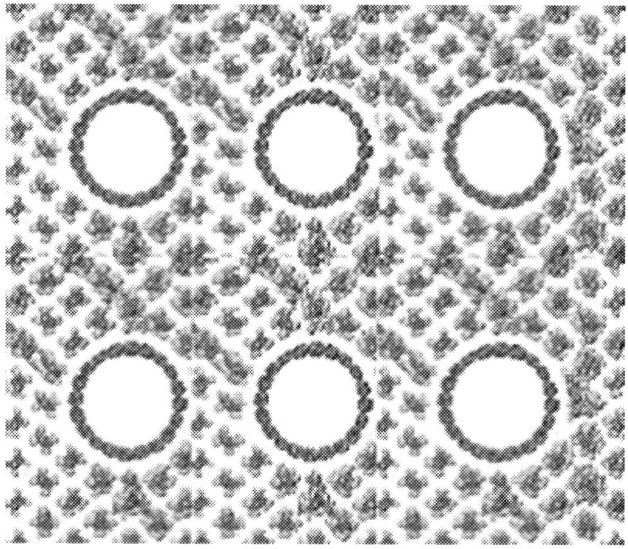

Figure 8.4
Atomic lattice of the crystalline polyethylene polymer and two arrays of the aligned (10,10) carbon nanotubes in the periodic nanoscale structure of the carbon nanotube-reinforced polymer matrix nanocomposites in the molecular dynamics simulations (after Frankland and Harik, 2003).

Periodic boundary conditions were used to replicate the unit cell in all three dimensions (Fig. 8.4). An equivalent block of polymer is also simulated of the same size without any carbon nanotube. In this nanocomposite, the PE density is 0.71 g/cm^3.

Interfacial interactions between the polymer matrix and carbon nanotube can be modeled by the relatively weak van der Waals forces described by the Lennard-Jones potential:

$$U = 4\varepsilon_{\mathrm{LJ}}\left(\frac{\sigma_{\mathrm{LJ}}^{12}}{r^{12}} - \frac{\sigma_{\mathrm{LJ}}^{6}}{r^{6}}\right),$$

where U is the potential energy of interactions between a pair of atoms, r is the distance between a pair of atoms, $\varepsilon_{\mathrm{LJ}}$ is the well depth, and σ_{LJ} is the van der Waals separation distance. Interfacial interactions between the carbon atoms of the (10,10) carbon nanotube and the polymer $-CH_2-$ units are described by the Lennard-Jones potential parametrized with $\varepsilon_{\mathrm{LJ}}$ equal to 0.4492 kJ/mol and σ_{LJ} of 3.825 A. The PE chains can be simulated with a molecular mechanics type force field, in which the $-CH_2-$ units of the PE chains are separated by the bond length of 1.53 A using the SHAKE algorithm.[2] The SHAKE

[2] The MD simulations were carried out by using DL-POLY, a large-scale MD simulation package available from Daresbury Laboratory. This package was adapted to include the carbon potential for the carbon nanotubes and to apply strain to the unit cell of a nanocomposite. All of the simulations have been carried out at 300K with a 2-fs time-step.

algorithm is based on a constraint dynamics method, which keeps the bond length constant with the user-defined tolerance. The angle bending forces can be modeled with a harmonic valence angle potential having an equilibrium angle of 112.813° and the barrier of 520 kJ/mol, and a torsional potential $\phi(\alpha)$ of the form:

$$\phi(\alpha) = A_1(1 + \cos\alpha) + A_2(1 + \cos 2\alpha) + A_3(1 + \cos 3\alpha)$$

was used where α is the torsional angle around the $CH_2{-}CH_2$ bond, and $A_1 = -18.087$, $A_2 = -4.88$, and $A_3 = 31.8$ kJ/mol. The nonbonded interactions within the chains were described by the Lennard-Jones potential parametrized with ε_{LJ} equal to 0.4742 kJ/mol and σ_{LJ} of 4.28 A for the internal PE chains and interchain interactions.

Deformation of a unit cell can be applied at an increment, when a uniform strain is prescribed on the entire MD model and across the unit cell. For the longitudinal and transverse deformations, pure states of strain: ε_{11} and ε_{22}, respectively, were initially applied. The application of strain was accomplished by uniformly expanding the dimensions of the unit cell in the direction of the deformation and rescaling the new coordinates of the atoms to fit within the new dimensions. After the initial deformation, the MD simulation was continued and the atoms were allowed to equilibrate within the new MD cell dimensions. This process was carried out for the subsequent increments of deformation. The applied strain for each configuration in either the longitudinal or transverse direction was 2%, applied in two increments of 1% each, 1 ps (500 steps) apart. The corresponding strain rate is $1.0 \times 10^{10}\ s^{-1}$. For each configuration six increments of strain were applied up to ~ 12% total over a period of 72 ps. After each 2% increment of strain, the system was relaxed and data was taken about the system of atoms over an interval of 10 ps. A series of stress–strain curves (Fig. 8.5) can be generated using the MD simulation of nanocomposites with the long and short carbon nanotubes and for the pure polymer matrix subjected to the longitudinal strain (i.e., the axial one-direction) and the transverse deformation (i.e., the two- or three-direction).

In the MD simulations of the carbon nanotube-reinforced nanocomposites, the maximum strains are on the order of 0.1 after the unit cell had been subjected to stepwise increases of iso-strain. The applied strain leads to an overall displacement of 0.5 nm for each unit-cell edge in the longitudinal direction or 0.25 nm in the transverse direction. Furthermore, the form of the carbon potential used was not parametrized to accurately simulate the plastic yielding of the carbon nanotube. In the rule-of-mixture analysis, it is assumed that the volume fraction of a carbon nanotube is 0.08 and the (10,10) carbon nanotube has an effective modulus of 600 GPa. The stress–strain curve (Fig. 8.5) for the nanocomposite containing the long carbon nanotubes strained in the longitudinal direction shows the most significant enhancement in Young's modulus and the slope of its stress–strain curve relative to the polymer. The solid line in Fig. 8.5 is a result of the micromechanical rule of mixtures analysis, which will be discussed below.

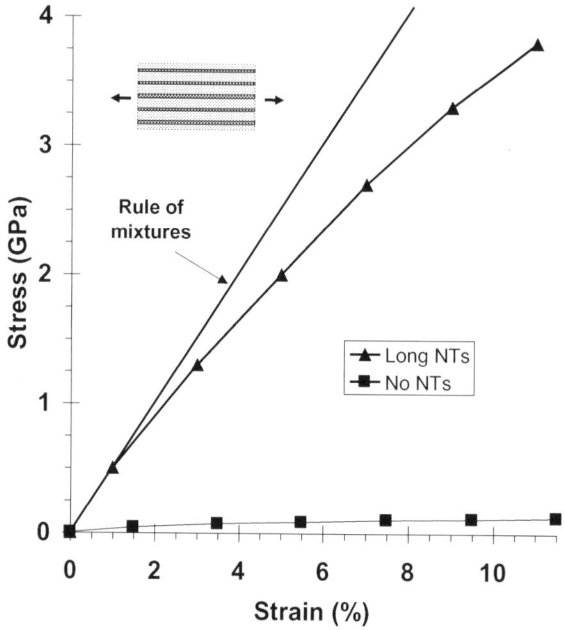

Figure 8.5

The stress–strain curves for the carbon nanotube-reinforced nanocomposites and the polymer matrix material subjected to the axial strain shown with the predictions based on the rule-of-mixtures (Frankland et al., 2003).

8.2.2 Nanoscale Rules of Mixtures for Nanocomposites

Material properties of the carbon nanotube-reinforced nanocomposites depend on the elastic modulus of carbon nanotubes and on their volume fraction in the polymer matrix, which can be estimated by the rule of mixtures (Tsai, 1988). The classical rule of mixtures is based on three assumptions: (1) the constituents of a composite are represented by a continuum media or the well-defined material, (2) the matrix is isotropic, and (3) there is a perfect interfacial bonding between the constituents. Analysis of nanocomposites requires that nanoscale effects should be properly taken into account. According to the first assumption, the entire volume of the unit cell is occupied by the composite constituents:

$$V_{NT} + V_m = 1, \tag{8.2a}$$

where V_{NT} is the volume fraction occupied by carbon nanotubes and V_m is the volume fraction associated with the polymer matrix. The volume fraction, V_{NT}, of a carbon nanotube is given by $\pi(R_{NT} + h_{NT}/2)^2/A_{cell}$, where h_{NT} is the effective thickness of a carbon nanotube such that $h_{NT} \leq h_{vdW}$, where h_{vdW} is the average van der Waals separation between a carbon nanotube and the matrix, and A_{cell} is the cross-sectional area of the unit cell. Note that the van der Waals separation depends on the nature of the polymer/nanotube interfacial interactions, while the effective thickness of a carbon nanotube may be affected by the direction of loading, because of the transversely semiisotropic structure of carbon nanotubes.

The rule of mixtures requires that the smallest element (or a material point) of a constituent must have the same material properties as the bulk material. To obtain an equivalent-continuum medium with the uniquely defined average properties, the polymer chains must consist of many repeat units (e.g., 1095 CH_2 beads). The nanoscale homogenization criteria are satisfied, since there are so many polymer units. The amorphous polymer is further assumed isotropic in accordance with assumption (2). Finally, the interfacial separation between the polymer and carbon nanotube due to the van der Waals forces can be divided equally between the volume associated with the polymer and that of the carbon nanotube (Frankland et al., 2003). Note that the average van der Waals separation may vary from one polymer matrix to another.

The nanoscale volume fraction analysis can be more accurate in order to explicitly include the volume of the van der Waals spacing, V_{vdW} (Harik, 2011, 2014):

$$V_{\mathrm{NT}} + V_{\mathrm{m}} = 1 - V_{\mathrm{vdW}}, \tag{8.2b}$$

which is the modified volume fraction analysis for nanocomposites. The nanoscale volume fraction analysis is based on the following assumptions: (1) the constituents of a nanocomposite are composed of the well-defined atomic structure; (2) the volumes of the constituents: V_{NT} and V_{m} satisfy the nanoscale homogenization criteria, and (3) the contour of the interfacial region is well-defined. According to Fig. 8.2, nanocomposites may have a noticeable interphase region; therefore the modified volume fraction analysis for nanocomposites can be also presented as follows:

$$V_{\mathrm{NT}} + V_{\mathrm{m}} = 1 - V_{\mathrm{vdW}} - V_{\mathrm{int}}, \tag{8.2c}$$

where V_{vdW} is the volume fraction of the van der Waals spacing and V_{int} is the volume fraction of the interphase region around the carbon nanotube (Fig. 8.2).

According to the rule-of-mixtures (Tsai, 1988), the longitudinal elastic modulus, E_1, of the composite with long fibers can be estimated by

$$E_1 = E_{\mathrm{NT}} V_{\mathrm{NT}} + E_{\mathrm{m}} V_{\mathrm{m}}, \tag{8.3a}$$

where E_{NT} and E_{m} are the effective elastic moduli of a carbon nanotube and of the polymer matrix, respectively. Evaluation of the elastic modulus of carbon nanotubes depends on their radius and their atomic lattice structure. According to Fig. 8.2, nanocomposites may have a recognizable van der Waals separation as well as a noticeable interphase region; therefore the modified rule of mixtures for nanocomposites can be presented as follows:

$$E_1 = E_{\mathrm{NT}} V_{\mathrm{NT}} + E_{\mathrm{m}} V_{\mathrm{m}} + E_{\mathrm{vdW}} V_{\mathrm{vdW}} \tag{8.3b}$$

and

$$E_1 = E_{NT}V_{NT} + E_m V_m + E_{vdW}V_{vdW} + E_{int}V_{int} \tag{8.3c}$$

depending on the nanoscale structure of a nanocomposite. Here, V_{vdW} is the volume fraction of the van der Waals spacing (or the van der Waals radius), V_{int} is the volume fraction of the interphase, E_{int} is the elastic modulus of the interphase, which is typically size-dependent and ill-defined, and E_{vdW} is the elastic modulus associated with the van der Waals interactions:

$$E_{vdW} \approx \sum_{i=1}^{n_{vdW}} \frac{1}{V_{vdW,i}} \left| \left(\frac{\partial^2 U}{\partial \varepsilon^2} \right)_{\varepsilon=0,i} \right|_{vdW}$$

approximated as the rate of changes of the potential energy, U, with respect to infinitesimal strain, ε, associated with the van der Waals forces (Harik, 2011, 2014). In the case of the covalent bonding along the interface, the elastic modulus of the interfacial region would also depend on the density of the covalent bonding per unit area in the van der Waals spacing.

The rule of mixtures[3] also allow to estimate the transverse elastic modulus of nanocomposites, E_2, as follows:

$$\frac{1}{E_2} = \frac{V_{NT}}{E_{2NT}} + \frac{V_m}{E_m}, \tag{8.4a}$$

where E_{2NT} is the effective elastic modulus of a carbon nanotube in transverse direction, which is highly sensitive to their radius. The modified rule of mixtures for nanocomposites may include the van der Waals separation, as well as the interphase region. Consequently, the modified rule of mixtures for nanocomposites can be presented as follows:

$$\frac{1}{E_2} = \frac{V_{NT}}{E_{2NT}} + \frac{V_m}{E_m} + \frac{V_{vdW}}{E_{2vdW}} \tag{8.4b}$$

and

$$\frac{1}{E_2} = \frac{V_{NT}}{E_{2NT}} + \frac{V_m}{E_m} + \frac{V_{vdW}}{E_{2vdW}} + \frac{V_{int}}{E_{int}}, \tag{8.4c}$$

[3] In the case of transverse loading, one may assume that the stress is distributed uniformly in a carbon nanotube and the matrix and the van der Waals spacing with or without the covalent bonds: $\sigma_\infty = \sigma_{NT} = \sigma_m = \sigma_{vdW}$. The overall strain in nanocomposite is distributed between the constituents of the nanocomposite as follows: $\varepsilon_c = \varepsilon_{NT} \cdot V_{NT} + \varepsilon_m \cdot V_m + \varepsilon_{vdW} \cdot V_{vdW}$. The overall elastic modulus, E_2, of the nanocomposite is then given by $E_2 = \sigma_\infty / \varepsilon_c = \sigma_{NT} / (\varepsilon_{NT} \cdot V_{NT} + \varepsilon_m \cdot V_m + \varepsilon_{vdW} \cdot V_{vdW})$, or $E_2 = (V_{NT}/E_{2NT} + V_m/E_m + V_{vdW}/E_{2vdW})^{-1}$, since $\sigma_{NT} = E_{2NT} \cdot \varepsilon_{NT}$ and $\sigma_m = E_m \cdot \varepsilon_m$, according to Hooke's law, and a similar approximation $\sigma_{vdW} = E_{2vdW} \cdot \varepsilon_{vdW}$.

where E_{2vdW} is the transverse modulus of the van der Waals interactions across the interface with or without the covalent bonding. The nanoscale rules of mixtures (8.3a) and (8.4a) can take into account the effect of atomic structure of carbon nanotubes, as the aforementioned geometric and material parameters can change for different classes of carbon nanotubes.

8.2.3 Halpin–Tsai Relation for Nanocomposite Materials

Mechanical properties of the polymer matrix nanocomposites reinforced with carbon nanotubes can be described by the Halpin–Tsai relation, which relates Young's modulus, E_m, of the polymer matrix and the reinforcing carbon nanotubes, E_{NT}, to Young's modulus of the composite material, E, through the volume fraction, V_{NT}, of the randomly oriented nanotubes and other important parameters. The length, L_{NT}, of carbon nanotubes, their diameter, d_{NT}, and, especially their aspect ratio, d_{NT}/L_{NT}, affect the stiffness of nanocomposites or their Young's modulus. The elastic moduli of nanocomposites are proportional to the ratio of the elastic moduli, E_{NT}/E_m, and the volume fraction of the reinforcing nanofibers, V_{NT}, as given by the Halpin–Tsai relation (Chou, 1992):

$$E = E_m \left[\frac{3}{8} \frac{1 + 2(L_{NT}/d_{NT})\eta_{E_L} V_{NT}}{1 - \eta_{E_L} V_{NT}} + \frac{5}{8} \frac{1 + 2\eta_E V_{NT}}{1 - \eta_E V_{NT}} \right], \tag{8.5a}$$

where the nondimensional parameters, η_E, and η_{E_L}, are based on the ratio of the elastic moduli, E_{NT}/E_m:

$$\eta_E = \frac{E_{NT}/E_m - 1}{2 + E_{NT}/E_m} \quad \text{and} \quad \eta_{E_L} = \frac{E_{NT}/E_m - 1}{2(L_{NT}/E_m) + E_{NT}/E_m}. \tag{8.5b}$$

Here, the volume fraction, V_{NT}, of a carbon nanotube is given by $\pi(R_{NT} + h_{NT}/2)^2/A_{cell}$, where h_{NT} is the effective thickness of a carbon nanotube such that $h_{NT} \leq h_{vdW}$, where h_{vdW} is the average van der Waals separation between a carbon nanotube and the matrix material, and A_{cell} is the cross-sectional area of the unit cell (Fig. 8.2). Note that the effective thickness of a carbon nanotube may be affected by the direction of loading, because of the transversely semiisotropic structure of carbon nanotubes. Young's modulus of carbon nanotubes depends on their radius and their atomic lattice structure.

The Halpin–Tsai relation (8.5a) shows that Young's modulus of a composite material increases with the volume fraction, V_{NT}, of carbon nanotubes and their aspect ratio, L_{NT}/d_{NT}. As a result, the applied loads have to be higher in order to achieve the same mechanical response, e.g., the degree of strain. It has been noted that the Halpin–Tsai model underestimates the stiffness of a composite for high volume fractions in glass polymers. The Halpin–Tsai relation (8.5b) is based on micromechanical analysis and therefore does not take into account the nanoscale effects and the strength or the weakness of the nanotube–polymer interfaces. The strength and structure of nanoparticles and their interfaces in nanocomposites may vary, as well as the resulting strength and stiffness of nanocomposite materials.

8.2.4 Radial Stiffness of Carbon Nanotubes in Nanocomposites

Material properties of carbon nanotubes can influence evaluation of the average material properties of nanocomposites. The effective thickness of SWCNTs and the number of internal walls in MWCNTs determine whether these carbon nanotubes behave as the thin lattice shells or as the thick lattice shells under transverse loads. It has been noted that the definition of the thin and the thick carbon nanotubes depends on the thickness-to-radius ratio, h_{NT}/R_{NT}. In the case of MWCNTs, the definition of their thickness is intimately connected with the number of internal lattice shells in their structure. The transverse elastic modulus of carbon nanotubes is much lower for the thin lattice shells resulting in the "thin-shell" reduction of their radial stiffness. This "thin-shell effect" can influence the average elastic modulus of nanocomposites. In order to evaluate the radial stiffness of SWCNTs and MWCNTs, their radius, R_{NT}, has to be compared with their typical thickness in a nanocomposite, i.e., 3.4 Å, as demonstrated in Table 8.3. The "thin-shell" reduction in the value of radial stiffness of carbon nanotubes needs further analysis.

8.2.5 Functionalization of Carbon Nanotubes in Nanocomposites

Functionalization of carbon nanotubes can improve the strength of the polymer–nanotube interface in nanocomposites. The interfacial region in nanocomposites may have either van der Waals bonding or covalent bonding (Fig. 8.6). Carbon nanotubes can be functionalized by either $\pi - \pi$ bonding interactions with polymer modifiers, including the aromatic rings, or by covalent bonding. In polymer matrices with long chain molecules of higher molecular weight, carbon nanotubes can be wrapped around their circumference. The strength of the interfacial region affects the elastic properties of the carbon nanotube-based nanocomposites and the carbon nanotube-modified polymers. The Halpin–Tsai relation can be used to

Table 8.3: Classes of the thin and thick SWCNTs and MWCNTs

Estimates of Thickness, h_{NT}	Class II-s/II-m Thin Shells[a] (R_{NT})	Class II-sb/II-mb Thick Shells (R_{NT})
Single graphene shell, 3.4 Å (with two van der Waals layers)	3.7 nm–...	3.4–34 Å
Double-wall carbon nanotubes (with four van der Waals layers)	7.5 nm–...	3.4–68 Å
Triple-wall carbon nanotubes (with six van der Waals layers)	11.2 nm–...	3.4–102 Å
MWCNTs (four walls) (with eight van der Waals layers)	15.0 nm–...	3.4–136 Å

SWCNT, single-wall carbon nanotube; *MWCNT*, multiwall carbon nanotube.
[a]A transitional range of 10% is assumed.

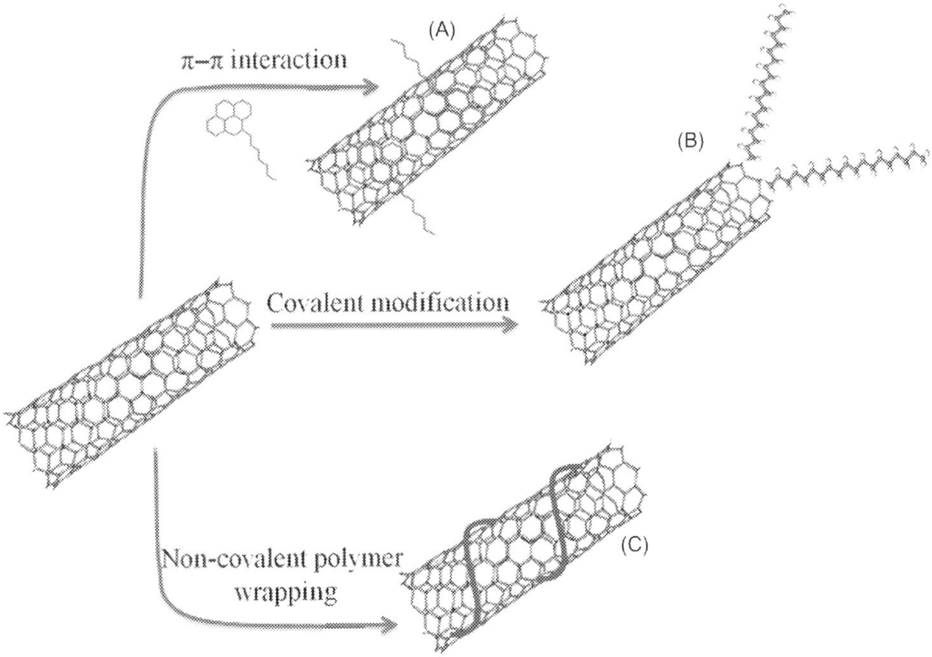

Figure 8.6
A schematic of possible methods of functionalization of carbon nanotubes in nanocomposites: (A) $\pi - \pi$ interactions, (B) covalent bonding modification, and (C) noncovalent polymer wrapping (Choudhary and Gupta, 2011).

predict the elastic properties of polymer matrix composites for the glassy polymers and often for the semicrystalline polymers. The strength of the interfacial bonding affects the strength of nanocomposites due to the increased interfacial stress transfer.

The structure of interfacial regions in nanocomposites depends on the molecular structure of the surrounding polymer, as well as on the molecular structure of the functionalizing nanostructures added to the polymers as modifiers (Fig. 8.6A). The thickness of the interfacial region, δt, can be evaluated with respect to the size or the radius of a nanostructure, e.g., a group of aromatic rings or the molecule, l_{m}, added as a polymer modifier. The thickness of the enriched interface, δt, is thin, when its normalized thickness, $\delta t/l_{\mathrm{m}}$, is such that $\delta t/l_{\mathrm{m}} < 1$, and thick, when $\delta t/l_{\mathrm{m}} > 1$ (see Fig. 8.6A). Here, the radius of the molecule can be replaced by the size of planar molecules made of aromatic rings. When the molecular structure of the interfacial region is complex, it actually can form an interphase region with the unique material properties and unique molecular structure which is distinct from the bulk polymer material. In order for a nanocomposite to have an interphase region, the normalized thickness of the interfacial region, $\delta t/l_{\mathrm{m}}$, should be large enough as compared to the smallest structural element, a_i, i.e., the size of a carbon ring, in the molecular structure of the interfacial region, δt (Fig. 8.2).

The weak interfacial regions found in the material systems with interfacial absorbents (Fig. 8.7), which have fused aromatic rings and the likely $\pi - \pi$ bonding with the nanotube surface (Fig. 8.6). The $\pi - \pi$ bonding through the van der Waals interfaces between adjacent carbon rings is stronger than other van der Waals interfaces between different atomic layers or different material phases. Nanocomposites with covalent bonding associated with the functionalized carbon nanotubes (Fig. 8.7) have stronger interfacial bonding and, hence, better stress transfer and higher elastic moduli of nanocomposites. Nanocomposites with strong interfaces are better for structural applications.

The molecular structure of the interfacial region can be analyzed by using the so-called scaling analysis (Harik, 2001a,b, 2002). Analysis of the key parameters describing the nanostructure of the interfacial region can be used to formulate a condition for the existence of the interphase with distinct material properties in the interfacial region in nanocomposites:

> the necessary and sufficient condition for the existence of the interphase region in a nanocomposite is that the nondimensional homogenization parameter for the molecular structure, $H = \delta t/a_i$ characterizing the relation between the thickness of the interface, δt,

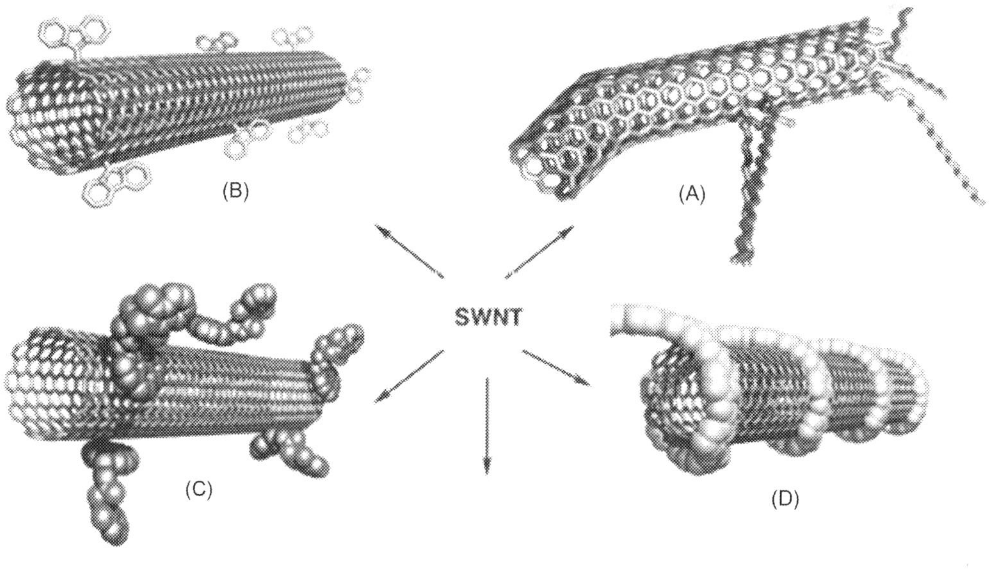

Figure 8.7
Atomic lattice structure of carbon nanotubes shown with the molecular structure associated with the different methods of functionalization of carbon nanotubes in nanocomposites by the covalent functionalization of defects (A), the covalent surface functionalization (B), and the noncovalent exohedral functionalization (C) with surfactants, as well as polymer wrapping (D) proposed by Jeon et al. (2011).

and the size of the smallest structural element, a_i, in its molecular structure is much greater than unity, that is $H \gg 1$.

The nondimensional homogenization parameter, H, for the molecular structures has some useful characteristics. First, it mathematically defines when and under what conditions a thick interface, $H > 1$, can become an interphase, that is, $H \gg 1$. This is a first step in material homogenization for any molecular structure. Second, it characterizes the size and the complexity of the molecular structure of various functional groups around a nanostructure. When the parameter is $H \approx 1$ or smaller, then the molecular structure of the interface region is not complex. When the parameter H is 3 or 5 and higher, then the molecular structure of the interface region can be complex. When the parameter H is large, that is, $H \gg 1$, then the molecular structure of the interphase is not only complex, but it is fully developed into a macroscopic material with the unique and well-defined material properties. It also means that when the parameter H is very large, that is, $H \geq 10$, then the interphase satisfies the nanoscale homogenization criteria. Therefore, when the parameter H is not very large, that is, $H < 10$, then the material properties of the interphase are size-dependent and their homogenization is not size invariant.

8.3 Conclusions

Analysis of the carbon nanotube-reinforced polymer matrix nanocomposites has been discussed, along with their unique material properties and the length scales in their atomic structure. Multiscale homogenization of nanocomposites was presented, along with the models of the RVEs for the multiscale analysis methods. MD simulation of the carbon nanotube-reinforced nanocomposites has been discussed in order to demonstrate their large molecular structure and their stress–strain dependence. MD simulation results for the stress–strain response of nanocomposites have been compared with the predictions of the classical rule of mixtures. Different methods of functionalization of carbon nanotubes have been discussed, along with their effect on the structure of the interfacial regions.

The modified rules of mixtures for the analysis of the carbon nanotube-reinforced nanocomposites have been presented, along with the effects of the matrix/nanotube interphase and the van der Waals interface. The formation of interphase regions in nanocomposites has been examined by the scaling analysis of their molecular structure. MD simulations of the carbon nanotube-reinforced polymer nanocomposites have been presented. MD results compare well with those of the rules of mixtures for small strains of the carbon nanotube-reinforced polymer nanocomposites. The Halpin–Tsai relation has been adapted for the carbon nanotube nanocomposites. The "thin-shell" effect in the reduction of the radial stiffness of carbon nanotubes in polymer nanocomposites has been discussed.

Sample Problems

1. Find and list a few examples of the carbon nanotube-reinforced nanocomposites and describe their structure.
2. Describe the length scales in the atomic lattice structure of the carbon nanotube-reinforced nanocomposites.
3. Make a PowerPoint viewgraph with an illustration of the atomic lattice structure of the carbon nanotube-reinforced nanocomposites.
4. Make a PowerPoint viewgraph with an illustration of the length scales in the atomic lattice structure of the carbon nanotube-reinforced nanocomposites.
5. Find examples of carbon nanotube-reinforced nanocomposites that include different classes of carbon nanotubes.
6. Describe the atomic structure of the carbon nanotube-reinforced nanocomposites, which include different classes of carbon nanotubes.
7. Describe the transversely (semi)isotropic material model for carbon nanotubes.
8. What is the physical basis associated with the values of the effective thickness, h_{NT}, of 3.4 Å for the carbon nanotube-reinforced nanocomposites?
9. What are the nanoscale homogenization criteria for the carbon nanotube-reinforced nanocomposites?
10. Describe the nanoscale rules of mixtures modified for the carbon nanotube-reinforced nanocomposites.

References

Brenner, D.W., 1990. Empirical potential for hydrocarbons for use in simulating the chemical vapor deposition of diamond films. Phys. Rev. B: Condens. Matter 42, 9458–9471.

Chou, T.W., 1992. Microstructural Design of Fiber Composites. Cambridge University Press, Cambridge.

Choudhary, V., Gupta, A., 2011. Polymer/carbon nanotube nanocomposites. In: Yellampalli, S. (Ed.), Carbon Nanotubes–Polymer Nanocomposites. InTech, Rijeka, Croatia.

Frankland, S.J.V., Harik, V.M., 2003. Carbon nanotube pull-out from a polymer matrix. Surf. Sci. Lett. 525, L103.

Frankland, S.J.V., Harik, V.M., Odegard, G.M., Brenner, D.W., Gates, T.S., 2003. The stress–strain behavior of polymer–nanotube composites from molecular dynamics simulation. Compos. Sci. Technol. 63, 1655–1661.

Harik, V.M., 2001a. Ranges of applicability for the continuum beam model in the mechanics of carbon nanotubes and nanorods. Solid State Commun. 120 (7–8), 331–335.

Harik, V.M., 2001b. Ranges of applicability for the continuum beam model in the constitutive analysis of carbon nanotubes: nanotubes or nano-beams? NASA/CR-2001-211013. NASA Langley Research Center, Hampton, VA.

Harik, V.M., 2002. Mechanics of carbon nanotubes: applicability of the continuum-beam models. Comput. Mater. Sci. 24 (3), 328–342.

Harik, V.M., 2011. Mechanics of Carbon Nanotubes. Nanodesigns Press, Newark, DE.

Harik, V.M. (Ed.), 2014. Trends in Nanoscale Mechanics. Springer, Dordrecht.

Jeon, I.-Y., Chang, D.W., Kumar, N.A., Baek, J.-B., 2011. Functionalization of carbon nanotubes. In: Yellampalli, S. (Ed.), Carbon Nanotubes–Polymer Nanocomposites. InTech, Rijeka, Croatia.

Park, C., Ounaies, Z., Watson, K.A., Crooks, R.E., Smith Jr., J., Lowther, S.E., et al., 2002. Dispersion of single wall carbon nanotubes by in situ polymerization under sonication. Chem. Phys. Lett. 364 (3–4), 303–308.

Schadler, L.S., Giannaris, S.C., Ajayan, P.M., 1998. Load transfer in carbon nanotube epoxy composites. Appl. Phys. Lett. 73 (26), 3842–3844.

Thostenson, E.T., Chou, T.-W., 2002. Aligned multi-walled carbon nanotube-reinforced composites: processing and mechanical characterization. J. Phys. D: Appl. Phys. 35 (16), L77–L80.

Tsai, S.W., 1988. Composites Design. Think Composites Press, Dayton, OH.

CHAPTER 9

Nanoscale Interfaces in Nanocomposites and Friction Laws

Chapter Outline

9.1 Interfaces Around Carbon Nanotubes in Nanocomposites

Carbon nanotubes and the matrix materials form different types of interfaces in the polymer, metal, or ceramic nanocomposite materials (Harik, 2011). Atomic lattice structure of carbon nanotubes is associated with higher values of elastic modulus and strength compared to that of the polymer, metal, and ceramic matrix materials. Atomic and molecular structures of the matrix materials are adjusted to the atomic lattice of carbon nanotubes depending on the processing conditions, polymer modifiers and additives, as well as the size and curvature of nanotubes. Functionalization of carbon nanotube surfaces can also change the polymer–nanotube interface and how polymer chains adjust to the local atomic structure. Surfactants and absorbents can also influence the stoichiometry at the polymer–nanotube interface and the resulting molecular structure of the interface region.

Geometry of carbon nanotubes can influence the size of the polymer–nanotube interface, since larger values of the radius of cylindrical shells are associated with longer

Mechanics of Carbon Nanotubes.
DOI: https://doi.org/10.1016/B978-0-12-811071-3.00009-3

© 2018 Elsevier Inc. All rights reserved.

circumferences and larger surface areas. Uniform dispersion of carbon nanotubes in polymer matrix is important for the increase of the total surface area exposed to the polymer matrix in nanocomposites. Better dispersion of carbon nanotubes results in larger interfacial area and more uniform distribution of the reinforcement in the matrix. Dispersion and orientation of carbon nanotubes can be controlled by the flow-induced alignment in the extrusion flows, which can affect the formation of the interfacial structures and their distribution in the matrix. Long high aspect ratio single-wall carbon nanotubes (SWCNTs) and multiwall carbon nanotubes (MWCNTs) have larger surface areas and, hence, an extensive interfacial surface (Fig. 9.1), which can significantly improve the stress transfer from the matrix material into the reinforcing carbon nanotubes.

Carbon nanotubes with the higher aspect ratios (Fig. 9.1) have a larger surface area and, hence, significant interfacial surface, which is proportional to their radius ($R_{NT} + h_{NT}/2$), and their length, L_{NT}, as well as their product: $2\pi(R_{NT} + h_{NT}/2)L_{NT}$, h_{NT} is the effective thickness of carbon nanotubes. Good interfaces in the carbon nanotube-reinforced nanocomposites provide a better load-carrying capability due to their better ability to assist the transfer of the external loads into the matrix reinforcement. The size of interfacial surface can be characterized by a nondimensional parameter: $2\pi(R_{NT} + h_{NT}/2)(L_{NT}/ah_{NT})$, where a is the size of a carbon ring. The strength of the carbon nanotube–polymer interface and the energy required to cause the interfacial debonding and failure are also proportional to the size of the interfacial region. It means that the critical shear stress, τ_{cr}, which characterizes the debonding strength of the polymer–nanotube interface, has the following functional dependence $f(.)$ on the aforementioned nondimensional parameter:

$$\tau_{cr} \propto 2\pi f\left(\frac{L_{NT}}{a}, \frac{R_{NT}}{h_{NT}}\right), \tag{9.1}$$

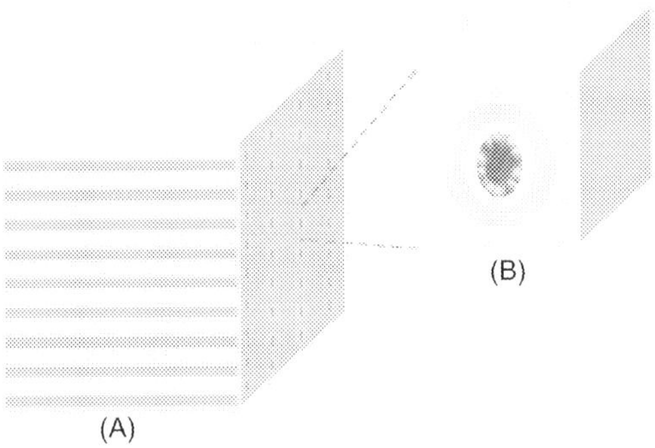

Figure 9.1
Long carbon nanotubes with interfacial regions (yellow or light) in the carbon nanotube-reinforced polymer matrix nanocomposites (A) and an example of the interphase region in the polymer matrix around the (5,5) carbon nanotube with van der Waals spacing (B).

where the radius-to-thickness ratio, R_{NT}/h_{NT}, has been used to describe the classes of the thin and thick SWCNTs and MWCNTs. The normalized length, L_{NT}/a, of carbon nanotubes is also associated with the characterization of the long SWCNTs and MWCNTs. It should be noticed, however, that the functional dependence (9.1) of the critical shear stress, τ_{cr}, does not include the aspect ratio, d_{NT}/L_{NT}, or its inverse, but a more complex dependence involving the familiar thickness-to-radius ratio, $(R_{NT}/h_{NT})^{-1}$, and the normalized length, L_{NT}/a, of carbon nanotubes. The strength of the interfacial region affects the elastic properties of the carbon nanotube-based nanocomposites and the carbon nanotube-modified polymers. The work and the energy required for the debonding of the whole carbon nanotube along its interface and the work associated with the pullout of a carbon nanotube from a polymer matrix have the same functional dependence described by Eq. (9.1).

The volume of carbon nanotubes, which have been pulled out from the polymer matrix, is proportional to the energy directed at the volume of a nanocomposite during an impact. An impact can be a low velocity or a high velocity impact associated with different failure events. This volume of the fractured carbon nanotubes can be controlled by the optimal design of the material structure of nanocomposites and by improving the strength of their interface (or interphase) regions. The volume of the carbon nanotubes pulled out of the matrix also depends on the size of the potential fracture surfaces,[1] which can be controlled by improving the interfacial strength and using the so-called fail-safe designs of composite materials with the controlled failure patterns. The strength of the polymer–nanotube interfaces can be used by optimizing the structure of the carbon nanotubes used for the reinforcement. Classification of carbon nanotubes identifies 20 main classes of SWCNTs and MWCNTs, which can be used for the optimization of material designs guided by the functional dependence (9.1) for improving the critical strength of interfaces. Each class of carbon nanotubes is characterized by a particular type of geometry associated with the specific ranges of geometric parameters, which can be used to improve the critical strength of the interfacial regions in nanocomposites, their load-carrying capability, and the efficiency of the load transfer.

9.1.1 Length Scales in the Atomic Structure of Functionalized Interfaces

The strength of interfaces in the carbon nanotube-reinforced nanocomposites can be improved by different methods. The noncovalent bonding of the polymer–nanotube interface with the van der Waals spacing can be improved by choosing the polymer material with aromatic rings in the molecular structure of the polymer chains. Aromatic rings have an

[1] The potential exposure to the high aspect ratio carbon nanotubes, which may be toxic as long MWCNTs, can be minimized by controlling the size of a potential fracture, the extent of the fracture surfaces, and the pullout of carbon nanotubes. It is possible to minimize the potential risks by optimizing the structure of nanocomposites and their use in a particular product. Carbon nanotubes with moderate aspect ratios can also provide enough reinforcement for polymer matrices, while their size can be optimized to reduce the potential health risks.

electronic structure that is capable of the $\pi - \pi$ interactions with the surface of carbon nanotubes. The ability of the polymer matrix material to have the $\pi - \pi$ interactions can be also improved by modifying the polymer with molecules, which are composed of the aromatic rings, as well as a short polymer chain (Fig. 9.2) for the better mixing with the polymer. Mixtures of molecules with aromatic rings can be added to carbon nanotubes as surfactants (Fig. 9.2), which can also improve the $\pi - \pi$ interactions at the interface and improve bonding with the polymer with the short polymer chains. These polymer modifiers can be also used as the interface modifiers. Their effectiveness in the forming good $\pi - \pi$ bonding along the polymer–nanotube interface and the formation of a strong interfacial region in the polymer depends on the atomic structure of the interfacial region and its material components, as well as the length scales (Table 9.1) of their atomic and molecular structures.

The length scales in the atomic structure of the interfacial surface of a carbon nanotube involve several important parameters: the length of the covalent bonds, l_{C-C}, the size of a carbon ring, a, the radius, R_{NT}, of carbon nanotubes, the effective thickness, h_{NT}, of their atomic lattice, the van der Waals spacing, h_{vdW}, the size of the surface surfactants, l_s, or their aromatic rings (Table 9.1), the size of the smallest elements in the molecular structure of polymer chains, l_m, and in the crystalline structure of a polymer, L_m, the thickness of the interface (or interphase) region, $h_{int} = R_{int,o} - R_{int,i}$, evaluated as the difference between their outer, $R_{int,o}$, and inner radii, $R_{int,i}$. The length, L_h, of the homogenization of carbon nanotubes, and their normalized length, L_{NT}/a, are also important for the characterization of the interfacial surface. Table 9.1 presents typical values of the important parameters describing the atomic structure of interfacial regions and their length scales.

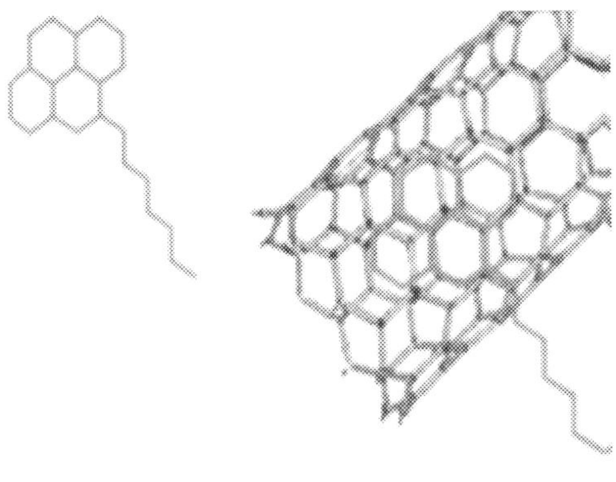

Figure 9.2
Atomic structure of a zigzag carbon nanotube with a molecule made of aromatic rings and a short polymer chain (also shown on a side) interacting through the $\pi - \pi$ electron interactions on the surface of the carbon nanotube with the van der Waals spacing (after Choudhary and Gupta, 2011).

Table 9.1: Length scales in the atomic structure of interfaces in nanocomposites

Parameters	Typical Values
Length of a C–C bond, l_{C-C}	1.41–1.42 Å
Size of a carbon ring, a	2.46 Å
van der Waals distance, d_{vdW}	3.4 Å
Thickness of electron layers, $2h_e$	1.82 Å
Thickness of nanotubes, h_{NT}	0.66–3.4 Å (single wall)
Size of (CH_2) zigzag chain, l_m	~3 Å
Size of surfactants, l_s	~3–7 Å
Crystalline unit in polymer, L_m	~7 Å
Radius of SWCNTs, R_{NT}	~2 Å–1.5 nm
Radius of MWCNTs, R_{NT}	~15 Å–45 nm
Homogenization length, L_h	~10a (2.5 nm)

SWCNT, single-wall carbon nanotube; *MWCNT*, multiwall carbon nanotube.

9.1.2 Scaling Analysis of Complex Interfaces in Nanocomposites

Functionalization of carbon nanotubes increases the complexity of the atomic structure of the polymer–nanotube interface in nanocomposites. The interfacial region in nanocomposites may have noncovalent bonding with the van der Waals spacing or covalent bonding (Fig. 9.3). Carbon nanotubes can be functionalized by small or large molecules through the $\pi - \pi$ bonding interactions with the molecules of surfactants or polymer chains, both of which include aromatic rings (Fig. 9.3C and D). The strength of their $\pi - \pi$ bonding interactions depends on the structure of a surfactant, which may include one or two rings and a short tail (Fig. 9.3C). Polymer matrix materials of higher molecular weight consist of the long chain molecules, which can wrap carbon nanotubes around their circumferences (Fig. 9.3D). The strength of the $\pi - \pi$ electron interactions between a long polymer chain, which is wrapping the surface of the carbon nanotube (Fig. 9.3D), depends on the number of aromatic rings in its structure and the period of its wrapping. The size of surfactant molecules, l_s (Fig. 9.3C), affects the molecular structure and the size of the interface region. The size and structure of the interphase region also depends on the atomic structure of different molecules and aromatic rings attached to SWCNTs by covalent bonding (Fig. 9.3B).

A nondimensional ratio defined as the radius-to-thickness ratio, R_{NT}/h_{NT}, is important in the evaluation of the size of interfacial region in the carbon nanotube-reinforced polymer matrix nanocomposites. Carbon nanotubes with large radii may have the interfacial region with the unique and the well-defined material properties, if their normalized circumference, $2\pi R_{NT}/a$, is such that $2\pi R_{NT}/a \gg 1$, or $R_{NT} > 5a/\pi$. Most of the carbon nanotubes (SWCNTs and MWCNTs) satisfy this condition. The larger value of the radius-to-thickness ratio, R_{NT}/h_{NT}, is associated with the larger size of the interfacial surface and that of the polymer–nanotube interface. Larger surface area of the interface allows

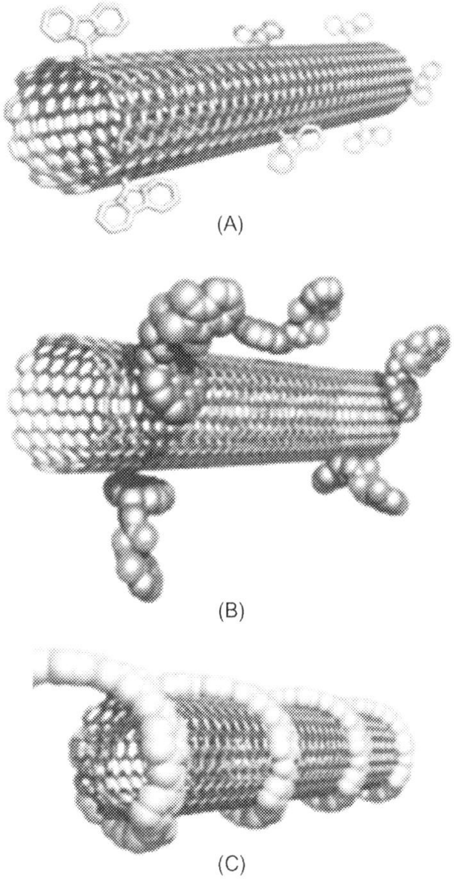

Figure 9.3
Atomic structure of carbon nanotubes with different molecules made of aromatic rings attached by covalent bonding (A) or the $\pi - \pi$ bonding interactions [a surfactant with a ring and a short tail (B)], or interacting through the $\pi - \pi$ electron interactions by wrapping the surface of the carbon nanotube (C) with the van der Waals spacing (after Jeon et al., 2011).

stronger bonding interactions and increases the strength of the interface. The strength of the interfacial bonding affects the strength of nanocomposites, due to the increased interfacial stress transfer. Nanocomposites with covalent bonding associated with the functionalized carbon nanotubes have stronger interfacial bonding and, hence, better stress transfer and higher elastic moduli. Nanocomposites with strong interfaces are better for structural applications.

The molecular structure of the interfacial region in nanocomposites depends on the atomic and molecular structure of the functionalizing molecules and the surrounding polymer. The complexity of the material structure of the interface also depends on the length scales associated with the atomic lattices of material constituents forming the interface. The number of the length scale parameters describing the atomic and molecular structure of the interface is large. Therefore, scaling analysis should be used for determining the critical nondimensional ratios describing the relation between different length scales in the atomic structure of the interface.

The thickness of the interfacial region, h_{int}, can be compared with respect to the size of the functionalizing molecules, e.g., a group of aromatic rings, l_s, added as a polymer modifier. The thickness of the enriched interface, h_{int}, is thin, when its normalized thickness, h_{int}/l_s, is such that $h_{\text{int}}/l_s < 1$, and thick, when $h_{\text{int}}/l_s > 1$. When the molecular structure of the interfacial region is complex, it actually can form an interphase region with the unique material properties and unique molecular structure, which is distinct from the bulk polymer material. In order for a nanocomposite to have an interphase region, the normalized thickness of the interfacial region, h_{int}/l_s, should be large enough as compared to the smallest structural element, a_i, i.e., the size of a carbon ring, in the molecular structure of the interfacial region, l_s.

The aforementioned characterization of the molecular structure and the normalized thickness of the interfacial region, h_{int}/l_s, is analyzed by using a simple example of the scaling analysis (Harik, 2001a,b, 2002). Analysis of the key parameters describing the nanostructure of the interfacial region can be used to formulate a condition for the existence of the interphase with distinct material properties in the interfacial region in nanocomposites (Harik, 2011, 2017):

> *the necessary and sufficient condition for the existence of the interphase region in a nanocomposite is that the nondimensional homogenization parameter for the molecular structure, $H = h_{int}/l_s$ characterizing the relation between the thickness of the interface, h_{int}, and the size of the smallest structural element, l_s, in its molecular structure is much greater than unity, that is $H \gg 1$.*

The nondimensional homogenization parameter, H, for the molecular structure of the polymer–nanotube interface has some useful characteristics. First, this parameter defines when and under what conditions a thick interface, $H > 1$, can become an interphase, that is, $H \gg 1$. This is a first step in the nanoscale homogenization of a material with any molecular structure. Second, this parameter characterizes the size and the complexity of the molecular structure of various functional groups around a nanostructure. This parameter can equal to unity: $H \approx 1$ or be smaller; then, the molecular structure of the interface region is not complex. When the parameter H is 3 or 5 and higher, then the molecular structure of the interface region can be complex. When the parameter H is large, that is, $H \gg 1$, then the molecular structure of the interphase is not only complex, but it is fully developed into a macroscopic material with the unique and well-defined material properties. It also means that when the parameter H is very large, that is, $H \geq 10$, then the interphase satisfies the nanoscale homogenization criteria. Therefore when the parameter H is not very large, that is, $H < 10$, then the material properties of the interphase are size dependent and their homogenization is not size invariant.

9.2 Analysis Methods for Carbon Nanotube Interfaces in Nanocomposites

Analysis of interfacial interactions in the carbon nanotube-reinforced nanocomposites requires understanding of the length scales in the atomic structure of the polymer–nanotube interface. Scaling analysis of the length scale parameters describing the atomic and molecular structure of the interfacial region can provide useful nondimensional parameters that can indicate the important relations between the smallest structural elements, the thickness of the interface, and the radius of carbon nanotubes. Multiscale structure of the interfacial region in nanocomposites can be investigated by the finite element models, which reflect the material structure of a representative volume element (RVE), including a finite interface and the property gradients across the interface and, possibly, in the adjacent region of the polymer interphase. It should be noted that all parts of RVE in a finite element model should satisfy the nanoscale homogenization criteria.

Molecular dynamics (MD) simulations of interfacial interactions in nanocomposites can predict not only their effect on the material properties of RVEs in nanocomposites, but also can characterize the strength of the polymer–nanotube interface and interfacial friction mechanisms between atomic lattices (Frankland and Harik, 2003). MD simulations can also verify the nanoscale rules of mixtures for nanocomposite materials, as was shown in the previous chapter, and evaluate their modification for the inclusion of the interface effects. The atomic lattice structures and the periodic unit cells used in MD modeling have to satisfy the nanoscale homogenization criteria for their size in order to represent material properties with the unique and well-defined values, which are size-invariant. In order to properly resolve interfacial phenomena associated with the interfacial interactions of two atomic lattices at nanoscale level, MD simulations also have to satisfy certain criteria for the nanoscale homogenization (or averaging) of their results. Computational results from MD simulations require appropriate multiscale approaches for their analysis and the use for the nanoscale characterization of interfacial interactions and interfacial friction mechanisms between atomic lattices.

9.2.1 Molecular Dynamics Simulations of Interfaces in Nanocomposites

The atomic structure of the carbon nanotube interfaces in polymer matrix nanocomposites consists of the atomic lattice of SWCNTs and the interfacial segments of the molecular structure of the polymer matrix. The atomic lattice structure of carbon nanotubes has to satisfy the nanoscale homogenization criteria for its size. Interfacial atomic interactions in nanocomposites can be simplified by considering the crystalline polymer matrix such as in the polyethylene (PE), which have a regular arrangement of molecular chains. Furthermore, the noncovalent bonding of carbon nanotubes can be considered for the analysis of interfacial interactions between two regular atomic lattices. The nonbonding interactions between two adjacent lattices can be approximated by the van der Waals interactions represented by the Lennard-Jones (LJ) potential. Interactions between carbon atoms in the atomic lattice of carbon nanotubes can be described by the Tersoff–Brenner potential (Brenner, 1990).

Analysis of interfacial regions in the carbon nanotube-reinforced nanocomposites can be further simplified by taking into account the periodicity of the atomic and molecular structures along the interface. While a segment of a carbon nanotube can include a large, but finite, number of carbon atoms, the molecular structure of a polymer material is composed of thousands of atoms. The multiscale modeling method based on a RVE for the periodic molecular structure carries out MD simulations for a periodic unit cell with a carbon nanotube and a limited number of polymer chains (Fig. 9.4). The size of the interfacial region shown in Fig. 9.4 is smaller than the size of the unit cell in MD simulations (Frankland and Harik, 2003), which has to satisfy the nanoscale homogenization criteria.

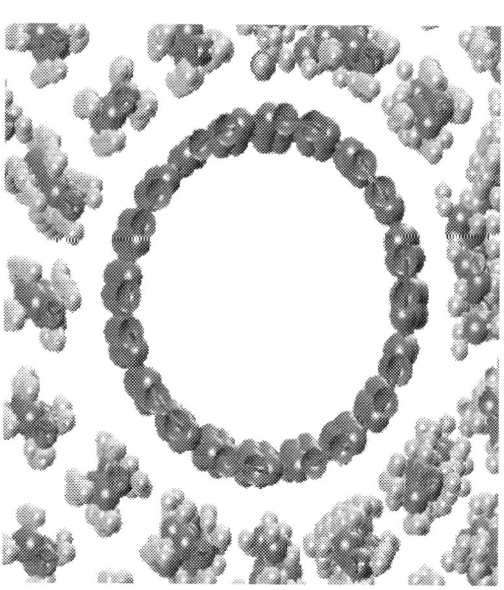

Figure 9.4
Molecular structure of the interface region around the (10,10) carbon nanotube with the van der Waals spacing (white) in the carbon nanotube-reinforced polymer matrix nanocomposite (after Frankland and Harik, 2003).

The atomic structure of a unit cell, which is shown in part in Fig. 9.4, is composed of a 52.9-Å long conductive (10,10) carbon nanotube of the radius, $R_{NT} = 6.8$ Å, surrounded by 15 zigzag $[CH_2-CH_2-CH_2]_n$ PE chains with the CH_2 asperities associated with a crystalline part of the PE polymer material. The typical size of the crystalline parts of the semicrystalline PE material is about 80–200 Å, while the size of its noncrystalline parts is about 50–300 Å. The length of a C–C bond in the PE chain is 1.53 Å, the zigzag bond angle is 112°, and the circumferential distance between the two adjacent PE chains is about 2.55 Å, which is slightly larger than the size of a carbon ring, *a*. The spacing between the two adjacent PE chains is associated with the internal PE-polymer structure and its orthorhombic crystal unit cell having the following lengths of its axes: $a = 7.42$ Å, $b = 4.95$ Å, and $c = 2.55$ Å. The RVE considered in this analysis is about 60 × 60 × 53 Å encompassing about 23,000 atoms with 880°C atoms in the lattice of carbon nanotube and 178 PE chains of 42 units long around the embedded in equilibrium at 300K. The considered material volume element is periodic in three dimensions.

The atomic structure of the 52.9-Å long (10,10) SWCNT is composed of 21.5 carbon rings along its axial direction. Therefore its length, L_{NT}, satisfies the nanoscale homogenization criterion: $L_{NT}/a \geq 10$, where a is the width of a carbon ring (Fig. 9.5). The next upscaling level, $L_H \geq 10a$, is valid for the averaging (or homogenization) of interlayer interactions. Below this homogenization length scale, L_H, the nanoscale effects have intrinsically discrete nature and the interfacial interlocking mechanisms increasingly reveal their dependence on

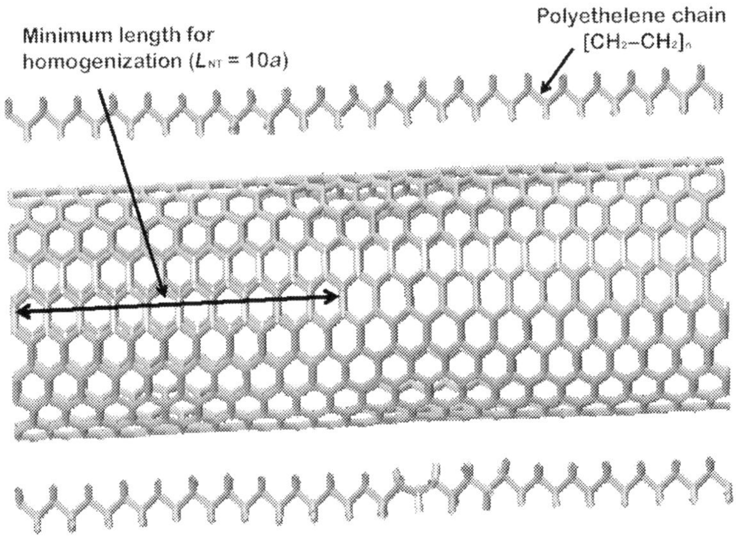

Figure 9.5
Atomic lattice structure of the armchair (10,10) carbon nanotube ($R_{NT} = 6.8$ Å and $L_{NT} = 52.89$ Å) and two $(CH_2)_n$ zigzag polyethylene chains above and below the nanotube along the interface with the van der Waals distance of 3.4 Å (after Frankland and Harik, 2003).

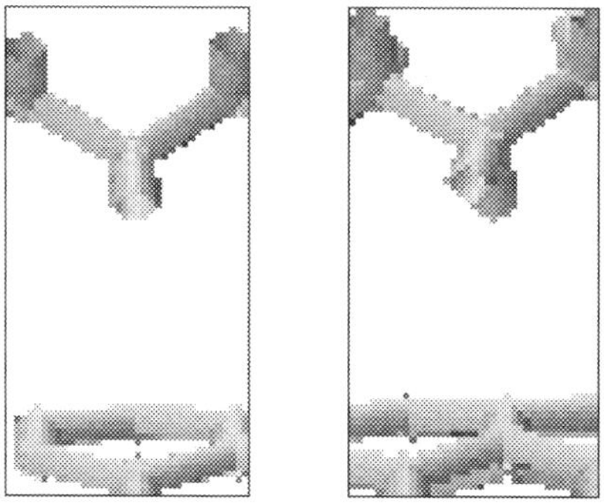

Figure 9.6
An asperity of a fragment of the $(CH_2)_n$ zigzag polyethylene chain above a carbon ring in the O-registry (A) and in the X-registry (B) over the C–C bond belonging to the two carbon rings in the atomic lattice structure of the armchair carbon nanotube along the interface with the van der Waals distance of 3.4 Å. © *2011 by Nanodesigns Consulting (reprinted with a permission).*

the local atomic-lattice structures and their registry configurations. The two zigzag PE chains, which have 21 CH_2 asperities along the lattice of the (Fig. 9.5), display two distinct registry configurations, i.e., the O-registry configuration (Fig. 9.6A) with the CH_2 asperity positioned above the center of the carbon ring, hence, the O-registry; and the X-registry configuration (Fig. 9.6B) with the CH_2 asperity positioned right above the transverse C–C bond of the two consecutive carbon rings, i.e., the X-registry.

All of the CH_2 asperities in the two zigzag chains shown in Fig. 9.5 have either O- or the X-registry configurations, or a different registry in between these two configurations. The O- and X-registry configurations satisfy the following lattice-translation relation: If the nth CH_2 asperity of the PE zigzag chain is in the O-registry configuration, the $n+9$ CH_2 asperity is in the X-registry configuration. Hence, the $n+18$ CH_2 asperity is also is in the X-registry configuration, and $N=18$ is the registry unit cell for the PE and SWCNT lattices. Therefore the RVE encompasses the entire registry unit cell. The distance between the CH_2 asperities and the lattice of the SWCNT is smaller than the van der Waals distance, $h_{vdW}=3.4$ Å, because the length of the two C–H bonds is 1.09 Å, which is close to the length of the C–C bond, 1.53 Å, in these zigzag asperities.

9.2.2 Molecular Dynamics Simulations of Interfacial Registry Barriers

Interfacial interactions at the carbon nanotube–polymer interface in nanocomposites include the atomic lattice of SWCNTs and the interfacial molecular structure of the polymer matrix. At the nanoscale level, the fundamental origins of sliding friction between two atomic lattices involve the surface-to-surface interactions, such as the interlocking and registry effects between the lattice of SWCNTs and the surrounding polymer surface with its morphology of asperity distribution, local bending and buckling of the SWCNT, the sliding induced excitation of atomic lattice vibrations, interaction of phonons propagating along the sliding surfaces, electrostatic interactions, electron motion, and the electron interactions akin to the $\pi - \pi$ bonding interactions, especially those involving aromatic ring structures. Some of these interfacial friction mechanisms have been studied to some extent by a quartz-crystal microbalance, scanning force microscopy, and the nanotube pullout and experiments with atomic force microscopes (AFM), as well as by theoretical modeling, including MD simulations.

MD simulation of the polymer–nanotube interface in the carbon nanotube-reinforced nanocomposites has demonstrated the complex atomic structure of the adjacent atomic lattices separated by the van der Waals spacing and their commensurability characterized by different atomic registries. The atomic lattice structure of the interfacial region is intimately connected with the nanoscale mechanisms of adhesion between two adjacent surfaces. Local variations in the atomic structure affect the strength of van der Waals interactions, as well as the local commensurability and the specific registry between two interfacial lattices. After the details of the atomic structure of the interfacial lattices have been determined, along with the degree of their commensurability, the mechanisms of interfacial interactions can be examined in order to investigate the nanoscale origins of the interfacial adhesion.

The mechanisms of interfacial interactions include the interlocking between carbon rings of the carbon nanotube and the zigzag asperities in the molecular chains. The degree and the type of such interlocking can be characterized by different atomic registries described earlier. The ring-asperity interlocking greatly affects the interfacial adhesion of the atomic lattice of carbon nanotubes and the interfacial strength of adhesion, τA_S, characterized by the interlocking-induced shear stress, τ, and the contact area of the sliding surface, A_S, which are linearly proportional to the nanotube radius, R_{NT}, the length, L_{NT}, and, obviously, the surface area. The ring-asperity interlocking can be measured by the potential registry—barriers for interfacial displacement between two atomic lattices and the force, F_B, needed to overcome a particular registry barrier associated with the commensurability between two lattice and the specific registry. The coupling of two or more interfacial interaction mechanisms such as strong registry effects, asperity interlocking, or nonstandard stick-slip behavior with generation of lattice phonons can be challenging.

MD simulations of interfacial interactions of the zigzag CH_2 PE chains with the atomic lattice of a carbon nanotube can be described by the LJ potential and an atomic potential describing the torsion of the chain (Frankland and Harik, 2003). The angle bending forces can be modeled with a harmonic valence angle potential having an equilibrium angle of 112.813° and the barrier of 520 kJ/mol, and a torsional potential $\phi(\alpha)$ of the form

$$\phi(\alpha) = A_1(1 + \cos\alpha) + A_2(1 + \cos 2\alpha) + A_3(1 + \cos 3\alpha),$$

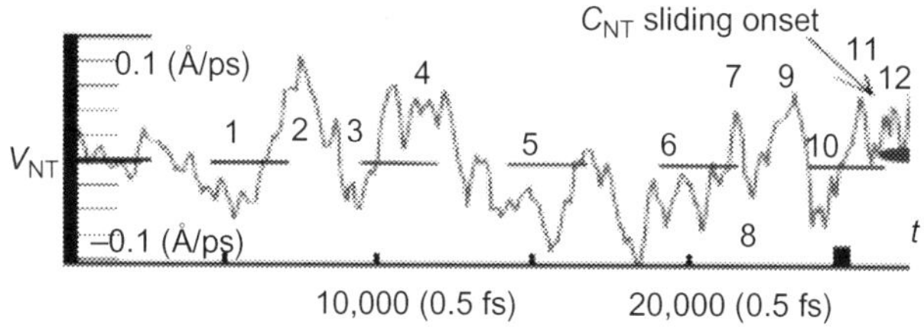

Figure 9.7

Axial center-of-mass velocity, V_{NT}, of the carbon nanotube varies from 0.027 Å/ps to ± 0.1 Å/ps with forward ($V_{NT} > 0$) and backward ($V_{NT} < 0$) motion during the five and half intervals of 5000 time steps (0.5 fs each) under slowly increasing distributed axial force, F_{NT}, from 0 to 0.12 nN (additional analysis of data from Frankland and Harik, 2003). *© 2011 by Nanodesigns Consulting (reprinted with a permission).*

where α is the torsional angle around the $CH_2{-}CH_2$ bond, and $A_1 = -18.087$, $A_2 = -4.88$, and $A_3 = 31.8$ kJ/mol. The nonbonded interactions within the chains were described by the LJ potential parametrized with ε_{LJ} equal to 0.4742 kJ/mol and σ_{LJ} of 4.28 Å for the internal PE chains and interchain interactions.

In MD simulations, the interfacial interactions between two atomic lattices and their multiatom configurations are described on the femto- and the picosecond timescales. These timescales are associated with small relative displacements between the segments of the carbon nanotube (Fig. 9.7) and the polymer interface, as well as the generation of the lattice waves or phonons in the atomic structure of the interface (Harik, 2011). The strength of the registry effects can be characterized by the potential barriers between different atomic registries associated with the changes in the commensurability between two atomic lattices. The strength of the potential barriers can be examined by monitoring changes in the asperity interlocking associated with the strongest registries and describing the nonstandard stick-slip behavior of the carbon nanotube lattice. Changes in the axial center-of-mass velocity, V_{NT}, of the carbon nanotube (Fig. 9.7) describe the gradients in the potential barriers associated with the changes in the atomic registry and the degree of commensurability between adjacent atomic lattices. In MD simulations (Frankland and Harik, 2003), the axial center-of-mass velocity, V_{NT}, of the carbon nanotube varies from 0.027 to ± 0.1 Å/ps, with forward ($V_{NT} >$ 0) and backward ($V_{NT} <$ 0) motion during the five and half intervals of 5000 time steps (0.5 fs each) under slowly increasing distributed axial force, F_{NT}, from 0 to 0.12 nN. The magnitude of the force characterizes the potential barrier gradients associated with the atomic registries along the axial displacement of the carbon nanotube.

9.2.3 Nanoscale Effect of the Spatial Exclusion of Electrons at the Interface

Interfacial interactions between atomic lattices are associated with the van der Waals forces acting across the interface. However, the potential barriers associated with the local changes in the registry and the degree of commensurability between the atomic lattices are also connected to the physical origins of different components contributing to the van der Waals forces. At the initial thermodynamic equilibrium, the lattice of the SWCNT relaxes into and oscillates around a more favorable position within the initial lattice-asperity registry well[2] with the axial center-of-mass velocity, $V_{NT} = 0.027$ Å/ps (Fig. 9.7), under the ambient registry potential akin to the capillary effect equivalent to the axial force of about 0.01 nN. During the oscillation period (1) (Fig. 9.7), the lattice of the SWCNT actually moves backwards ($V_{NT} < 0$) under the uniformly distributed force of about 0.018 nN, simply because of the prior repulsion of a few or more CH_2 asperities encountered by the carbon atoms and their electron distributions.

This seemingly paradoxical backward lattice motion and the nature of the asperity-induced repulsion is easily understood just by taking into account the onset of the X-registry configuration (Fig. 9.6), which is associated with the need of the π-electrons of the SWCNT lattice to overcome the asperity-induced registry potential, and the Coulomb repulsion stemming mainly from the closest electrons of the two C–H covalent bonds. Since the electron-electron repulsive influence is proportional to $1/r^{12}$, where r is the distance, the electron repulsion of the π-electrons of the SWCNT lattice greatly increases as $r \approx 0$, when they are pushed against the electrons of the CH_2 asperities (Fig. 9.6). This effect is associated with the electrons spatial exclusion (ESE) zones for the π-electrons at the X-registry configuration of the two lattices, which amounts to a nanoscale analog of the Pauli exclusion-principle for the π-electrons associated with the local Coulomb repulsion of the carbon-based lattices. This is the nanoscale effect of the spatial exclusion of electrons (ESEE) associated with the transient $\pi - \pi$ interactions along the interface between two atomic lattices. The repulsive influence of the neighboring, but more remote, electrons of the C–C bonds in the PE chain also has some effect.[3] The ESEE and the associated Coulomb repulsion of the π-electrons at the

---

[2] The onset of the X-registry configuration (Fig. 9.6) is associated with the need of an electron to overcome the registry potential of the C–C bond and the associated Coulomb repulsion within the ESE zone. The size of the ESE zone depends on the local atomic lattice configuration, the registry potential barriers, the nanoscale Coulomb repulsion proportional to $1/r^2$, and the nanoscale Pauli repulsion proportional to $-1/r^{12}$. The combined effect results in the ESE effect. The nanoscale Pauli repulsion for the electrons stems from the quantum Pauli principle for the identical particles with the spin 1/2 (fermions); the two identical particles cannot occupy the same energy state, as their combined wave function, ψ, is antisymmetric. The nanoscale Pauli repulsion and the quantum Pauli principle for the electrons both affect the precise dimensions of the ESE zone.

[3] The total energy of the interacting electrons involved in the ESEE in the electron–electron repulsion between two lattices is described by a system of the Schrodinger equations for the wave function $\Psi(r,t)$: $i\hbar(\partial/\partial t)\Psi(r,t) = \left(-\left(\hbar^2/2m\right)\nabla^2 + V(r)\right)\Psi(r,t)$, where i ($i^2 = -1$) is the imaginary unit, $\hbar$ is the reduced Planck's constant: $\hbar = h/2\pi$, with the Planck's constant $h = 4.135667516(91) \times 10^{-15}$ eV s, m is the rest state mass, $-\left(\hbar^2/2m\right)\nabla^2\Psi$ is the kinetic energy of an electron, $V(r)$ is the time-independent potential within which an electron is positioned and described by the Hamiltonian, $H\Psi(r,t)$: $H\Psi(r,t) = \left(-\left(\hbar^2/2m\right)\nabla^2 + V(r)\right)\Psi(r,t)$ for an electron in the potential $V(r)$ with the corresponding potential energy: $V(r)\Psi(r,t)$.

polymer–nanotube interface are the cause of the decreasing axial center-of-mass velocity prior to the onset of the oscillation periods 1, 3, 5, 6, and 10 (Fig. 9.7), and during the initial part of these axial oscillation periods, when the lattice of the SWCNT begins to move repeatedly backwards ($V_{NT} < 0$) under the slowly increasing axial force, which pushes the nanotube lattice against the CH_2 asperities.

The effect of interfacial asperities is surprising at the nanoscale, since the initial axial oscillations of the lattice of the SWCNT do not progressively lead to the weakening of adhesion and debonding of the nanotube under the increasing axial force, F_{NT}, as indicated in Fig. 9.7 through the repeated pushbacks ($V_{NT} < 0$) by the asperities and the greater registry interlocking afterwards with lower V_{NT} during the oscillation periods 11 and 12. Therefore these results indicate that there exists a sequence of adjustments of the atomic lattice of the SWCNT within the distributed potential barriers associated with the atomic registry between the SWCNT and the polymer interface. Therefore the nanoscale mechanisms of the adhesion between two atomic lattices separated by the van der Waals spacing are more complex, since the adhesion has its own transient, but stable states. The length of this sequence of adjustments of the atomic lattice of the SWCNT is an additional characteristic of the adhesion between the two surfaces and the strength of adhesion. In fact, this sequence of adjustments in the atomic registry can be used to increase the interfacial strength and to monitor the safety of the interface prior to the onset of the critical debonding of the interface and its failure. The critical decohesion stress, τ_{shear}, of the lattice of the SWCNT is assessed to be about 4 MPa, which is similar to the existing results: 2 MPa (Harik, 2011). The nanoscale ESEE can be used to monitor and control the interfacial strength in the carbon nanotube-reinforced nanocomposites.

9.2.4 Molecular Dynamics Simulation of the Carbon Nanotube Pullout

Computational simulation of the carbon nanotube pullout from the polymer matrix can characterize not only the interfacial interactions at the carbon nanotube–polymer interface at different stages of the pullout process, but also the strength of the interfacial adhesion during deformation of nanocomposites and their failure. Initially, the atomic lattice of the embedded SWCNT has to overcome the potential registry berries associated with its adhesion to the interfacial molecular structure of the surrounding polymer matrix. The critical force, F_{cr}, associated with the onset of the pullout process of the carbon nanotube out of the polymer matrix determines the critical shear stress, τ_{cr}, which characterizes the strength of the polymer–nanotube interface in nanocomposites.

After the details of interfacial interaction mechanisms between two atomic lattices have been examined before the onset of interfacial sliding of the SWCNT, the mechanisms of interfacial friction can be investigated in order to characterize the nanoscale interfacial friction of SWCNTs. The extended pullout process of a SWCNT out of the polymer matrix can demonstrate the origins of the mechanisms of the interfacial friction. MD simulation of the pullout process can provide new insights about the fundamental origins of sliding friction between two atomic lattices of the SWCNT and the polymer interface. MD simulation results can indicate the nanoscale effects of the atomic lattice structure on the surface-to-surface interactions, including the sequential interlocking and registry barriers during the interfacial sliding of the lattice of SWCNT through the surrounding polymer surface with its complex morphology of asperity distribution. Interfacial sliding of the SWCNT can induce vibrations of the atomic lattices, interaction of phonons propagating along the sliding surfaces, electrostatic interactions, and the electron interactions akin to the transient $\pi - \pi$ bonding interactions.

MD simulation of the carbon nanotube pullout from the PE matrix is carried out for the (10,10) SWCNT. The periodic unit cell in MD simulations contains about 23,000 atoms, including 880 carbon atoms in SWCNT and 178 chains of 42 methylene units with the hydrogen atoms explicitly included (Frankland and Harik, 2003). A unidirectional force is applied to each atom of the SWCNT along the axis. The pullout force is increased incrementally over time. The simulation is run for ~100,000 MD time steps of 0.5 fs. The MD simulation captures the full NT pullout process. Initially, after a sufficient amount of force is applied to the SWCNT, interfacial sliding begins. The observed variations in the axial velocity of the SWCNT (Fig. 9.8) and the entire pullout process can be divided into several stages.

Prior to the onset of interfacial sliding, the SWCNT experiences thermal vibrations and interacts with the adjacent polymer chains in equilibrium at 300K for the first 27,000 time steps (up to time of pullout, t_{pullout}) as the applied force is being gradually increased to 0.1 nN. Beyond this critical force, the SWCNT starts to slide through the polymer. Following the details of MD simulation data presented in Fig. 9.8, as well as the established conceptual framework for viscoplastic interfacial yielding (Harik, 2011), the entire pullout process can be described by four stages:

1. Initially, the equilibrium exists between the SWCNT and the surrounding polymer for the time period: $t < t_{\text{pullout}}$.
2. For an applied force of about 0.1 nN, accession of the energy/force barrier occurs, initiating the slippage between the SWCNT and the polymer. This onset of slippage is marked by a noticeable increase in the axial velocity of the SWCNT, which is observed at $t = t_{\text{pullout}}$.
3. Slippage of the SWCNT in the polymer cavity fully develops during a transition stage between t_{pullout} and $t_{ss} \approx 50{,}000$ time steps, where t_{ss} refers to the onset of steady sliding.

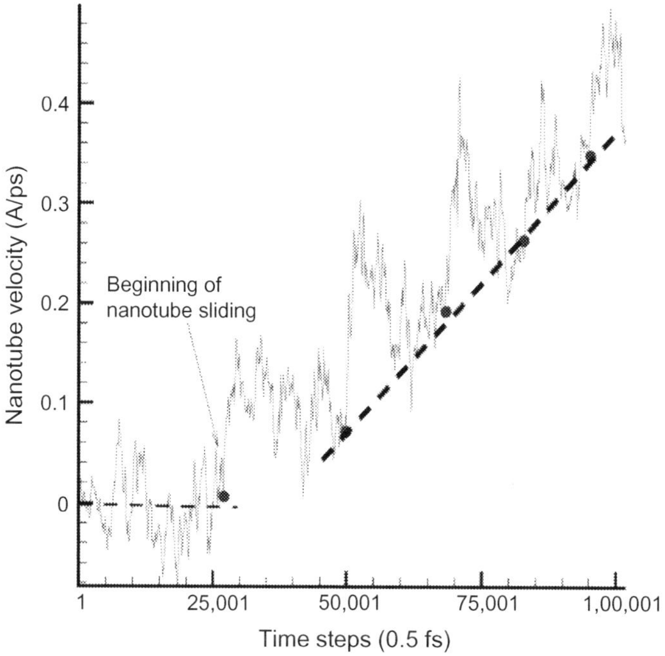

Figure 9.8
Axial sliding velocity, V_{NT}, of the center-of-mass of the carbon nanotube prior to debonding, at the onset of the sliding (marked by a dot) and during the subsequent periodic increases in the sliding velocity (similar accelerations are marked by dots within similar potential-registry-barrier wells and a joining line) (Frankland and Harik, 2003).

4. The steady sliding of the SWCNT (for $t > t_{ss}$) is characterized by semiperiodic variations in the sliding velocity of the SWCNT as it accelerates under the step-wise increases in the applied force. The onset of periodic accelerations in the sliding of the SWCNT is marked by dots in Fig. 9.8.

In Fig. 9.8, the dots on the velocity trace mark the onset of accelerations which are related to the ascent of local energy barriers at the beginning of the carbon ring period, 2.46 A. The resulting displacement of the SWCNT is affected by these periodic accelerations in the axial velocity. The interfacial sliding of the SWCNT shows periodic increases in the sliding velocity with the period directly connected to the size of a carbon ring. The results for the interfacial sliding velocity of the SWCNT can be used to characterize the interfacial friction and the friction mechanisms associated with the tribological properties of the two atomic lattices.

9.2.5 Nanoscale Analog of Newton's Friction Law for Interfaces

Interfacial friction between two atomic lattices separated by the van der Waals spacing can be described by a relation between a pulling force and the sliding velocity. The normal force applied during the interfacial sliding along the nanotube–polymer interface is determined by the van der Waals forces described by the LJ potential:

$$U_{LJ}(r) = \varepsilon\left(\frac{r_0^{12}}{r^{12}} - \frac{r_0^6}{r^6}\right), \tag{9.2}$$

where ε is the depth of the potential well and r_0 is the distance between two lattices at which the potential, U_{LJ}, is zero. The nanoscale friction between the SWCNT and the molecular structure of the surrounding polymer can be described by the nanoscale analog of the Newton's friction law (Frankland and Harik, 2003; Harik, 2011, 2014), according to which:

> *the velocity, $\langle V_{NT}\rangle$, of the moving nanotube is, on average, linearly proportional to the average applied force, $\langle f\rangle$, with the coefficient of proportionality, χ_{eff}, describing the effective viscosity of the nanotube–polymer interface.*

This nanoscale friction law is described by the following force–velocity dependence:

$$\langle f\rangle = \chi_{\text{eff}}\langle V_{\text{NT}}\rangle, \tag{9.3}$$

which is a nanoscale analog of the Newton's friction law (Frankland and Harik, 2003). This nanoscale law of friction has been used to describe the friction between the atomic lattices of two adjacent graphene sheets either in graphite, nanocomposites or in nanodevices. The same friction law describes the friction between the adjacent internal walls in MWCNTs during the sliding of an inner wall inside of a larger lattice shell.

At a length scale longer than that of the carbon rings, a nanoscale friction law can be developed as follows. The average shear stress, $\langle\tau_{rx}\rangle$, for planar sliding interactions can be defined as $\langle\tau_{rx}\rangle \approx \langle f\rangle_{pull}/(A_s)$, where $\langle f\rangle_{pull}$ is the average force applied to the SWCNT, and A_s is the common interfacial area of steady sliding between the SWCNT and the polymer matrix. The average strain rate, $\langle\dot{\gamma}\rangle$, can be estimated by $\langle\dot{\gamma}\rangle \approx \partial\langle V_{NT}\rangle/\partial r$, where $\langle V_{NT}\rangle$ is the average velocity in the axial x-direction and r is the radial distance. Then, the nanoscale analog of Newton's law for steady interfacial sliding is

$$\langle\tau_{\text{rx}}\rangle = \mu_{\text{eff}}\frac{\partial\langle V_{\text{NT}}\rangle}{\partial r}, \tag{9.4}$$

where μ_{eff} is an effective viscosity for the interfacial sliding. The effective viscosity coefficient, μ_{eff}, is connected with the momentum transfer between the adjacent atomic lattices through the interfacial friction.

9.2.6 Friction Model for the Carbon Nanotube Pullout from a Polymer

The nanoscale analog of Newton's friction law can be used to develop an interfacial friction model for the entire sliding process of the SWCNT, which is given by

$$\langle f \rangle_{\text{pull}} = f_0 + \chi_{\text{eff}} \langle V_{\text{NT}} \rangle, \tag{9.5}$$

where f_0 is the critical force associated with the onset of interfacial sliding of the carbon nanotube (Frankland and Harik, 2003). According to this nanoscale friction law:

> *after the applied force exceeds the critical force, f_0, associated with the strength of interfacial adhesion, the velocity, $\langle V_{NT} \rangle$, of the moving nanotube is, on average, linearly proportional to the average applied force, $\langle f \rangle$, with the coefficient of proportionality, χ_{eff}, describing the effective viscosity of the nanotube–polymer interface.*

The second nanoscale friction law (9.5) involves an initial process of overcoming the initial registry barrier described by the critical force, f_0, which also characterizes the strength of interfacial bonding and adhesion. The particular registry barrier and the value of the critical force, f_0, depends on a particular lattice registry, which can be described by an appropriate registry matrix (Harik, 2015). The two nanoscale friction laws (9.3) and (9.5) have the atomic scale limit of their applicability, which is associated with the size of the carbon ring, a, or the size of the smallest structural element of any atomic lattice, in general. The nanoscale homogenization length, L_h, identifies the length scale, $L_h \gg a$, at which these nanoscale friction laws apply. At the homogenization length scale, the average applied force, $\langle f \rangle$, is related to the shear stress by the force balance:

$$\langle f \rangle = \langle \tau_{\text{rz}} \rangle A_{\text{ss}}.$$

The average shear stress, τ, is related to the interfacial velocity gradient through the effective viscosity, μ_{eff}, of the two atomic lattices of the polymer–nanotube interface. The interfacial force interactions along the adjacent atomic lattices of the nanotube–polymer interface are described by the LJ potential (9.2) and its gradient. As a result, interfacial debonding and frictional sliding of the SWCNT along the polymer–nanotube interface can be described by the Bingham-type model for the viscoplastic shear debonding of the interface, which relates the shear stress, τ, to the strain rate, $\dot{\gamma}$, by the Newton's friction law: ($\tau = \mu\dot{\gamma}$) after the critical stress, τ_0, is reached:

$$\tau = \tau_0 + \mu_m \dot{\gamma}. \tag{9.6}$$

Here, μ and μ_m are the viscosities of a fluid and a viscous matrix material, respectively. The Newton's friction law is defined for the adjacent layers of a continuum fluid that is subjected to the shearing. At the nanometer level, an analog of Newton's friction law can be developed, which relates the applied forces and the sliding velocity, to determine an effective interfacial viscosity. The stresses and strain rates can be presented in terms of the average forces and the velocity gradients derived from the MD simulations.

In order to model the sliding between the SWCNT and the surrounding polymer interface on a length-scale level larger than the carbon rings, multiple carbon ring lengths have to be considered for sufficiently meaningful averaging of physical parameters characterizing the interfacial sliding. The length, L_{NT}, should be sufficiently long, that is, $L_{NT}/a > 10$, where a is the width of a carbon ring. The perimeter of the interface and the radius, R_{NT}, should satisfy an inequality: $2\pi\left(\left(R_{NT} + \overline{h}_{vdW}\right)/2\right)/a > 10$, where R_{NT} is the radius and $\overline{h}_{vdW}$ is an average surface separation caused by the van der Waals forces.

As a result, an interfacial friction model for the entire pullout process can be developed for the total shear stress, $\langle \tau_{rx} \rangle_{pull}$, which, in addition to the sliding component $\langle \tau_{rx} \rangle$, includes an initial threshold pullout stress, τ_0:

$$\langle \tau_{rx} \rangle_{pull} = \tau_0 + \mu_{eff} \frac{\partial \langle V_{NT} \rangle}{\partial r}. \tag{9.7a}$$

After approximating the change in velocity across the interfacial separation, h_{vdW}, as the average velocity, $\langle V_{NT} \rangle$, Eq. (9.7a) can be rewritten as

$$\langle \tau_{rx} \rangle_{pull} = \tau_0 + \mu_{eff} \frac{\langle V_{NT} \rangle}{\overline{h}_{vdW}}, \tag{9.7b}$$

where the strain rate is $\partial \langle V_{NT} \rangle / \partial r \approx \langle V_{NT} \rangle / \overline{h}_{vdW}$. The viscosity coefficient, μ_{eff}, can be related to the effective viscosity χ_{eff} via

$$\mu_{eff} = \frac{\chi_{eff} \overline{h}_{vdW} / L_{NT}}{2\pi\left(\left(R_{NT} + \overline{h}_{vdW}\right)/2\right)}. \tag{9.8}$$

The effective viscosity of the interfacial sliding, χ_{eff}, and the effective viscosity of the interfacial media, μ_{eff}, are proportional to the magnitude of the force required to move an atomic lattice at the specific velocity during the interfacial sliding of two atomic lattices.

9.2.7 *Intrinsic Viscosity of the Spatially Distributed Electrons at the Interface*

The effective viscosity of the interfacial friction between two atomic lattices depends on their atomic registry and the degree of commensurability, which in turn affect the strength of the potential registry barriers and the strength of the transient $\pi - \pi$ interactions, as well as the resulting momentum transfer between the lattices. The nanoscale analog of Newton's friction law (9.3) includes a proportionality coefficient, χ_{eff}, between the interfacial sliding velocity and the force applied to the atomic lattice. This proportionality coefficient, χ_{eff}, characterizes the effective viscosity of the interfacial sliding between two atomic lattices and the distributions of π-electrons along their interfaces. The value of this coefficient can be determined from the molecular dynamic simulations of the interfacial sliding between two atomic lattices. The value of the effective viscosity coefficient, χ_{eff}, is estimated as $\chi_{eff} \approx 1.5$ nN ps/Å.

The effective viscosity coefficient, χ_{eff}, is related to the intrinsic viscosity of the spatially distributed electrons by the relation: $\mu_e \approx \chi_{\mathrm{eff}} \cdot h_{\mathrm{vdW}}/2\pi R_{\mathrm{NT}} L_{\mathrm{NT}}$, where the value of the interfacial perimeter should be adjusted to $R_{\mathrm{NT}} + h_{\mathrm{NT}}/2$, as noted above (Harik, 2011). The intrinsic viscosity of the spatially distributed π-electrons is estimated as $\mu_e \approx 0.2$ cP (centipoise). Note that the higher values of both the effective viscosity coefficient, χ_{eff}, and the intrinsic sliding viscosity of the electrons involved, μ_e, are directly affected by the increases in the effective density of the activated electrons along the asperity-rich polymer–nanotube interface. The intrinsic viscosity of the interfacial electrons, μ_{eff}, is proportional to the van der Waals spacing and the magnitude of the force required for moving an atomic lattice at the specific velocity.

The intrinsic viscosity of the spatially distributed electrons is increased by the effect of the spatial exclusion of electrons (or the SEE effect). It can be further evaluated by using the above nanoscale analog of Newton's friction law. The axial oscillations of the SWCNT described by the variations in the axial velocity in Fig. 9.7 can be used to evaluate the intrinsic viscosity of the π-electrons by using a nanoscale analog of Newton's friction law: $\langle F_{\mathrm{NT}} \rangle = \chi_{\mathrm{eff}} \langle V_{\mathrm{NT}} \rangle$, where χ_{eff} is the effective viscosity coefficient ($\chi_{\mathrm{eff}} \approx 0.5$ nN ps/Å) related to the intrinsic sliding viscosity of the spatially distributed electrons, $\mu_e \approx 0.07$ cP (centipoise). This oscillation-based value compares favorably with the higher value of the effective interfacial sliding "viscosity" of the asperity-rich polymer–nanotube interface, i.e., 0.2 cP, which is close to the viscosity of pentane at 0.22 cP. The above result demonstrates that the intrinsic viscosity of the spatially distributed electrons is, indeed, increased by the effect of the spatial exclusion of electrons and the transient $\pi - \pi$ interactions, the strength of which is affected by the commensurability of the lattices. The above results also demonstrate the nanoscale phenomena underlying the effective interfacial sliding "viscosity" of the polymer–nanotube interface. A more in-depth influence of the SEE effect on the intrinsic viscosity, μ_e, of the spatially distributed electrons can be further evaluated by using Newton's friction law for the analysis of interfacial sliding of graphene sheets, as well as other evaluations of the effective viscosity coefficient χ_{eff} for different atomic lattices.

9.2.8 Modified Halpin–Tsai Relation for Nanocomposites With Interfaces

The Halpin–Tsai relation (Tsai, 1988; Chou, 1992) can be used to predict the average elastic properties of the carbon nanotube-reinforced polymer matrix nanocomposites, as was discussed in the previous chapter. This relation can be modified in order to take into account the finite thickness of the nanoscale interfaces in nanocomposites. The strength and the form of the interfacial bonding and its density can affect the strength of nanocomposites. The weak interfacial regions found in nanocomposites with nonfunctionalized carbon nanotubes have lower stress transfer through the van der Waals interface. The strength of interfacial regions affects the mechanical properties of nanocomposites along with Young's modulus, E_m, of the polymer matrix and the reinforcing carbon nanotubes, E_{NT}.

The average Young's modulus of nanocomposites, E, depends on the volume fraction, V_{NT}, of the randomly oriented carbon nanotubes and the volume fraction of the interfacial regions. The radius, R_{NT}, of carbon nanotubes, their diameter, d_{NT}, and, hence, their aspect ratio, d_{NT}/L_{NT}, are also affected by the thickness of the interface. The elastic modulus of nanocomposites is proportional to the ratio of the elastic moduli, E_{NT}/E_m, and the volume fraction of the reinforcing nanofibers, V_{NT}, as given by the modified Halpin–Tsai relation:

$$E = E_m \left[\frac{3}{8} \frac{1 + 2\left(L_{NT}/R_{NT}\left(2 + h_{NT}/R_{NT}\right)\right)\eta_{E_L} V_{NT}\left(1 + (1/2)h_{NT}/R_{NT}\right)^2}{1 - \eta_{E_L} V_{NT}\left(1 + (1/2)h_{NT}/R_{NT}\right)^2} \right.$$
$$\left. + \frac{5}{8} \frac{1 + 2\eta_E V_{NT}\left(1 + (1/2)h_{NT}/R_{NT}\right)^2}{1 - \eta_E V_{NT}\left(1 + (1/2)h_{NT}/R_{NT}\right)^2} \right], \tag{9.9a}$$

where the nondimensional parameters, η_E, and η_{E_L}, are based on the ratio of the elastic moduli, E_{NT}/E_m:

$$\eta_E = \frac{E_{NT}/E_m - 1}{2 + E_{NT}/E_m} \quad \text{and} \quad \eta_{E_L} = \frac{E_{NT}/E_m - 1}{2\left(L_{NT}/R_{NT}\left(2 + h_{NT}/R_{NT}\right)\right) + E_{NT}/E_m}. \tag{9.9b}$$

Here, the volume fraction, V_{NT}, of a carbon nanotube is adjusted by a correction factor $\left(1 + (1/2)h_{NT}/R_{NT}\right)^2$, since the volume of a nanoscale fiber depends on the effective thickness of the atomic lattice wall: $\pi(R_{NT} + h_{NT}/2)^2/A_{cell}$, where h_{NT} is the effective thickness of a carbon nanotube such that $h_{NT} \leq h_{vdW}$, where h_{vdW} is the average van der Waals separation between a carbon nanotube and the matrix material, and A_{cell} is the cross-sectional area of the unit cell. The nanoscale correction in the modified Halpin–Tsai relation is based on the normalized thickness of the SWCNTs, or the thickness-to-radius ratio, h_{NT}/R_{NT}, which is an important nondimensional parameter. The aspect ratio, L_{NT}/d_{NT}, in the modified Halpin–Tsai relation is also adjusted in order to include the effective thickness of a carbon nanotube. The nanoscale correction to the aspect ratio is also based on the thickness-to-radius ratio, h_{NT}/R_{NT}. Note that the effective thickness of a carbon nanotube may be affected by the direction of loading, because of the transversely semiisotropic structure of carbon nanotubes. In the case of MWCNTs, their thickness depends on the number of walls in their structure. Young's modulus of carbon nanotubes also depends on their radius and their atomic lattice structure.

9.3 Conclusions

This chapter describes the nanoscale interfaces in the carbon nanotube-reinforced nanocomposites and the length scales of their atomic structure. Nanoscale homogenization of the material structure of the polymer interphases at the surface of carbon nanotubes has been discussed, along with an example of the scaling analysis and homogenization criteria for MD modeling. MD simulations of the atomic structure of interfaces in nanocomposites, interfacial registry barriers between adjacent atomic lattices, and the pullout process of a carbon nanotube out of a polymer matrix have been carried out. The atomic registry between the zigzag PE chains and the surface of a carbon nanotube has been analyzed. New nanoscale mechanisms associated with the transition between stable states in the semitransient adhesion of the polymer–nanotube interface have been discussed. Nanoscale effect of the spatial exclusion of π-electrons at the interfaces between atomic lattices has been presented for the polymer–nanotube interface.

Computational data from MD simulations have been analyzed in order to demonstrate their dependence on the size of a carbon ring and the atomic structure of the interface. MD results for the axial velocity of a carbon nanotube have been presented for the case of interfacial sliding through the polymer matrix in nanocomposites. Dependence of the periodic increases in the sliding velocity of the pulled carbon nanotube on the size of the carbon ring in its lattice structure has been demonstrated. A nanoscale analog of Newton's friction law for the polymer–nanotube interfaces has been presented, along with a friction model for the carbon nanotube pullout from a polymer matrix. MD data have been used to estimate the intrinsic viscosity of the spatially distributed electrons during interfacial sliding of two adjacent atomic lattices. It has been shown that the intrinsic viscosity of the spatially distributed π-electrons depends on the effect of the spatial exclusion of electrons and the transient $\pi - \pi$ interactions. The modified Halpin–Tsai relation has been presented with the nanoscale corrections for the carbon nanotube nanocomposites.

The nanoscale effect of the spatial exclusion of π-electrons (SEE) is associated with the interfacial sliding and friction, and the need of π-electrons to overcome the transient $\pi - \pi$ bonding, the registry potential of the opposing covalent bonds with the associated Coulomb repulsion within the spatial exclusion zone (SEZ) for electrons (Harik, 2011, 2014). The size of the SEZ depends on the local atomic lattice configuration (e.g., orientation of the C–C bonds) and the registry potential barriers including the boundary for the available transient $\pi - \pi$ interactions. Note that the nanoscale Coulomb repulsion is proportional to $1/r^2$, while the nanoscale repulsion is proportional to $1/r^{12}$. The nanoscale repulsion results in part from the spatially distributed *SEE effect*, which is a nanoscale analog of the Pauli exclusion-repulsion for electrons stemming from the quantum Pauli principle for the identical electrons, i.e., particles with the spin 1/2 (fermions) prohibiting them to occupy the same energy level. The two identical particles cannot occupy the same energy state, because their combined wave function, ψ, is antisymmetric. The nanoscale analog of Pauli repulsion and the quantum Pauli principle for the electrons both affect the precise dimensions of the SEZ for interfacial electrons. The spatial exclusion of π-electrons along the nanoscale interfaces between atomic lattices affects their interfacial interactions under both dynamic and quasistatic conditions at low temperatures.

Sample Problems

1. Find and list a few examples of the atomic structure of nanoscale interfaces between carbon nanotubes (SWCNTs or MWCNTs) and other materials, and describe their structure.
2. Describe the length scales in the atomic structure of nanoscale interfaces between carbon nanotubes (SWCNTs or MWCNTs) and other materials in problem 1.
3. Make a PowerPoint viewgraph with an illustration of the atomic structure of nanoscale interfaces between carbon nanotubes (SWCNTs or MWCNTs) and other materials.
4. Make a PowerPoint viewgraph with an illustration of the length scales in the atomic structure of nanoscale interfaces between carbon nanotubes (SWCNTs or MWCNTs) and other materials.
5. Find examples of the atomic structure of nanoscale interfaces which involve different classes of carbon nanotubes.
6. Describe the atomic structure of the polymer–nanotube interfaces which involve different classes of carbon nanotubes.
7. What are the nanoscale homogenization criteria for the carbon nanotube-reinforced nanocomposites?
8. Describe the nanoscale analog of Newton's friction law for the polymer–nanotube interfaces.
9. Describe the nanoscale effect of the spatial exclusion of electrons.
10. Describe the effective viscosity of the spatially distributed electrons during interfacial sliding between atomic lattices.
11. Describe the new nanoscale corrections to the modified Halpin–Tsai relation for the carbon nanotube-reinforced nanocomposites.

References

Brenner, D.W., 1990. Empirical potential for hydrocarbons for use in simulating the chemical vapor deposition of diamond films. Phys. Rev. B: Condens. Matter 42, 9458–9471.

Chou, T.W., 1992. Microstructural Design of Fiber Composites. Cambridge University Press, Cambridge.

Choudhary, V., Gupta, A., 2011. Polymer/carbon nanotube nanocomposites. In: Yellampalli, S. (Ed.), Carbon Nanotubes-Polymer Nanocomposites. InTech, Rijeka, Croatia.

Frankland, S.J.V., Harik, V.M., 2003. Carbon nanotube pull-out from a polymer matrix. Surf. Sci. Lett. 525, L103.

Harik, V.M., 2001a. Ranges of applicability for the continuum beam model in the mechanics of carbon nanotubes and nanorods. Solid State Commun. 120 (7−8), 331−335.
Harik, V.M., 2001b. Ranges of Applicability for the Continuum Beam Model in the Constitutive Analysis of Carbon Nanotubes: Nanotubes or Nano-Beams? NASA/CR-2001-211013, NASA Langley Research Center, Hampton, Virginia.
Harik, V.M., 2002. Mechanics of carbon nanotubes: applicability of the continuum-beam models. Comput. Mater. Sci. 24 (3), 328−342.
Harik, V.M., 2011. Mechanics of Carbon Nanotubes. Nanodesigns Press, Nanodesigns Consulting, Newark, DE.
Harik, V.M. (Ed.), 2014. Trends in Nanoscale Mechanics. Springer, Dordrecht.
Harik, V.M., 2015. Lecture notes: a short course, Nanodesigns Consulting Mechanics of Carbon Nanotubes and Graphene. Nanodesigns Press, Wilmington, DE.
Harik, V.M., 2017. Mechanics of carbon nanotubes and nanomaterials. In: Gogotsi, Yu (Ed.), Nanomaterials Handbook. CRC Press, Boca Raton, FL, Chapter 6.
Jeon, I.-Y., Chang, D.W., Kumar, N.A., Baek, J.-B., 2011. Functionalization of carbon nanotubes. In: Yellampalli, S. (Ed.), Carbon Nanotubes-Polymer Nanocomposites. InTech, Rijeka, Croatia.
Tsai, S.W., 1988. Composites Design. Think Composites Press, Dayton, OH.

CHAPTER 10

Constitutive Modeling of Multifunctional Nanocomposites: Electromechanical, Thermomechanical and Elastomagnetic Properties

Chapter Outline

10.1 Electromagnetic Properties of Multifunctional Nanocomposites

Carbon nanotubes in the polymer matrix nanocomposites change their material properties by strengthening their load-carrying capabilities, their elastic modulus, as well as their electric conductivity originating from the high electric conductivity of the metallic lattices of the armchair carbon nanotubes. Addition of single wall carbon nanotubes (SWCNTs) to the metal materials improves the durability of these materials in radiation environments. Carbon nanotube-based nanocomposites can incorporate the sensing and actuating capabilities of SWCNTs for different environments including temperature, chemical vapors, stresses, and strains (Frankland et al., 2003), as well as electromechanical, electrochemical, and electrical stimuli. Electric properties of SWCNTs can be used to improve their alignment in polymer matrices in order to enhance both the elastic and electric properties of nanocomposites.

Mechanics of Carbon Nanotubes.
DOI: https://doi.org/10.1016/B978-0-12-811071-3.00010-X

© 2018 Elsevier Inc. All rights reserved.

During the processing of polymer nanocomposites, SWCNTs can be aligned by the field-induced dipolar interactions between the dispersed nanotubes under an AC electric field in a liquid matrix material, followed by their immobilization by photopolymerization with the application of an electric field (Park et al., 2006). Alignment of SWCNTs can be controlled by changing the magnitude, frequency, and duration of the applied electric field. During the in-field processing, SWCNTs can align along the direction of the applied field and migrate laterally to form thick percolation columns between the electrodes. The aligned SWCNTs improve the electrical and dielectric properties of multifunctional nanocomposites, in addition to enhancing their mechanical properties and their load-carrying capability.

The dielectric constants of the pristine urethane-hexanediol polymer matrix is 5.2 (Park et al., 2006), while that of the carbon nanotube-reinforced nanocomposite with the above matrix is 25.2 for 0.03 wt.% of unaligned SWCNTs. The dielectric constant of nanocomposites increases with better alignment of SWCNTs at higher electric fields. For lower strengths of the applied electric field (up to 50 V_{p-p}), there was no noticeable change in the alignment of SWCNTs during processing and in the resulting values of the dielectric constant of nanocomposites. For larger magnitudes of the applied electric field (up to 100 V_{p-p}), there are significant increases in the alignment of SWCNTs and in the resulting values of the dielectric constant of nanocomposites, i.e., over 600 at 1 Hz, and further to around 1500 for 150 V_{p-p} (Park et al., 2006).

Frequency of the electric field applied during the processing of nanocomposites also strongly influenced the alignment of SWCNTs and the dielectric properties of nanocomposites. There was no significant alignment of SWCNTs for the frequency of 0.01 Hz of the applied electric field, which resulted in little increase in the dielectric constant as compared to the nanocomposite with the unaligned SWCNTs. At the higher frequency of 1 Hz, there was a better alignment of SWCNTs and noticeable increase in the dielectric constant (up to 900). The degree of alignment has not changed at higher frequencies (above 100 Hz) similarly to the dielectric constant of 1000 (Park et al., 2006). These experiments demonstrate that the dielectric constant of the carbon nanotube-reinforced polymer nanocomposites can be changed over three orders of magnitude by controlling the frequency and strength of the AC field used for the alignment of SWCNTs.

10.1.1 Conductivity of the Carbon Nanotube-Reinforced Nanocomposites

The conductivity of the pristine resins such as a photopolymerizable monomer, a blend of urethane dimethacrylate and 1,6-hexanediol dimethacrylate, is about 1.4×10^{-13} S/cm, and that of the unaligned nanocomposite is about 3.3×10^{-13} S/cm at 1 Hz, which indicates less than an order of magnitude increase in conductivity of the polymer matrix by adding 0.03 wt.% of SWCNTs (Park et al., 2006). The electric conductivity of such polymer nanocomposites increases linearly with the frequency on a logarithmic scale across most of the frequency range. Such dependence between the conductivity of a nanocomposite and the frequency of the applied field points to the insulative properties and a lack of percolation for the volume fraction of SWCNTs at 0.03 wt.% loading. Electric conductivity of nanocomposites with the aligned SWCNTs parallel to the applied electric field (Park et al., 2006) did not noticeably increase until the strength of the electric field reached 50 V_{p-p} during the alignment of SWCNTs in the liquid resin, and then increased significantly at 100 V_{p-p} from 1.4×10^{-12} to 3.0×10^{-9} S/cm. Further increases were observed for stronger electric fields used for the alignment of SWCNTs up to the saturation at around 150 V_{p-p}, when the conductivity of the nanocomposite is on the order of 10^{-7} S/cm. Electric conductivity of the carbon nanotube-reinforced nanocomposites showed metallic behavior for the samples aligned by the electric field with the strength above 100 V_{p-p}. The metallic behavior was marked by the relatively constant conductivity at low frequencies (e.g., below 102 Hz) with a slight increase at higher frequencies (e.g., above 102 Hz). These results (Park et al., 2006) indicate that the conductivity of the carbon nanotube-reinforced nanocomposites can be changed over the range of six orders of magnitude by controlling the strength of the electric field used for aligning SWCNTs.

10.1.2 Volume Fraction of Nanotubes and Conductivity of Nanocomposites

Experimental data about the conductivity of the carbon nanotube-reinforced nanocomposites can be used to describe the effect of the volume fraction of SWCNTs on the improvement of electric conductivity of the polymer matrix. Conductivity of the SWCNT- and multiwall carbon nanotubes (MWCNT)-reinforced nanocomposites depends on the distribution of carbon nanotubes in the polymer matrix, their network within the nonconductive polymer, and the number of random conductive paths made of the loosely connected SWCNTs. The theory about the influence of the volume fraction of SWCNTs and the associated network of the dispersed SWCNTs on the conductivity of nanocomposite material is called the percolation theory. The percolation theory describes the electric conductivity of nanocomposite materials filled with the disordered network of SWCNTs (Stauffer and Aharony, 1994). The critical volume fraction, V_{NTcr}, of SWCNTs or MWCNTs associated with the onset of the conductive behavior of the carbon nanotube-reinforced nanocomposites, is called the percolation threshold. As a result, the conductivity, G, of the polymer nanocomposite is approximated by the following expression:

$$G \propto (V_{NT} - V_{NTcr})^{\beta}, \qquad (10.1)$$

where the exponent β describes the scaling relation between the volume fraction, V_{NT}, and the conductivity, G, of the SWCNT- and MWCNT-reinforced nanocomposite.

The percolation relation (10.1) can be further examined in the framework of the scaling analysis, which results in the following normalized percolation relation:

$$G \propto \left(\left(\frac{V_{NT}}{V_{NTcr}} \right) - 1 \right)^{\beta} (V_{NTcr})^{\beta}, \qquad (10.2)$$

where the exponent β now describes the scaling relation between the critical volume fraction, V_{NTcr}, the normalized volume fraction, V_{NT}/V_{NTcr}, and the conductivity, G, of the SWCNT-based nanocomposites. The normalized percolation relation (10.2) emphasizes the importance of the critical volume fraction, V_{NTcr}, at the threshold percolation and underscores that the conductivity, G, is proportional to the ratio, V_{NT}/V_{NTcr}, and not the difference between the two volume fractions. It also indicates that the conductivity, G, is proportional to the normalized scaling of the characteristics associated with the critical volume fraction, V_{NTcr}, and not just simply on the value of the volume fraction, V_{NT}, as a number. The percolation relation (10.1) and the normalized percolation relation (10.2) are useful for the analysis of experimental data on conductivity of nanocomposites (Fig. 10.1).

Experimental data (Chiolerio et al., 2011) for the conductivity of the PVB nanocomposites does not show a clear transition between the insulating and conductive behavior (Fig. 10.1). Dependence of electric conductivity of the PVB nanocomposites on the volume fraction of

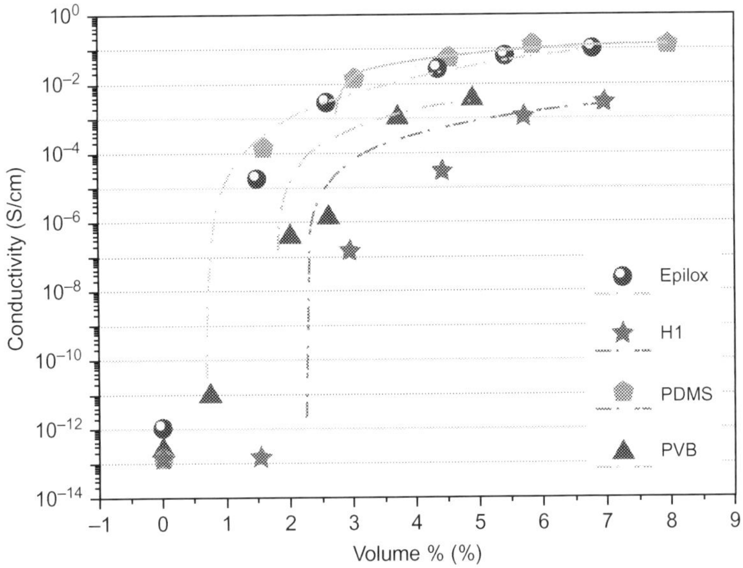

Figure 10.1
Experimental data for the DC conductivity of several polymer matrix nanocomposites with different matrices (listed) approximated by the percolation relation above the critical volume fraction of MWCNTs (Chiolerio et al., 2011).

MWCNTs seems linear in the semi-logarithmic scale, although there is some lack of data as well. Nanocomposites with the commercial H1 polymer matrix display clear insulating behavior even after the addition of more than 1 wt.% of MWCNTs, which may also indicate poor dispersion of conductive MWCNTs in this polymer. Nanocomposites with the commercial Epilox and PDMS matrices display a better transition. The values of the parameters used for the extrapolation of the experimental data are listed in Table 10.1. Nanocomposites with the Epilox matrix have the lowest percolation threshold: 0.2 $\pm$ 0.1 wt.%, which is followed by that of the PDMS nanocomposites: 0.7 $\pm$ 0.2 wt.%. The critical exponent β is close to 2.0 for all nanocomposites considered (Chiolerio et al., 2011).

Experimental data for the conductivity, G, of the carbon nanotube-reinforced nanocomposites can be also approximated by a model (Kilbride et al., 2002), which takes into account the high aspect ratio of MWCNTs and the higher probability of a contact between the uniformly dispersed MWCNTs:

$$\ln G = -A(V_{\mathrm{NT}})^{-1/3}, \tag{10.3}$$

where A is the constant of proportionality with the values listed in Table 10.2 for several nanocomposites, which were examined above and approximated with the new relation in Fig. 10.2. The high aspect ratio of the homogeneously dispersed MWCNTs results in a lower percolation threshold as compared to those with the low aspect ratios. The scaling relation (10.3) is a result of an interesting simplification of the fluctuation-induced tunneling for constant temperature (Kilbride et al., 2002), which depends on the mean distance between MWCNTs determining the potential barrier height in nanocomposites.

Table 10.1: Percolation parameters for nanocomposites

Matrix	V_{NTcr} (%)	β
Epilox	0.2 $\pm$ 0.1	2.0 $\pm$ 0.2
H1	2.0 $\pm$ 0.1	2 $\pm$ 1
PDMS	0.7 $\pm$ 0.2	2.7 $\pm$ 0.5
PVB	2.0 $\pm$ 0.1	2 $\pm$ 1

Table 10.2: Percolation parameters for nanocomposites

Matrix	V_{NTcr} (%)	R^2	A	R^2
Epilox	0.2 $\pm$ 0.1	0.97	11 $\pm$ 1	0.96
H1	2.0 $\pm$ 0.1	0.90	31 $\pm$ 2	0.99
PDMS	0.7 $\pm$ 0.2	0.91	9 $\pm$ 1	0.93
PVB	2.0 $\pm$ 0.1	0.84	17 $\pm$ 2	0.97

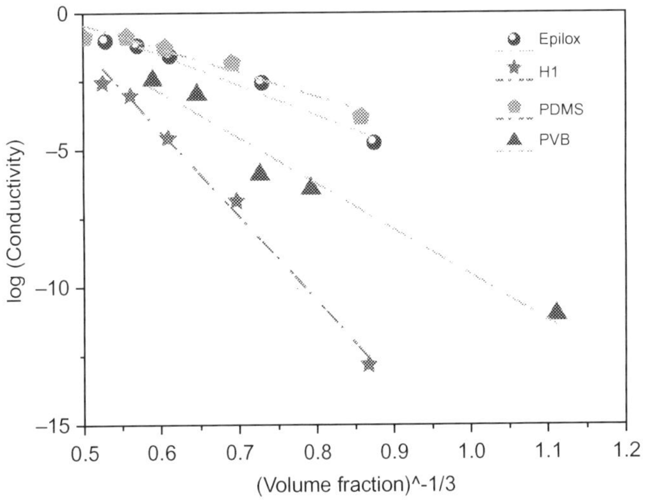

Figure 10.2
Experimental data for the DC conductivity of several polymer matrix nanocomposites with different matrices (listed) and increasing volume fractions of MWCNTs approximated by a scaling relation with a power of " − 1/3" of the volume fraction (Chiolerio et al., 2011).

The scaling relation (10.3) can be replaced by an equivalent scaling relation, which is based on the normalized volume fraction, V_{NT}/V_{NTcr}, as follows:

$$\ln G = -A_0\left(\frac{V_{NT}}{V_{NTcr}}\right)^{-1/3}, \tag{10.4}$$

where A_0 is the constant of proportionality, which can be related here to the parameter A through the normalization relation: $A_0 = A\ (V_{NTcr})^{-1/3}$, involving the critical volume fraction, V_{NTcr}. The scaling relation (10.4) with the normalized volume fraction, V_{NT}/V_{NTcr}, is physically more robust, while its data points should be more informative. For example, the critical volume fraction, V_{NTcr}, associated with the threshold percolation corresponds to the normalized volume fraction, $V_{NT}/V_{NTcr} = 1$. Therefore the normalized volume fraction, V_{NT}/V_{NTcr}, defines a nondimensional percolation parameter, $P = V_{NT}/V_{NTcr}$. The percolation parameter, P, defines when the percolation occurs, that is when: $P = 1$, and links the conductivity of the SWCNT- and MWCNT-reinforced nanocomposites with the degree of percolation: $P > 1$. The lack of percolation naturally corresponds to $P < 1$. This nondimensional parameter just reveals the physical nature of the normalized volume fraction, V_{NT}/V_{NTcr}. Therefore these two parameters are interchangeable.

The values of the parameters used in the relation (10.3) and in the extrapolation of experimental data in Fig. 10.2 are presented in Table 10.2. The mean distance, R^2, between

MWCNTs extrapolated from the relations (10.1) and (10.3) and the experimental data in Fig. 10.2 is also presented in Table 10.2. The results for the mean distance, R^2, between MWCNTs listed in the two columns indicate that the percolation relation (10.1) is less accurate (the first column of values for R^2) than the scaling relation (10.3), since the second column of values for R^2 has higher values. The results for the mean distance, R^2, between MWCNTs in PDMS polymer matrix indicate that the probability of the matrix affecting the dispersion geometry is the lowest, as compared to other nanocomposites. This corresponds to the case of the lowest effect on affinity between the polymer and MWCNTs.

10.2 Analysis Methods for Multifunctional Nanocomposites

Analysis of constitutive behavior and deformation response of the nanotube-reinforced nanocomposites requires understanding of their material properties, which can be challenging if boron nitride (BN) nanotubes are used for their piezoelectric, pyroelectric, or photogalvanic capabilities. The length scales in the atomic structure of BN nanotubes and the BN nanotube-based nanocomposites are similar to those of carbon nanotubes and their nanocomposites, respectively, both of which have been discussed in earlier chapters (Harik, 2011, 2017). Multifunctional nanocomposites are made of the nanotube-enhanced polymers and the particle-based polymer modifiers. The nanotube-modified polymers and multifunctional polymer films can be electrically conductive, when the metallic SWCNTs or MWCNTs have been added. Electric conductivity of the multifunctional nanotube-based nanocomposites depends on the volume fraction of nanotubes, their weight ratio, the dispersion, the percolation length, and the aspect ratio. The material properties of multifunctional nanocomposites involving the high aspect ratio piezoelectric BN nanotubes also depend on their volume fraction, weight ratio, the dispersion, and the aspect ratio, as well as the intrinsic piezoelectric properties of BN nanotubes. The dependence of mechanical response of piezoelectric materials on the applied electric field is described by the constitutive relations for their electromechanical coupling. The macroscopic form of these constitutive relations is quite complex, as are their nanoscale analogs.

10.2.1 Constitutive Modeling of Electromechanical Properties

Mechanical behavior of multifunctional nanocomposites in electric fields is affected by their electromechanical properties, i.e., electric conductivity, electric permittivity, dielectric constants, and piezoelectric susceptibility, in addition to the elastic, shear, and bulk moduli. The relation between the mechanical stresses described by the stress tensor, σ_{ij}, and the mechanical strains expressed by the strain tensor, ε_{ij}, is described by the Hooke's law:

$$\sigma_{ij} = C^{E}_{ijkl}\varepsilon_{kl}, \tag{10.5}$$

where C^{E}_{ijkl} is the elastic stiffness matrix and the indices i, j, k, and l take the following values: 1, 2, and 3. Values of the stiffness matrix coefficients depend on the values of the elastic modulus, E, the shear modulus, G, and Poisson's ratio, ν. At the nanoscale level, the value of these material parameters depend on the atomic lattice structure of the nanostructured materials and the atomic lattice of nanotubes. The nanoscale analogs of the macroscopic rules of mixtures for composite materials and their effective longitudinal and transverse moduli were presented earlier.

Materials' susceptibility to respond to the electric field is governed by the nanoscale and sub-micron carriers of the electric dipoles. The constitutive relation describing the electromechanical coupling (Harik, 2011) in materials is given by

$$\sigma_{ij} = C^{E}_{ijkl}\varepsilon_{kl} - e_{ijk}E_{k}, \tag{10.6}$$

with the electric field, E_k, and the 27-component piezoelectric tensor, e_{ijk}, in the reduced form

$$e_{ijk} = \begin{pmatrix} e_{111} & \underline{e_{122}} & e_{133} & e_{123} & \underline{e_{113}} & e_{112} \\ e_{211} & e_{222} & e_{233} & \underline{e_{223}} & e_{213} & e_{212} \\ \underline{e_{311}} & e_{322} & \underline{e_{333}} & e_{323} & e_{313} & e_{312} \end{pmatrix},$$

where the typical nonzero coefficients are underlined. Eq. (10.14) predicts the mechanical stresses generated by the boundary displacements (or tractions) and the applied electric field.

Mechanical strains and the applied electric fields may also induce the displacement of electric dipoles (Harik, 2011) in the piezoelectric materials:

$$D_i = e_{ikl}\varepsilon_{kl} - \kappa^{\varepsilon}_{ij}E_j, \tag{10.7}$$

where D_i is the electric displacement and $\kappa^{\varepsilon}_{ij}$ is electric permittivity or dielectric tensor, depending on the materials dielectric constants. Dependence of these material parameters on the atomic lattice structure of the nanostructured materials and their sizes has been investigated only to some extent for the material components of the microelectromechanical systems (MEMS). The constitutive relations (10.6) and (10.7) can be presented in the fully-coupled lamped matrix form:

$$\begin{pmatrix} \sigma \\ D \end{pmatrix} = \begin{pmatrix} C^{E} & -e^{T} \\ e & \kappa^{\varepsilon} \end{pmatrix} \begin{pmatrix} \varepsilon \\ E \end{pmatrix}, \tag{10.8}$$

which is typically used in the modeling of MEMS devices (e.g., the finite element code *ANSYS Multiphysics*).

10.2.2 Constitutive Modeling of Electromechanical and Thermal Coupling

The macroscopic values of the continuum (i.e., volume-averaged and quasistatic) mechanical stresses include contributions due to the thermal and the electric field effects (Harik, 2011) in the following form:

$$\sigma_{ij} = C^E_{ijkl}\varepsilon_{kl} - e_{ijk}E_k - \alpha_{ij}\Delta T, \tag{10.9}$$

where α_{ij} is the thermal expansion matrix and ΔT is the temperature difference: $T - T_0$. The thermal and the electric field effects on the electric displacements, D_i, of the electric dipoles (Harik, 2011) are described by the pyroelectric vector, $p_i(m)$, and the electric permittivity tensor, $\kappa^{\varepsilon}_{ij}$, by

$$D_i = e_{ikl}\varepsilon_{kl} - \kappa^{\varepsilon}_{ij}E_j - p_i(m)\Delta T. \tag{10.10}$$

At the nanoscale level, the mechanical stresses, σ_{ij}, can be described by the formulae presented in earlier chapters.

10.2.3 Magnetoelastic, Electromechanical and Thermal Coupling

Mechanical behavior of multifunctional nanocomposites in magnetic, electric and electromagnetic fields is affected by their electromechanical and magnetic properties, i.e., electric conductivity, electric and magnetic permittivities, and the dielectric and magnetic permeability tensors in addition to the elastic moduli. The relation between the mechanical stresses, σ_{ij}, and strains and the electric and the magnetic fields, E_k and H_k, is described by the constitutive relation for the piezoelectric and piezomagnetic coupling (Harik, 2011):

$$\sigma_{ij} = C^E_{ijkl}\varepsilon_{kl} - \alpha_{ij}\Delta T - e_{ijk}E_k - q_{ijk}H_k, \tag{10.11}$$

where e_{ijk} is the piezoelectric tensor and q_{ijk} is the piezo-magnetic tensor. The effect of the magnetic field, H_k, on the electric displacements, D_i, is described by

$$D_i = e_{ikl}\varepsilon_{kl} - p_i(m)\Delta T - \kappa^{\varepsilon}_{ij}E_k - a_{ik}H_k, \tag{10.12}$$

where $p_i(m)$ is the pyroelectric vector, $\kappa^{\varepsilon}_{ij}$ is and the electric permittivity tensor, and a_{ik} is the tensor of magnetoelectric coefficients (Harik, 2011). The coupling between the mechanical stresses, σ_{ij}, and the electric displacements, D_i, in the electric and magnetic fields, E_k and H_k, can be modeled by the constitutive matrix relation for the piezoelectric and piezomagnetic coupling:

$$\begin{pmatrix} \sigma \\ D \end{pmatrix} = \begin{pmatrix} C^E & -e^T & -q^T \\ e & -\kappa^{\varepsilon} & -a \end{pmatrix} \begin{pmatrix} \varepsilon \\ E \\ H \end{pmatrix}, \tag{10.13}$$

where the thermal effects due to the temperature difference are excluded for simplicity, but can be easily incorporated. The electric and magnetic fields[1] can also generate the magnetic field flux, B_i, in the multifunctional nanocomposites with the electric and the magnetic susceptibilities:

$$B_i = q_{ikl}\varepsilon_{kl} - m_i\Delta T - a_{ik}E_k - \mu_{ik}H_k, \tag{10.14}$$

where q_{ikl} is the piezomagnetic tensor, the thermal coefficients for the magnetic flux m_i are the aforementioned pyroelectric coefficients, a_{ik} is the tensor of magnetoelectric coefficients, and μ_{ik} is the magnetic permeability tensor. The above constitutive equations for the electromechanical and the magnetomechanical coupling in the nanotube-enhanced multifunctional nanocomposites help the fundamental understanding of the nanoscale multiphysics[2] phenomena.

10.3 Conclusions

New developments in nanomechanics of nanodevices and nanoprobes, piezoelectric sensors, and actuators require constitutive models for the electromechanical and thermal coupling, as well as combine magnetoelastic, electromechanical, and thermal effects in advanced materials. These constitutive models are presented in order to describe new nanoscale physical phenomena and material properties of multifunctional components of various devices and composite structures. New applications in nanotechnology for aerospace, automotive, and computer industries require novel developments in the multiscale modeling of nanostructured materials, carbon nanotubes, and the carbon nanotube-reinforced polymer nanocomposites. Scaling analysis of carbon nanotubes and nanocomposites, the nanoscale homogenization of their atomic structures, and their molecular dynamics models provide a framework for the development of multiscale modeling methods and the multiphysics modeling of new nanoscale phenomena in multifunctional materials.

---

[1] Maxwell's electromagnetic field equations include the electric field equation: $\nabla \cdot \mathbf{E} = q(t)/\varepsilon_0$, where the electric field gradient is equal to the density of the electric charges and the ε_0 is the dielectric constant for the vacuum; and the magnetic field equation: $\nabla \cdot \mathbf{B} = 0$, where the gradient of the magnetic flux, $\mathbf{B}$, is zero. Maxwell's formulation of the Faraday's law of induction is given by $\nabla \times \mathbf{E} = -d\mathbf{B}/dt$, where the changes in the magnetic field flux can generate an electric field. The Ampere–Maxwell law of induction: $\nabla \times \mathbf{B} = \mu_0\varepsilon_0 d\mathrm{E}/dt + \mu_0\mathbf{I}$, where ε_0 and μ_0 are the electric and magnetic permittivities of the vacuum and the changes in the electric field, and an electric current, I, can generate a magnetic field flux.

[2] Multiphysics implies the presence of electromechanical or magnetomechanical interactions, such as in the microelectromechanial systems or nanoscale electromechanical systems, as well as possible microfluidic phenomena in charged liquids.

Sample Problems

1. Calculate the volume fraction of the armchair (10,10) carbon nanotube in a unit cell of the size: 16 × 16 × 80 nm and the effective longitudinal and the transverse Young's moduli of the corresponding nanocomposites.
2. Calculate the difference between the values of the longitudinal and the transverse Young's moduli of the nanocomposite in Problem 1 predicted by the macroscopic rules of mixtures.
3. Describe the difference between the O-registry and the X-registry of the atomic lattice of the carbon nanotube and a zigzag asperity of the polyethylene polymer.
4. Calculate the lattice-unit-sliding frequency, f_{unit}, and the so called "washboard frequency" for the interfacial sliding of a carbon nanotube by using the formula: $\omega = 2\pi f_{\text{unit}} = 2\pi V_{\text{NT}}/a$, peaks in the molecular dynamics data for the sliding velocity in Chapter 8, Carbon Nanotubes in Nanocomposites. This frequency provides an estimate for the frequency, ω_{ph}, of propagating phonons.
5. The matrix form of the linear elastic stress tensor, σ_{ij}, and the elastic strain tensor, ε_{ij}, can be written as

$$\begin{pmatrix} \sigma_{11} & \sigma_{12} \\ \sigma_{21} & \sigma_{22} \end{pmatrix}, \text{ or} \begin{pmatrix} \sigma_{xx} & \sigma_{xy} \\ \sigma_{xy} & \sigma_{yy} \end{pmatrix}, \text{ and} \begin{pmatrix} \varepsilon_{11} & \varepsilon_{12} \\ \varepsilon_{21} & \varepsilon_{22} \end{pmatrix}, \text{ or} \begin{pmatrix} \varepsilon_{xx} & \varepsilon_{xy} \\ \varepsilon_{xy} & \varepsilon_{yy} \end{pmatrix},$$

   where $\sigma_{12} = \sigma_{21}$ or $\sigma_{xy} = \sigma_{yx}$ and $\varepsilon_{12} = \varepsilon_{21}$ or $\varepsilon_{xy} = \varepsilon_{yx}$. Write out the terms of the product of these two tensors.
6. Write out the scalar product $(\nabla \cdot \mathbf{E})$ of the gradient vector: $\nabla = \begin{pmatrix} \dfrac{\partial}{\partial x_1} & \dfrac{\partial}{\partial x_2} & \dfrac{\partial}{\partial x_3} \end{pmatrix}$ and the vector of the electric field, E_k: $\begin{pmatrix} E_1 & E_2 & E_3 \end{pmatrix}$.
   Write down the corresponding Maxwell's equation.
7. Identify the coefficients, constants, and variables in the expression for the dipole–dipole interaction used in the Lennard–Jones potential for the van der Waals forces:

$$V(r) = -\frac{1}{r^6}\frac{1}{(4\pi\varepsilon_0)^2}\frac{2\mu_1^2\mu_2^2}{3kT} \quad \text{and} \quad V(r) = -\frac{1}{r^6}\frac{1}{(4\pi\varepsilon_0)^2}\left(\mu_1^2\alpha_1 + \mu_2^2\alpha_2\right).$$

8. Write out the tensor form of the piezoelectric tensor and the electric permittivity tensor.
9. Write out the tensor form of the piezomagnetic tensor, the tensor of magnetoelectric coefficients, and the magnetic permeability tensor.

References

Chiolerio, A., Castellino, M., Jagdale, P., Giorcelli, M., Bianco, S., Tagliaferro, A., 2011. Electrical properties of CNT-based polymeric matrix nanocomposites. In: Yellampalli, S. (Ed.), Carbon Nanotubes-Polymer Nanocomposites. InTech, Rijeka, Croatia, pp. 215–230.

Frankland, S.J.V., Harik, V.M., Odegard, G.M., Brenner, D.W., Gates, T.S., 2003. The stress–strain behavior of polymer–nanotube composites from molecular dynamics simulation. Compos. Sci. Technol. 63, 1655–1661.

Harik, V.M., 2011. Mechanics of Carbon Nanotubes. Nanodesigns Press, Newark, Delaware.

Harik, V., 2017. Mechanics of carbon nanotubes and nanomaterials. In: Gogotsi, Yu (Ed.), Nanomaterials Handbook. CRC Press, Boca Raton, Florida, Chapter 6.

Kilbride, B.E., Coleman, J.N., Fournet, P., Cadek, M., Drury, A., Hutzler, S., et al., 2002. Experimental observation of scaling laws for alternating current and direct current conductivity in polymer-carbon nanotube composite thin films. J. Appl. Phys. 92, 4024–4030.

Park, C., Wilkinson, J., Banda, S., Ounaies, Z., Wise, K.E., Sauti, G., et al., 2006. Aligned single wall carbon nanotube polymer composites using an electric field. J. Polymer Sci. B: Polymer Phys. 44 (12), 1751–1762.

Stauffer, D., Aharony, A., 1994. Introduction to Percolation Theory. Taylor and Francis, London.

CHAPTER 11

Carbon Nanotube Length and Criteria for Potential Toxicity

Chapter Outline

11.1 Introduction: Similarity Between Carbon Nanotubes and Asbestos Fibers

Hazardous properties of microscopic fibrous particles such as asbestos fibers have been known for decades, especially in the field of fiber toxicology (Kane et al., 1986; Moalli et al., 1987; Hill et al., 1995; Donaldson et al., 2010). Inhalation of these particles (see Fig. 11.1) has been shown to cause different diseases of the lung and its lining tissue (Kane et al., 1986; Viallat et al., 1986; Moalli et al., 1987; Cook, 2004; Donaldson et al., 2010). Most of these diseases prove to be fatal, as the harmful particles can reach the gas-exchanging part of the lung and most of them cannot be readily removed by macrophages (Kane et al., 1986; Moalli et al., 1987; Hill et al., 1995; Donaldson et al., 2010). The human immune system has little or no defenses against such hazards.

Natural or man-made fibers do not have to be asbestos to be toxic (see Fig. 11.1) as shown by Jaurand et al. (2009) and others. Toxicity of fibers is associated in part with their high aspect ratios and length. The size and shape of asbestos fibers, especially their length, have been singled out as the key characteristics affecting their potential toxicity (Kane et al., 1986; Viallat et al., 1986; Moalli et al., 1987; Hill et al., 1995; Cook, 2004; Donaldson et al., 2010). Cook (2004) has included the size and shape of microscopic fibers

Mechanics of Carbon Nanotubes.
DOI: https://doi.org/10.1016/B978-0-12-811071-3.00011-1

© 2018 Elsevier Inc. All rights reserved.

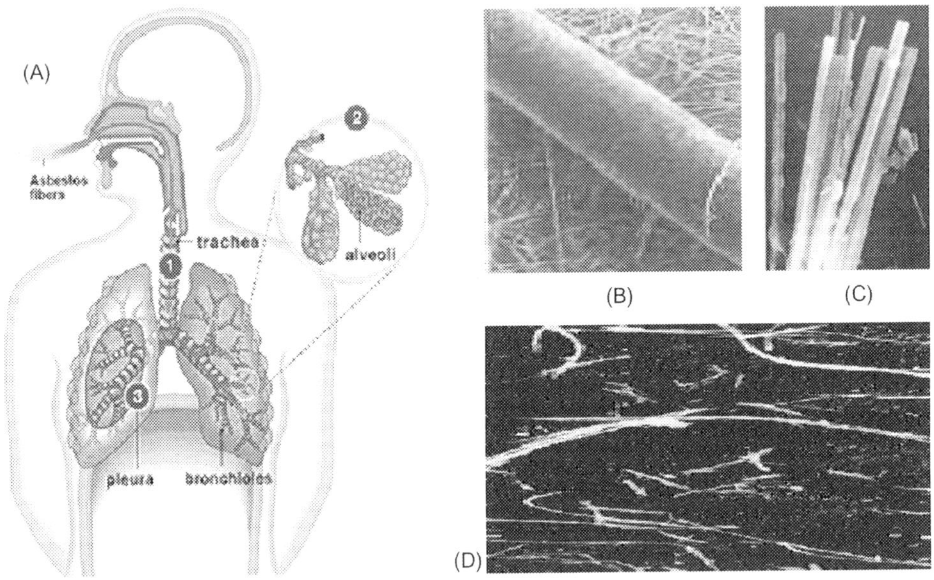

Figure 11.1
A schematic of asbestos and carbon nanotube transport (A) in the human lungs with (1) trachea, (2) alveoli, and (3) pleura. Amosite asbestos fibers (B) are shown below a human hair, next to a bundle of the rod-like crocidolite asbestos fibers (C) along with the bundles (D) and fibrils of chrysotile asbestos fibers (after Cook, 2004).

(and carbon nanotubes) into the criteria for the asbestos-like ability to cause pathologies, and has stated a hypothesis about the nanotube–asbestos similarity (Cook, 2004; Harik, 2005). Geometric or structural characteristics of microscopic fibers and carbon nanotubes have been rigorously linked to their ability to cause the asbestos-like pathologies through the scaling analysis of geometric parameters of asbestos fibers and carbon nanotubes by using the laws of similitude (Harik, 2001) in support of Cook's hypothesis about the nanotube–asbestos similarity (Harik, 2005, 2014). The structure–activity paradigm (Donaldson et al., 2010) has been also used to analyze the toxicity of the high aspect ratio fibers having the microscopic elongated size. Harik's scaling analysis of geometric parameters of asbestos fibers and carbon nanotubes (Harik, 2005, 2014) provides a mathematical framework for the structure–activity paradigm (Donaldson et al., 2010), and a mathematical extrapolation of potential toxic effects for different material fibers including carbon nanotubes.

Donaldson et al. (2010) have presented a review of toxicological studies, where they synthesize new data about the behavior of multiwall carbon nanotubes (MWCNTs) in the lung and presented a hypothesis about their retention as long fibers in the parietal pleura, which results in inflammation and pleural pathology such as mesothelioma. They have studied the difference in the behavior of short and long fibers in the lung. Donaldson's team pointed out similarity between long asbestos fibers and long MWCNTs in the context of their interaction with macrophages in the lung during the engulfment process and the efficiency of phagocytosis. Donaldson et al. (2010) have presented experimental data that support Cook's hypothesis about the nanotube–asbestos similarities (Cook, 2004; Harik, 2005, 2014).

A review of toxicological data, their analysis, and synthesis is valuable for understanding of origins and mechanisms of inflammation processes and the resulting pathologies. In the present review of toxicological studies, analysis of the structure–activity paradigm for fibrous materials is further developed in order to identify the key geometric parameters, as well as an important nondimensional parameter, which characterize the efficiency of phagocytosis for different fibers, including carbon nanotubes. Different sets of geometric, morphological, and behavioral criteria have been employed to evaluate the toxicity of inhaled particulates (Cook, 2004; Harik, 2005; Maynard et al., 2011). A quantitative extrapolation of the potential toxic effects can be performed by the scaling analysis of geometric parameters of various fibers (Harik, 2005, 2014).

The structure–activity paradigm (Donaldson et al., 2010), Cook's hypothesis about the similarity of asbestos fibers and carbon nanotubes (Cook, 2004), and Harik's proof of the Cook's hypothesis by using the scaling analysis and the laws of similitude (Harik, 2005, 2014) provide a methodology for the analysis of toxic effects of carbon nanotubes that connects their geometric and structural properties with their potential toxicity (Harik, 2014). Other researchers such as Stone and Donaldson (2006), Donaldson et al. (2006), Poland et al. (2008), Kostarelos (2008), Jaurand et al. (2009), Murphy et al. (2014) have also recognized similarities between carbon nanotubes and asbestos fibers: "Both CNTs [some MWCNTs] and asbestos particles share fibrous morphology, and their dimensions are in the same range" (Jaurand et al., 2009). The laws of geometric similitude (Harik, 2005, 2014) can extend geometric similarity between carbon nanotubes and asbestos fibers beyond the range of the same dimensions noted above, and provide a rigorous mathematical framework for the extrapolation of potential toxic effects of long carbon nanotubes.

The nature of the asbestos–nanotube similarities can be also explained by the fiber "pathogenicity paradigm developed in the 1970s–80s" and "a robust structure–toxicity relationship that enables the prediction of the pathogenicity of fibers depending on their length, thickness and biopersistence" (Donaldson et al., 2013). Fibers which "are sufficiently long," biopersistent, and "deposit in the lungs can cause oxidative stress and inflammation." In order to determine what "sufficiently long" means for fibrous materials, and to further develop more quantitative characterization of the structure–toxicity relationship, this review examines an important nondimensional parameter, $M_{\text{ph}} = L_M/L_f$, which links the size of macrophages, L_M, the length of fibers, L_f, the mass of carbon nanotubes, and the efficiency of phagocytosis, when $M_{ph} > 1$ (Harik, 2014). As a result, the critical values for the fiber length and the nondimensional parameter, M_{ph}, can be determined for various degrees of the efficiency of phagocytosis (for different $M_{ph} > 1$), its total inefficiency (for $M_{ph} < 1$), and the associated onset of various pathologies.

11.2 The Small Size Matters

The small size of nanomaterials, and fibrous nanomaterials in particular, should be considered in the analysis of their potential toxic effects, as was noted by British Royal Society, Royal Academy of Engineering (2004), Donaldson et al. (2004), Harik (2005), and Oberdorster et al. (2005). The joint British Royal Society and Royal Academy of Engineering report points out the potentially greater toxicity of nanoparticles (NP) due to their high surface area and, hence, the higher surface reactivity, as well as their ability to cross cell boundaries and membranes. NP have sizes that make them inhalable and allow them to penetrate deep into the lungs. Fine particles can be inhaled if their aerodynamic diameter, d_{ae}, which is about three times their actual diameter (Donaldson et al., 2013), is less than a few microns. The diameters of most NP are so small that they can be inhaled into the lungs and processed by macrophages. NP may evade the natural defenses of a human body and damage cells, cellular components and larger proteins. Hard NP with diameters that are smaller than cells can affect components of cells or cell surface. Some bodily defenses and the initial human response to the presence of aerosol particles also depend on the particle size. The mechanisms of interactions between NP and human cells or macrophages depend on their relative geometric dimensions, which can be analyzed by the scaling analysis (Harik, 2001, 2014) in order to evaluate the potential harm or their toxic effects.

It is well known that when the length of fibers or nanoparticle clusters is longer than about 20 μm, such geometry prevents the removal of such particles by the macrophages in lungs (Donaldson et al., 2010). The typical size of macrophages varies between 10 and 20 μm, therefore the critical size of fibers, L_f, that determines the limit in the efficiency of phagocytosis, when $M_{ph} = 1$, is at the limit of this range of about 10 μm (Harik, 2014). It should be noted that the typical size of NP (1–100 nm) is much smaller than the average size of macrophages and the critical size of fibers for phagocytosis, $L_{NT} < 10$ μm. Therefore the size of many NP will allow macrophages to process them through phagocytosis and avoid potential toxic effects.

Once NP are lodged in the lung tissue, i.e., their toxicity depends on the ability to initiate an inflammatory process or reaction, including interaction with macrophages and other defensive cells. Deposition of NP and nanofibers in human lung may result in lesions in the lung or scarring of the lung tissue (i.e., asbestosis), and even lung cancer, since the migration of asbestos fibers through the lung to its lining, the pleura, is known to result in the development of mesothelioma, a fatal tumor. The size of NP has been singled out as a key factor in a new field of nanotoxicology in an editorial in *Occupational and Environmental Medicine* by Donaldson et al. (2004):

> NP [nanoparticles] have greater potential to travel through the organism than other materials or larger particles. The various interactions of NP with fluids, cells, and tissues need to be considered, starting at the portal of entry and then via a range of possible pathways towards target organs. The potential for significant biological response at each of these sites requires investigation. In addition, at the site of final retention in the target organ(s), NP may trigger mediators which then may activate inflammatory or immunological responses. Importantly NP may also enter the blood or the central nervous system, where they have the potential to directly affect cardiac and cerebral functions. We therefore propose that a new subcategory of toxicology – namely nanotoxicology – be defined to address gaps in knowledge and to specifically address the special problems likely to be caused by nanoparticles.

In 2007, a new journal *Nanotoxicology* was launched. Nanotoxicology of NP can be linked with the toxicological studies of ultrafine particles (Oberdorster et al., 2005). Oberdorster et al. (2007), however, have questioned the uniqueness of nanoscale, and suggested that the toxicity of many nanomaterials is scalable and predictable from macroscopic materials. Quantitative analysis of the potential toxic effects of NP is important for the development of rigorous methodologies in nanotoxicology. The scaling analysis of NP (Harik, 2001), their sizes, structural and toxicological properties (Harik, 2005, 2014) can be carried out in the framework of the structure–activity paradigm (Donaldson et al., 2010) and its mathematical foundation, i.e., the dimensional analysis (Harik, 2005, 2014). This review presents further developments in the analysis of potential toxic effects of NP, especially carbon nanotubes, by using scaling analysis.

11.2.1 Nanoscale Corrections to the Dose and Exposure to Nanoparticles

A connection between the size of NP and various adverse health effects has been examined by Ferin et al. (1990), who have investigated the process of phagocytosis in alveoli of the lungs involving submicron particles of 20–30 nm in size. These nanoscale particles or their agglomerates were able sometimes to cause the death of macrophages, or a dysfunction that allowed translocation of NP into interstitial space. The size of these particles and their dynamics were the key aspects that affected their toxicity (Ferin et al., 1990; Song et al., 2009). Oberdorster (2000) has demonstrated that the inflammatory response to NP in lungs depends on their size, while "a size-invariant dose–response function" can describe the data normalized by the surface area. Normalization of various geometric parameters for carbon nanotubes has been instrumental in the classification of their mechanical behavior (Harik, 2001), the proof of Cook's hypothesis about the nanotube–asbestos similarity

(Harik, 2005, 2014), and the scaling analysis of the efficiency of phagocytosis (Harik, 2014). This study further develops the structure–activity paradigm (Donaldson et al., 2010) and nanoscale corrections for the ability of different geometric classes of carbon nanotubes to cause toxic effects.

Analysis of nanoscale corrections to the dose, concentration, and exposure to various NP depending on their size and their number per weight is useful for scalability and extrapolation of the potential toxic effects. The concept of *hazard* is defined as the potential of a substance, e.g., NP or carbon nanotubes, to cause harm. The content of this and other concepts, their meaning as well as the mechanisms of causing harm may, in fact, change at nanoscale. From this viewpoint, NP indeed have a higher total area of surface contact with biological tissues than their microscopic counterparts, and a greater ability to pierce through the cells and membranes. The greater area for material contacts in the body results in an increased ability of NP to take part in reactions with cells, including the tissue-damaging reactions. For example, the surface of the quartz crystal is highly reactive. It generates free radicals (reactive atoms), which lead to oxidative damage to the defensive cells. Nonreactive surfaces, however, can also generate oxidative stress on the cells. Other investigations have shown that weight-for-weight, the finely divided particles of a material, such as titanium dioxide or carbon black, are more toxic than larger particles of the same material. Significant increase in the total surface area and its reactivity at nanoscale level in relation to the finite size of macrophages seem to be fundamental to the potential toxicity of NP. Quantitative characteristics of the connection between increases in the surface area and the ability to cause harm are important for further development of the structure–activity paradigm for NP and nanofibers, in particular.

Exposure to a hazardous material is defined as the concentration of the substance in the relevant medium (air, food, water) multiplied by the duration of contact. Concentration of particles is defined as the number of particle per unit volume. At the same weight, the number of NP is much greater than that of microscopic particles, and, hence, their concentration would be considerably higher for the same weight-to-bio-volume ratio. For cylindrical fibers and microparticles, reduction in their length from 1 μm 10 (or 100) times without changing the total weight results in an increase in the number of NP of the size 100 nm (or 10 nm) and their concentration 10 (or 100) times. This ratio defines a dose as the amount of a substance that will reach a specific biological system or an organ, which can increase 10 or 100 times, depending on the decrease in the size of fibers. Consequently, the exposure and a dose will be both higher. The exposure may be further increased by the ability of NP to have a better interfacial contact and higher surface energy of adherence, and, hence, the longer duration of contact. Therefore determination of the nanoscale corrections to the dose and exposure is important in the analysis of scalability of the toxicity of NP from their macroscopic counterparts.

11.2.2 The Size of Carbon Nanotubes and Criteria for the Asbestos-Like Pathologies

Nanoscale particles have been shown to have intrinsically higher toxicity due to their smaller nanoscale size (Ferin et al., 1990; British Royal Society, Royal Academy of Engineering, 2004; Donaldson et al., 2004; Warheit et al., 2004; Harik, 2005). Safety concerns about the toxicity of NP, in general, and carbon nanotubes, in particular, are understandable, because their smaller size results in a larger surface area and higher concentration in addition to their high aspect ratios (Hill et al., 1995; Donaldson et al., 2004, 2006, 2010; Warheit et al., 2004; Harik, 2005, 2014; Poland et al., 2008; Maynard et al., 2011). Reduction in the length or the aspect ratio of 1-μm long carbon nanotubes 10 (or 100) times results in an increase in the number of 100 nm (or 10 nm) long nanotubes and their concentration exactly 10 (or 100) times if their total weight has not been changed. This is the nanoscale correction to the concentration of fibers based on the reduction of length.

The ability of carbon nanotubes to absorb various chemicals on their surface and exert oxidative stresses on the surrounding biological tissue is proportional to their geometric size and the surface area. Nanoscale corrections for the increases in the surface area of fibers can be estimated as follows. Reduction in the length or the aspect ratio of 1-μm long carbon nanotubes 10 (or 100, or n) times results in an increase in the surface area by

$$1 + 2(n-1)\left(1+2\frac{L_{\mathrm{NT}}}{d_{\mathrm{NT}}}\right)^{-1}$$

and 10 (or 100, or n) times more of 100 nm (or 10 nm) long nanotubes along with their concentration, if their total weight stays the same. These nanoscale corrections can support the structure–activity paradigm (Donaldson et al., 2010) for the toxicological studies and the analysis of toxicity of carbon nanotubes. Further development of the structure–activity paradigm tailored for carbon nanotubes can be achieved by increasing the understanding of how the structural properties of carbon nanotubes and their geometric parameters (Table 11.1) are linked with the nanoscale mechanisms of toxicity (Poland et al., 2008; Donaldson et al., 2010; Harik, 2014).

Table 11.1: [a]Size of carbon nanotubes and their potential health effects (in vivo studies)

Geometric Parameters	Dose and Exposure	Resulting Effects
Length, $L_{\mathrm{NT}} = 1–5$ μm Diameter, $d_{\mathrm{NT}} = 14.8$ nm	50 μg (24 h–7 days)	No inflammation reported
Length, $L_{\mathrm{NT}} = 40–50$ μm Diameter, $d_{\mathrm{NT}} = 84.9$ nm	50 μg (24 h–7 days)	Inflammation, granulomas
Length, $L_{\mathrm{NT}} = 20–100$ μm Diameter, $d_{\mathrm{NT}} = 165 \pm 5$ nm	50 μg (24 h–7 days)	Inflammation, granulomas

[a]This table is based on the data of toxicological studies presented by Poland et al. (2008).

Cook (2004) of the National Health and Environmental Effects Research Laboratory in Duluth, Minnesota, has proposed using the "size and shape" of microscopic asbestos fibers and their total "surface area" in his criteria for the asbestos-like pathologies for the fibers that may behave like asbestos, for example, carbon nanotubes:

- size and shape of particles that allow respiration and retention in lungs;
- size and shape of particles that allow translocation to pleura;
- durability of fibers, persistent contact in tissues;
- reactive surfaces, ability to induce oxidative stresses;
- high collective surface area; and
- propensity to split into thin fibers in vivo.

After presenting the aforementioned criteria, Cook (2004) has suggested a hypothesis about the similarities between carbon nanotubes and asbestos fibers. Cook's hypothesis about the asbestos–nanotube similarity has been analyzed and supported within the conceptual framework of scaling analysis (Harik, 2005, 2014) that deals with the characteristics of "the size and shape of particles" and their effects on different processes associated with their transport, translocation, and engulfment. Cook's hypothesis about the nanotube–asbestos similarity (Cook, 2004) has been proven both theoretically (Harik, 2005, 2014) and experimentally for long carbon nanotubes (see Table 11.1) by Poland et al. (2008).

In the toxicological study of long carbon nanotubes, Poland et al. (2008) have demonstrated (see Table 11.1) that the size of these nanoscale fibers is connected to the initiation of inflammatory processes and granulomas in the lungs. The size of carbon nanotubes, their length and diameter, which extend well into the microscopic range, satisfy Cook's criteria for the ability to cause asbestos-like pathologies (Cook, 2004; Harik, 2005), as well as the ability to affect phagocytosis (Harik, 2014). Analysis of experimental results of Poland et al. (2008) also points out (see Table 11.1) that there is a "critical length of carbon nanotubes" beyond which carbon nanotubes are able to cause inflammation and, sometimes, granulomas in the lungs. These experimental results support the structure–activity paradigm, and the hypothesis that the size of carbon nanotubes is particularly important in the ability to cause asbestos-like pathologies (Cook, 2004; Harik, 2005, 2014; Stone and Donaldson, 2006; Poland et al., 2008; Kostarelos, 2008; Donaldson et al., 2010). The length-based approach in the evaluation of key parameters in the structure–activity paradigm in the analysis of toxic effects of nanofibers and carbon nanotubes has resulted in the length-based criteria for the ability of nanofibers to cause asbestos-like pathologies (Harik, 2014). This review presents an analysis of geometric parameters of the structure of carbon nanotube and the potential mechanisms of toxicity, which can be attributed to the structure–activity paradigm for these nanoscale fibers.

This study demonstrates that the process of phagocytosis in the lungs and its effectiveness can be analyzed by using scaling analysis (Harik, 2001, 2005) and a scaling parameter, which is based on the length of carbon nanotubes and the size of macrophages (Harik, 2014). This method provides a quantitative framework for the analysis of experiments on phagocytosis and the engulfed NP, such as carbon nanotubes or other particles in the macrophages. The effectiveness of phagocytosis, the rate of engulfing process, and the probability of complete phagocytosis can be characterized within the framework of such scaling analysis.

The length-based approach in the analysis of the structure–activity paradigm for the evaluation of toxic effects of carbon nanotubes (Cook, 2004; Harik, 2005, 2014; Stone and Donaldson, 2006; Poland et al., 2008; Kostarelos, 2008; Donaldson et al., 2010) can provide not only the length-based criteria for the ability of nanofibers to cause asbestos-like pathologies (Harik, 2014), but also a new parametric analysis of phagocytosis for different classes of carbon nanotubes. After the scaling analysis of size effects in toxicological studies by Stone and Donaldson (2006), Poland et al. (2008), Kostarelos (2008), and Donaldson et al. (2010), Harik (2014) has proposed a quantitative hypothesis about the existence of a critical nondimensional parameter, which connects the length-based criteria for the ability of carbon nanotubes to cause harm, the size of macrophages, and the efficiency of phagocytosis. Carbon nanotubes, the length of which is clearly less than the size of macrophages (see Table 11.1, data in the first row) satisfy the criterion for the efficiency of phagocytosis, i.e.,

$$M_{\mathrm{ph}} > 1,$$

where the nondimensional parameter, $M_{\mathrm{ph}} = L_M/L_{\mathrm{NT}}$, links the size of macrophages, L_M, the length of carbon naotubes, L_{NT}, and the efficiency of phagocytosis with the required criterion: $M_{\mathrm{ph}} > 1$ (Harik, 2014). Carbon nanotubes, the length of which, L_{NT}, is clearly larger than the average size of macrophages (see Table 11.1, data in the second and third rows) do not satisfy the criterion for the efficiency of phagocytosis with $M_{\mathrm{ph}} < 1$. The following inequality

$$M_{\mathrm{ph}} < 1$$

is associated with the total deficiency of phagocytosis, therefore the experimental data in Table 11.1 indicate the corresponding inflammation and granulomas. The critical size of carbon nanotubes, L_{NT}, for the onset of phagocytosis is about 10 μm, i.e., $L_{\mathrm{NT}} < 10$ μm. Different values for the length of nanotube fibers and the nondimensional parameter, M_{ph}, can be also linked to various degrees of the efficiency or deficiency of phagocytosis (for different $M_{\mathrm{ph}} > 1$) not just to the total inefficiency of phagocytosis (for $M_{\mathrm{ph}} < 1$).

According to Esmen and Lippmann,[1] different sizes of asbestos fibers can be linked to different degrees in the ability to cause harm to human lungs: fibers with the length of 2 μm or more may cause asbestosis, asbestos fibers with the length of 5 μm or more may result in mesothelioma, and fibers with the length of 10 μm or more may cause lung cancer. Assuming that the average size of macrophages varies between 10 and 20 μm, asbestosis is associated with the following degree of efficiency of phagocytosis: $5 < M_{ph} < 10$. The lower efficiency of phagocytosis in the range of $2 < M_{ph} < 4$ can be linked to mesothelioma, while the nearly total deficiency of phagocytosis in the range of $1 < M_{ph} < 2$ is associated with lung cancer. These observations support theoretical predictions about the total deficiency of phagocytosis for $M_{ph} < 1$. Effective phagocytosis and, therefore, safety of nanotechnological processes can be achieved for the values of nondimensional number, M_{ph}, much higher than 10. Further studies in the quantitative evaluation of the efficiency of phagocytosis are needed to improve the understanding of different mechanisms and stages of phagocytosis. This study further develops the structure–activity paradigm (Donaldson et al., 2010) for different geometric classes of carbon nanotubes.

The early toxicological studies of carbon nanotubes in mammals (Warheit et al., 2004; Lam et al., 2004; Muller et al., 2005; Murr et al., 2005) can benefit from a methodology for the extrapolation of their assessment of nanotube toxicity in lung tissues of rats and human cells to other similar cases as defined by the scaling analysis and the so called laws of similitude (Harik, 2005, 2014). In other toxicological studies carried out at NIOSH (Shvedova et al., 2003), Du Pont and NASA (Warheit et al., 2004; Lam et al., 2004; Muller et al., 2005; Murr et al., 2005), nanotubes were instilled rather than inhaled in a natural manner. Nevertheless, these studies clearly demonstrate that carbon nanotubes can be toxic, thus, they support the scaling extrapolations of Harik (2005). The role of length of carbon nanotubes in the structure–activity paradigm and quantitative extrapolations of potential toxic effects can be further clarified by the analysis of the toxicological effects associated with the aspect ratios of carbon nanotubes.

11.3 The Length Matters: The Aspect-Ratio of Carbon Nanotubes and Safety

Hill et al. (1995) have noted the difference between the short and long asbestos fibers in their study of the differential release of superoxide anions by macrophages. Harik (2005) has used Cook's criteria of the size and shape of microscopic asbestos fibers (Cook, 2004) and the aspect ratio of carbon nanotubes in the analysis of the nanotube–asbestos analogy. Stone and Donaldson (2006) and Kostarelos (2008) have argued that the "size does matter"

[1] The following studies are recommended for the evaluation of size effects by S. Hirano of National Institute of Environmental Studies of Japan: Esmen, N.A., Environ. Health Perspect., 88:277–286,1990; and Lippmann, M., Environ. Health Perspect., 88:311–317,1990.

in their studies of "the long and short of carbon nanotube toxicity" and the engulfment of various nanotubes by macrophages in phagocytosis in the lungs. They have pointed out that the length of carbon nanotubes affects the process of phagocytosis. This study provides a quantitative analysis of the efficiency of phagocytosis, and identifies different ranges and degrees of the low efficiency phagocytosis by using a quantitative hypothesis about a critical nondimensional parameter. These theoretical results are supported by experimental data of Jaurand et al. (2009), who have examined the size effects on the toxicity of MWCNTs and their similarities with asbestos fibers. They have demonstrated that the size of carbon nanotubes, especially their length, affects the initiation of inflammation in the lungs. Moreover, Poland et al. (2008) have shown that the length of carbon nanotubes is linked to the asbestos-like pathologies in mice. Later, the same team Donaldson et al. (2010) confirmed these results in toxicological studies for pleura and attributed them to the structure–activity paradigm for the length-dependent toxicity of microscopic fibers. Yamashita et al. (2010) have also demonstrated that the long MWCNTs cause the most inflammation and DNA damage, as opposed to the short and thin MWCNTs. Park et al. (2014) have shown that the HIPCO single wall carbon nanotubes (SWCNTs), which tend to have high aspect ratios, exhibit toxic response in macrophages cells. Murphy et al. (2014) have found a common molecular signature in the lesions caused by the toxicity of the long carbon nanotubes and asbestos fibers (Fig. 11.1). Analysis of mechanisms behind the molecular signature in the structure of lesions can also benefit from the scaling analysis of the local molecular structure of both carbon nanotubes and biological tissue.

Variations in the diameter of carbon nanotubes, d_{NT}, such that $\delta d_{NT} < a/\pi$, can have a small effect on the intensity of local interactions between the surface of carbon nanotubes and biological fluids and tissues, and virtually no effect on the effectiveness of phagocytosis. Carbon nanotubes, which have such small variations in their diameters, can be considered to have virtually the same dimensions, as far as the effectiveness of phagocytosis is concerned. This conclusion is based on the scaling analysis of the distributed effects that the surface carbon atoms have on the surface-surface interactions between carbon nanotubes and biological fluids and tissues. In the case of MWCNTs, carbon nanotubes may have large diameters, d_{NT}, and large variations in the distribution of their diameters, such that $\delta d_{NT} \gg a/\pi$. Such variations may have a noticeable effect on the intensity of local surface-surface interactions between carbon nanotubes and biological tissues, as well as on the effectiveness of phagocytosis. Since the length and the high aspect ratios of carbon nanotubes and variations in their diameters can affect the efficiency of phagocytosis, it is important to examine the ranges of geometries of carbon nanotubes including MWCNTs.

11.3.1 Classification of Carbon Nanotubes and the Efficiency of Phagocytosis

In 2001, shortly after the National Nanotechnology Initiative, a research report from NASA Langley Research Center (Hampton, Virginia) has presented a classification of carbon nanotubes into four classes (Harik, 2001): the thin and thick lattice shells, the long high-aspect-ratio nanotubes, and the beam-like carbon nanotube crystals of small radii (Fig. 11.2). Different classes of carbon nanotubes have distinct geometries (i.e., length, shape, the aspect ratio, the wall thickness, and diameter), unique structural properties (i.e., thin or thick atomic lattice shells and large or small high aspect ratio beams) and different mechanical behavior that affect interactions with biological tissues (e.g., spontaneous Euler buckling, see Harik, 2011). The size and shape of carbon nanotubes have been identified as the key characteristics associated with the ability to cause asbestos-like pathologies in the development of Cook's hypothesis about the nanotube–asbestos similarity (Cook, 2004; Harik, 2005). The length of carbon nanotubes has been singled out as a critical geometric parameter, which is linked to the efficiency of phagocytosis through a critical nondimensional parameter, M_{ph} (Harik, 2014). Classification of carbon nanotubes can be also used to develop a quantitative framework for the analysis of different mechanisms of phagocytosis and the structure–activity paradigm.

A modified classification of carbon nanotubes is shown in Fig. 11.2 in order to identify nine classes of carbon nanotubes, including the long high-aspect-ratio nanotubes (SWCNTs

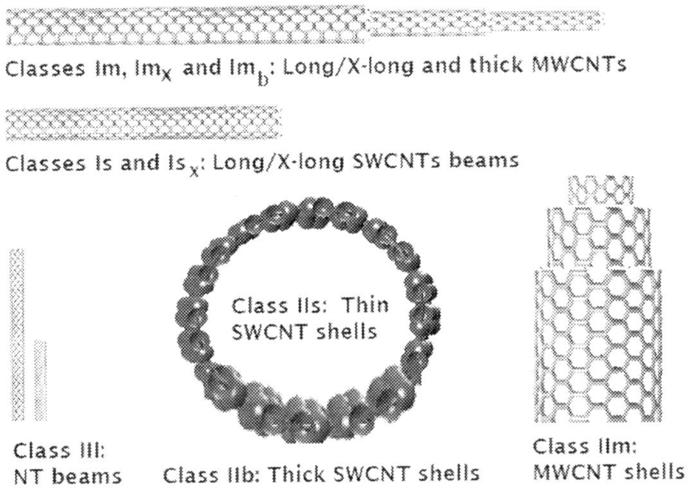

Figure 11.2
The modified classification of CNTs into different classes: *the long and X-long SWCNT shells* (class Ia, i.e., the high aspect ratio SWCNTs), *the long high aspect ratio MWCNT shells* (class Ib of long MWCNTs), *the thin SWCNT shells* (class IIa), *the thick MWCNT shells* (class IIb), and *CNT nanobeams* (class IIIa) and *thick SWCNT shells* (class IIIb). *CNT*, carbon nanotube; *SWCNT*, single wall carbon nanotube; *MWCNT*, multiwall carbon nanotube. Source: *This illustration is adapted from Harik (2017).*

and MWCNTs) and the extra-long carbon nanotubes (the three classes of the X-long SWCNTs, the X-long thin and thick MWCNTs). The extra-long SWCNTs and thin MWCNTs with the high aspect ratios are prone to the spontaneous Euler buckling, which promotes separation of nanotubes from the bundles and increases of the local surface area. The ability of fibers to split in vivo has been identified as an important factor for the increases in total surface area and higher potential toxicity of the inhaled fibers (Cook, 2004; Harik, 2005). The size of surface area of carbon nanotubes and their volume both depend on their geometric parameters, i.e., radius, R_{NT}, or diameter, d_{NT}, and the length, L_{NT}, as well as their aspect ratio, $d_{\mathrm{NT}}/L_{\mathrm{NT}}$. These geometric parameters are also important for the evaluation of the aforementioned nanoscale corrections to the dose, concentration, and exposure. For the same weight, different classes of carbon nanotubes will have different distribution of factors contributing to the potential toxicity of such nanoscale fibers.

The length of carbon nanotubes has been identified as one of the main parameters contributing to the toxicity of carbon nanotubes, as shown in the experimental toxicological studies (Poland et al., 2008; Jaurand et al., 2009; Donaldson et al., 2010) and the scaling analysis of the efficiency of phagocytosis (Harik, 2014). The long carbon nanotubes $\left(L_{\mathrm{NT}}/a \gg 1\right)$ normalized by the size of a carbon ring, a (Harik, 2001) and characterized by the high aspect ratios $(d_{\mathrm{NT}}/L_{\mathrm{NT}} \ll 1/10)$ have been shown to exhibit higher rates of inflammation (Hill et al., 1995; Stone and Donaldson, 2006; Kostarelos, 2008; Donaldson et al., 2010; Park et al., 2014), as well as structural similarities to the asbestos fibers (Cook, 2004; Harik, 2005; Poland et al., 2008; Jaurand et al., 2009; Donaldson et al., 2010; Murphy et al., 2014). The potential toxic effects of different classes of carbon nanotubes can be analyzed by using the scaling analysis of geometric and structural parameters, classification of carbon nanotube geometries (Harik, 2001, 2011), and the scaling parameters based on their geometric size or their surface area (Harik, 2005, 2014).

The five classes of long carbon nanotubes (Fig. 11.2) are characterized by one common geometric parameter, i.e., their high aspect ratio $(d_{\mathrm{NT}}/L_{\mathrm{NT}} \ll 1/10)$. When the aspect ratio, $d_{\mathrm{NT}}/L_{\mathrm{NT}}$, is so high that the length of carbon nanotube, L_{NT}, is greater than the size of the macrophages, the macrophages are unable to clear these NP from the lungs. This means that carbon nanotubes belonging to these five (out of nine) classes of nanotubes may have potential toxic effects. The probability of the effective phagocytosis can be evaluated for these five classes of carbon nanotubes by a nondimensional scaling parameter, M_{ph}, based on the ratio of the nanotube length, L_{NT}, and the average size of macrophages, L_m, i.e., $M_{\mathrm{ph}} = L_m/L_{\mathrm{NT}}$ (Harik, 2014). The ability of macrophages to perform an effective phagocytosis in the lungs for the inhaled carbon nanotubes can be described by the following inequalities:

$$M_{\mathrm{ph}} \gg 1 \quad \text{or} \quad d_{\mathrm{NT}}/L_{\mathrm{NT}} \gg d_{\mathrm{NT}}/L_m,$$

where the length of carbon nanotubes, L_{NT}, is such that macrophages of the length, L_m, are able to clear these nanoscale fibers from the lungs with the probability of the effective phagocytosis is being much greater than unity, i.e., $M_{ph} \gg 1$. Therefore the nondimensional parameter, M_{ph}, can be used to characterize the probability of the effective phagocytosis, P_{ph}, by the following relation:

$$P_{ph} = m_{ph} M_{ph},$$

where the proportionality coefficient, m_{ph}, characterizes the ability of macrophages to carry out an effective phagocytosis. The proportionality coefficient, m_{ph}, is proportional to the mass of macrophages and their volume, V_m. Therefore the aforementioned inequality for the efficiency of phagocytosis can be written for the elongated macrophages as

$$\frac{d_{NT}^2}{V_{NT}} \gg \frac{d_m^2}{V_m} \quad \text{or} \quad \frac{V_m}{V_{NT}} \gg \frac{d_m^2}{d_{NT}^2},$$

where V_{NT} is the volume of carbon nanotubes and d_m is an average size of the crossectional size of the elongated macrophages. This inequality demonstrates the importance of the volume of macrophages with respect to the volume of carbon nanotubes in relation to the ratio of the crossectional areas of macrophages and carbon nanotubes. The aforementioned inequalities provide further mathematical framework for the structure–activity paradigm in the analysis of toxicological studies of carbon nanotubes.

Donaldson et al. (2010) have conducted toxicological studies for pleura and stressed the importance of the structure–activity relations and the high aspect ratios in the length- and the aspect-ratio-dependent toxicity of various fibers. Following the scaling analysis (Harik, 2014) of different classes of carbon nanotubes, Table 11.2 presents a summary of different toxicological studies (Poland et al., 2008; Jaurand et al., 2009) in the form that demonstrates the role of aspect ratio on the initiation of inflammation during in vivo

Table 11.2: [a]Size of carbon nanotubes and their potential health effects (in vivo studies)

Aspect Ratio, d_{NT}/L_{NT}	Dose and Exposure	Resulting Effects
Between 1/8 and 1/12,500	1–7 mg/kg (1–90 days)	Inflammation
Between 1/14 and 1/35	2 mg/rat (3 and 60 days)	Inflammation (3 days) Granulomas (60 days)
About 1/20	40 g/mouse (4 h)	Inflammation
Between 1/83 and 1/1250	50 mg/mouse (24 days)	Inflammation, oxidative stress, coagulation
Between 1/121 and 1/606	50 mg/mouse (24 h, 7 days)	Inflammation
Between 1/125 and 1/250	40 mg/mouse (4 h)	Inflammation, oxidative stress, coagulation
Between 1/388 and 1/470	50 mg/mouse (24 h, 7 days)	Inflammation

[a]This table is based on the summary information presented in Lam et al. (2004) and Muller et al. (2005) without the aspect ratios.

experiments. Donaldson et al. (2010) have also emphasized the ability of the high aspect ratio particles (HARPs) to cause toxic effects [e.g., the extra-long high aspect ratio carbon nanotubes (HARCNTs) or see Fig. 11.2]. All these studies show that the long and relatively thick carbon nanotubes (among the five classes of group I), especially with aspect ratios above 100, are likely to cause inflammation and potentially toxic effects. Propensity of the long HARCNTs to split in vivo and behave similarly to microscopic asbestos fibers has been also pointed out in the extrapolation methodology based on the scaling analysis of nanotube geometry (Harik, 2005). The potential toxicity of SWCNTs was also noticed by Shvedova et al. (2003), Salvador-Morales et al. (2007), and Wan et al. (2013).

The presented developments of the structure–activity paradigm for the analysis of the potential toxicity of carbon nanotubes demonstrate that both the length of nanoscale fibers and their aspect ratio are important in their ability to cause the asbestos-like pathologies in the lungs. These geometric parameters have been also used for the classification of carbon nanotubes into different classes (Fig. 11.2), which have different mechanical properties, unique mechanical behavior, and distinct mechanical interaction mechanisms. The aforementioned inequalities show how the length of carbon nanotubes and their aspect ratio can affect the efficiency of phagocytosis, which is sensitive to the size effects described by those inequalities. It is possible to develop a parametric map, which can help to identify the range of geometric parameters associated with the different degrees of the efficiency of phagocytosis (Harik, 2014).

11.3.2 A Parametric Map for the Efficiency of Phagocytosis for Different Nanotube Geometries

The efficiency of phagocytosis of the inhaled nanoscale fibers depends on several important parameters that describe the structure of carbon nanotubes and the intensity of their interactions with biological tissues. The quantitative characteristics of the "size and shape" of carbon nanotubes have been employed in the scaling analysis, and its results have been utilized for the development of methodology for the extrapolation of the potential toxic effects associated with the inhaled SWCNTs (Harik, 2005, 2014). Dimensional analysis of the physical parameters describing the structure of carbon nanotubes can provide the critical nondimensional scaling parameters, which can help to identify the ranges of geometric parameters associated with the different degrees of the efficiency of phagocytosis (Harik, 2014). The laws of similitude (Harik, 2001) and the nondimensional scaling parameter such as the aspect ratio have been used to more rigorously demonstrate the geometric and functional similarities between asbestos fibers and carbon nanotubes, as well as their ability to cause potentially toxic effects (Harik, 2005, 2014). Experimental results shown in Table 11.2 support the use of the aspect ratio of carbon nanotubes in the structure–activity paradigm for the analysis of toxicological effects. These experimental results indicate that there is a wide range of the aspect ratios associated with the ability of carbon nanotubes to cause inflammation and harm. The laws of similitude of the dimensional analysis can help to identify the extent of the ranges of geometric parameters associated with the ability of carbon nanotubes to cause toxic effects (Harik, 2014). Such quantitative extrapolations from the scaling analysis (Harik, 2005, 2014) are supported by the toxicological studies of Stone and Donaldson (2006), Kostarelos (2008), Poland et al. (2008), Jaurand et al. (2009), Donaldson et al. (2010), and Park et al. (2014).

According to the laws of similitude (Harik, 2001, 2011) the research results on one type of phagocytosis for a group of similar carbon nanotubes extends to the entire class of carbon nanotubes (Fig. 11.2), to which that group belongs. The inhaled nanoscale particles belonging to the class of short or thin carbon nanotubes (e.g., the class III) are easily engulfed by macrophages and cleared from the lungs through the lymphatic system (Warheit et al., 2004; Murr et al., 2005). Kostarelos (2008) and Donaldson et al. (2010) have shown that the "size does matter," both for the ability and the inability of macrophages to engulf and process various nanotubes, as illustrated in Fig. 11.3. A nondimensional scaling parameter, M_{ph}, based on the ratio of the nanotube length, L_{NT}, and

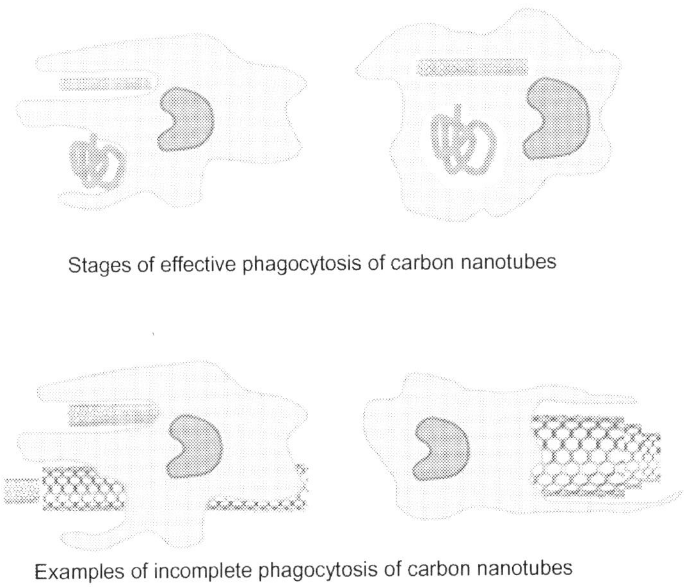

Figure 11.3
An illustration of two stages in the successful engulfment of short and the long curled carbon nanotubes in the process of an effective phagocytosis in macrophages (top) and two examples of an incomplete deficient phagocytosis of the long high aspect ratio carbon nanotubes (bottom/left) and thick multi wall carbon nanotubes (bottom/right) in macrophages (after Harik, 2014).

the average size of macrophages, L_m, i.e., $M_{ph} = L_m/L_{NT}$, can be also used to refine the boundaries of a parametric map for the evaluation of an effective phagocytosis in lungs. It was noted before that when the nondimensional number, M_{ph}, satisfies inequality: $M_{ph} \gg 1$, then the length of carbon nanotubes in a particular class is such that macrophages are able to easily clear these NP from the lungs. Therefore the probability of an effective phagocytosis is being much greater than 1. However, if carbon nanotubes have very high aspect ratios (as in the five classes of nanotubes mentioned before in Fig. 11.2), then $M_{ph} < 1$, and the probability of an effective phagocytosis is less than unity (see Fig. 11.3). It is not surprising that the region on the parametric map shown in Fig. 11.4 corresponding to the long HARCNTs shows the most ineffective phagocytosis [see Fig. 11.4 at the region, where the radius of nanotubes, R_{NT}, normalized by the size of a carbon ring, a, and the ratio $(d_{NT}/L_{NT})^{-1}$ are large]. The parametric map for the efficiency of phagocytosis shown in Fig. 11.4 represents different carbon nanotube geometries and the corresponding types of phagocytosis (Fig. 11.3). A distribution of carbon nanotube sizes or lengths can provide a distribution of probabilities for the effective phagocytosis in different parts of the lungs, depending on where different nanotubes settle according to their length, local penetrability, and transport processes. The large surface area of the engulfed carbon nanotubes needed to

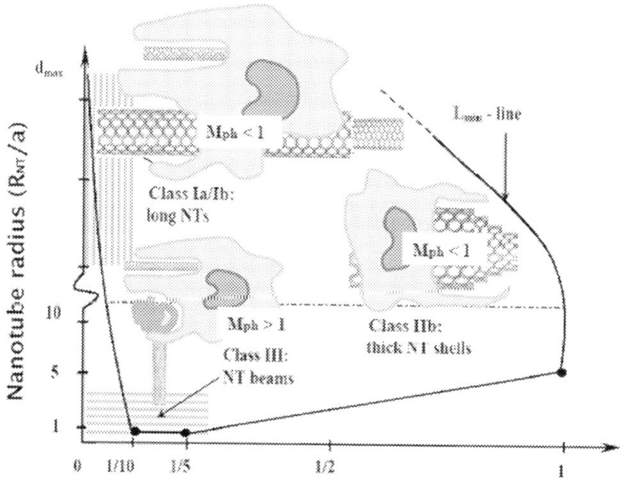

Figure 11.4

A revised parametric map of the carbon nanotube parameters showing different classes of carbon nanotubes and the corresponding type of incomplete phagocytosis (for large aspect ratios/the upper left corner, $M_{ph} < 1$, and thick MWCNTs, $M_{ph} < 1$) and effective phagocytosis (short or small nanotubes/the lower left corner, $M_{ph} > 1$). The engulfment of carbon nanotubes of distinct sizes and shapes by the macrophages is illustrated in the insets applicable within each class of carbon nanotubes (after Harik, 2014). *MWCNT*, multiwall carbon nanotube.

be covered by the macrophages in phagocytosis results in the oxidative stress and inflammation (see Table 11.1, where L_{NT} can be up to 500 μm).

The parametric map for the efficiency of phagocytosis shown in Fig. 11.4 is also supported by experimental results. For example, Donaldson et al. (2010) have found that the HARPs or the extra-long HARCNTs may cause potentially toxic effects. The region on the parametric map corresponding to these HARNTs corresponds to the most ineffective phagocytosis [see Fig. 11.4, where the ratios R_{NT}/a and $(d_{NT}/L_{NT})^{-1}$ are large]. Note that the nondimensional scaling parameter, M_{ph}, for such NP is very small, i.e., $M_{ph} \ll 1$, so the probability of the effective phagocytosis is much less than unity. These NP belong to the class I of nanotubes (Fig. 11.2). The process of phagocytosis of these and other carbon nanotubes is illustrated in Figs. 11.3 and 11.4. Yamashita et al. (2010) have shown that the long and thick MWCNTs results in more inflammation, as opposed to the short and thin MWCNTs or SWCNTs (Fig. 11.2). These results link different classes of carbon nanotubes (Fig. 11.2) with different degrees of inflammation. Further scaling analysis shows that some variations in the geometry of carbon nanotubes within one class (e.g., a class of the high aspect ratio SWCNTs) do not change the degree of inflammation.

The parametric map for the efficiency of phagocytosis (Fig. 11.4) can characterize the process of phagocytosis for other classes of carbon nanotubes as well. The thick carbon nanotubes belonging to the class II may also cause the incomplete phagocytosis and oxidative stresses (Figs. 11.3 and 11.4). In order to analyze experimental data in these cases, a scaling parameter, $(d_m/d_{NT})M_{ph}$, can be used. It is based on both the radial and longitudinal sizes of carbon nanotubes, and represents a nondimensional ratio of the surface area of macrophages and that of carbon nanotubes. Note that the process of phagocytosis satisfies the law of the conservation of mass, which is reflected in the illustrations of Fig. 11.3. The relevance of the nondimensional scaling parameter, $(d_m/d_{NT})M_{ph}$, to the process of engulfment of carbon nanotubes (Fig. 11.3) is also connected to the law of the conservation of mass, as it characterizes a relation between the merging surface areas. According to the laws of similitude (Harik, 2011) the results on the efficiency of phagocytosis for one group of similar carbon nanotubes extends to the entire class of these carbon nanotubes (Fig. 11.2).

The long and very thin carbon nanotubes are prone to the spontaneous buckling (Harik, 2011) and coiling, which may limit their linear size and result in a better phagocytosis than originally expected (see Fig. 11.4 and Fig. 11.5). In such cases, the scaling parameter, M_{ph}, should be based on the effective size of the newly formed nanotube bundles and not on their original total length. Nanotube bundles may deform in vivo as long nanotubes have to adjust, and often buckle under the ambient pressure and surface interactions, which depend on the type of biological cells and tissues being present. The value of the scaling parameter, M_{ph}, is also indicative of the rate of phagocytosis and the rate of engulfing carbon

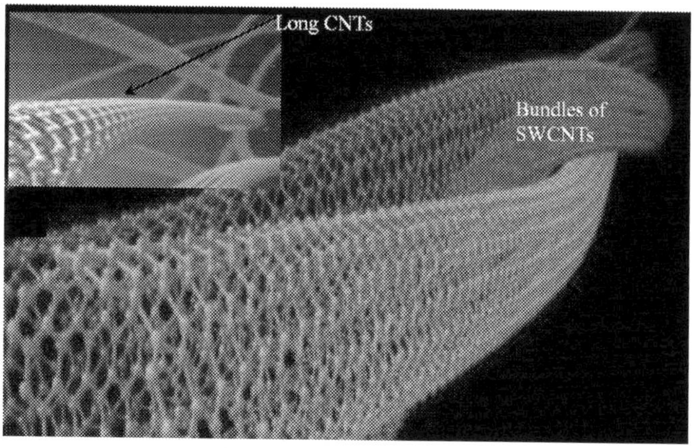

Figure 11.5
A bundle of long carbon nanotubes and a loose bundle of carbon nanotubes (the inset).

nanotubes depending on their size, i.e., the smaller the parameter, M_{ph}, the lower rate of the process of engulfing and, hence, the lower rater of phagocytosis. It is possible to show that $M_{ph} \approx 1$, for $L_{NT} \approx 10–20\ \mu m$, by using data from Donaldson et al. (2010). The size and the vitality of the local cells in the lung tissue directly affect the rate of phagocytosis. Moreover, in different parts of the lungs with distinct nanotube sizes being settled there the rate of phagocytosis will be different. Therefore the map of the values of parameter M_{ph} is useful for representing variations of the rate of phagocytosis across the lungs with different regions of biological tissues and varying sizes of cells and the carbon nanotubes inhaled. Another parametric map can be constructed for different classes of carbon nanotubes and their penetration into deep lung with the critical nanotube length, $L_{NT} \approx 3\ \mu m$, which allows such penetration. The scaling analysis also suggests that the normalized volume of inflammation of biological tissues, V_i, and the degree of inflammation, α_i, are proportional functions, f and ζ, say, of the aspect ratio, L_{NT}/d_{NT}, i.e.,

$$V_i = f\left(\frac{L_{NT}}{d_{NT}}\right), \quad \text{for} \quad M_{ph} < 4,$$

and

$$\alpha_i = \zeta\left(\frac{L_{NT}}{d_{NT}}\right), \quad \text{for} \quad M_{ph} < 4.$$

These expressions can be used for the extrapolation of experimental data. The size of macrophages is very critical for the evaluation of the efficiency of phagocytosis, therefore the inequality: $M_{ph} < 1$ is very important (and quite strong) in determining the probability of totally deficient phagocytosis.

11.3.3 Bundles of Carbon Nanotubes and Safety

In some experimental studies (Warheit et al., 2004), entanglement of carbon nanotubes has resulted in large clusters/clumps and a blockage of airways and intraairway fibrosis. Inhalation of small clumps or bundles may also lead to problems for normal lung defenses, including the possibility of increasing surface area during separation of single nanotubes or smaller bundles by the action of lung surfactants or cellular mechanisms, which is also due to the spontaneous Euler buckling (Harik, 2011). Groups of carbon nanotubes belonging to one class or distinct classes can form bundles (Fig. 11.5) which can be inhaled and transported to the lungs. Distribution of carbon nanotube sizes also may result in the formation of different bundles with varying shapes and sizes, which can be inhaled and undergo different processes of phagocytosis when separated. If a bundle is not separated (see the inset in Fig. 11.5), then different parts of a bundle may take part in distinct types of phagocytosis. However, the largest portion of the bundle determines the final result of phagocytosis. If the long HARCNTs are surrounded by short and thin nanotubes, then the ineffective phagocytosis will become even more difficult. This is a consequence of the law for the conservation of mass of the biological fluid in macrophages during phagocytosis.

The efficiency of phagocytosis for the carbon nanotube bundles can be also analyzed by using the aforementioned nondimensional scaling parameter, M_{ph}, based on the ratio of the largest nanotube length, L_{NT}, in a bundle, and the average size of macrophages, L_m, i.e., $M_{\text{ph}} = L_m/L_{\text{NT}}$. As was noted before, the nondimensional number, M_{ph}, should satisfy inequality: $M_{\text{ph}} \gg 1$, in order for the macrophages to easily clear such nanoscale bundles from the lungs. Volumetric effects of the carbon nanotube bundles can be analyzed by the aforementioned inequalities for the efficiency of phagocytosis:

$$\left(\frac{d_{\text{NT}}^2}{V_{\text{NT}}}\right)_{\text{bundle}} \gg \frac{d_m^2}{V_m} \quad \text{or} \quad \frac{V_m}{V_{\text{NT}_{\text{bundle}}}} \gg \frac{d_m^2}{d_{\text{NT}_{\text{bundle}}}^2},$$

where d_{NT} is the average diameter of the carbon nanotube bundle, V_{NT} is the volume of the carbon nanotube bundle, and d_m is an average size of the cross-sectional size of the elongated macrophages as before. These inequalities demonstrate the importance of the volume of macrophages with respect to the volume of carbon nanotube bundles, in relation to the ratio of the cross-sectional areas of macrophages and carbon nanotube bundles. In the case of MWCNTs (Fig. 11.6) and the bundles of MWCNTs, the inequalities shown above indicate that the larger volume of these carbon nanotubes will result in greater inflammation and significant inefficiency of phagocytosis. The aforementioned inequalities provide additional mathematical framework for the structure–activity paradigm in the analysis of toxicological studies of carbon nanotube bundles.

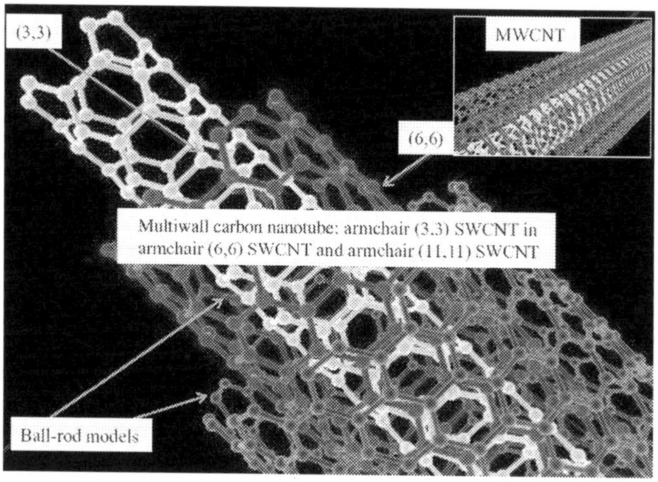

Figure 11.6
An example of the atomic structure of MWCNTs with small diameters and a few lattice shells. *MWCNT*, multiwall carbon nanotube (after Gollonet).

11.4 Conclusions

This chapter has presented a review of toxicological studies of potential toxic effects of carbon nanotubes and the development of quantitative approaches for the structure–activity paradigm in nanotoxicology. A methodology based on the structure–activity paradigm, the scaling analysis of size effects, and the classification of carbon nanotubes have been used in the analysis of toxicological studies of carbon nanotubes. Similarities between carbon nanotubes and asbestos fibers have been reviewed (Cook, 2004; Harik, 2005, 2014; Donaldson et al., 2010, 2013). It has been noted that more than a decade ago the nanotube–asbestos analogy was represented merely by Cook's hypothesis about their geometric similarities (Cook, 2004). Now Cook's hypothesis about similarity of asbestos fibers and carbon nanotubes has been proven both theoretically by the scaling analysis and by the laws of similitude of the dimensional analysis (Harik, 2005, 2014) and numerous experimental studies (e.g., Poland et al., 2008; Donaldson et al., 2010, 2013). The nanotube–asbestos analogy is now a well-established scientific fact, which is also reflected in the law.

In this chapter, the potential toxic effects of different classes of carbon nanotubes have been characterized by their normalized length, their aspect ratio, and other parameters related to their inhalability, engulfment by macrophages, and the effectiveness of phagocytosis. Several geometric scaling parameters and the classification of carbon nanotubes have been used to develop an updated parametric map for the extrapolation of the potential toxic effects resulting from the inhalation of long and short carbon nanotubes. An updated parametric map has been applied to the evaluation of the efficiency of phagocytosis involving distinct geometric classes of carbon nanotubes. An approximate critical value of an important nondimensional parameter that characterizes the efficiency of phagocytosis for different nanotubes is presented. The present evaluation of the potential toxicological effects of the HARCNTs is found to be in the agreement with other available studies and earlier scaling analyses.

References

British Royal Society, Royal Academy of Engineering, 2004. Nanoscience and Nanotechnologies: Opportunities and Uncertainties, Joint Report, the Royal Academy of Engineering and the British Royal Society, London, United Kingdom, August 2004.

Cook, P.M., 2004. Asbestos and related durable fibers: too ubiquitous, too persistent, too complex to put health risks to rest? Mid-Continent Ecology Division, National Health and Environmental Effects Research Laboratory, Duluth, MN.

Donaldson, K., Stone, V., Tran, C.L., Kreyling, W., Borm, P.J.A., 2004. Nanotoxicology. Occup. Environ. Med. 61, 727–728.

Donaldson, K., Aitken, R., Tran, L., Stone, V., Duffin, R., Forrest, G., et al., 2006. Carbon nanotubes: a review of their properties in relation to pulmonary toxicology and workplace safety. Toxicol. Sci. 92, 5–22.

Donaldson, K., Murphy, F.A., Duffin, R., Poland, C.A., 2010. Asbestos, carbon nanotubes and the pleural mesothelium: a review of the hypothesis regarding the role of long fibre retention in the parietal pleura, inflammation and mesothelioma. Part. Fibre Toxicol. 7 (5), 1–17.

Donaldson, K., Poland, C.A., Murphy, F.A., MacFarlane, M., Chernova, T., Schinwald, A., 2013. Pulmonary toxicity of carbon nanotubes and asbestos—similarities and differences. Adv. Drug Delivery Rev. 65, 2078–2086.

Ferin, J., Oberdorster, G., Penney, D.P., Soderholm, S.C., Gelein, R., Piper, H.C., 1990. Increased pulmonary toxicity of ultrafine particles 1. Particle clearance, translocation, morphology. J. Aerosol Sci. 21, 381–384.

Harik, V.M., 2001. Ranges of applicability of the continuum beam model in the mechanics of carbon nanotubes. Solid State Commun. 120, 331–335.

Harik, V.M., 2005. Nanotechnology of carbon nanotubes and safety issues. In: Proceedings of the 20th Technical Conference of American Society for Composites, 7–8 September 2005, Philadelphia, PA. pp. 44.

Harik, V.M., 2011. Mechanics of Carbon Nanotubes. Nanodesigns Press, Nanodesigns Consulting, Newark, DE.

Harik, V.M., 2014. Carbon nanotubes and safety. In: Harik, V. (Ed.), Trends in Nanoscale Mechanics. Springer, Dordrecht.

Harik, V.M., 2017. Geometry of carbon nanotubes and mechanisms of phagocytosis and toxic effects. Toxicol. Lett. 273, 69–85.

Hill, I.M., Beswick, P.H., Donaldson, K., 1995. Differential release of superoxide anions by macrophages treated with long and short fibre amosite asbestos is a consequence of differential affinity for opsonin. Occup. Environ. Med. 52, 92–96.

Jaurand, M.-C.F., Renier, A., Daubriac, J., 2009. Mesothelioma: do asbestos and carbon nanotubes pose the same health risk? Part. Fibre Toxicol. 6 (16), 1–14.

Kane, A.B., Macdonald, J.L., Moalli, P.A., 1986. Acute injury and regeneration of mesothelial cells produced by crocidolite asbestos fibers. Am. Rev. Respir. Dis. 133, A198.

Kostarelos, K., 2008. The long and short of carbon nanotube toxicity. Nat. Biotechnol. 28 (7), 774–776.

Lam, C.-W., James, J.T., McCluskey, R., Hunter, R.L., 2004. Pulmonary toxicity of single-wall carbon nanotubes in mice 7 and 90 days after intratracheal instilation. Toxicol. Sci. 77, 126–134.

Maynard, A.D., David, B., Warheit, D.B., Philbert, M.A., 2011. New toxicology of sophisticated materials: nanotoxicology and beyond. Toxicol. Sci. 120 (S1), S109–S129.

Moalli, P.A., Macdonald, J.L., Goodglick, L.A., Kane, A.B., 1987. Acute injury and regeneration of the mesothelium in response to asbestos fibers. Am. J. Pathol. 128, 426–445.

Muller, J., Huaux, F., Moreau, N., Misson, P., Heilier, J.F., Delos, M., et al., 2005. Respiratory toxicity of multi-wall carbon nanotubes. Toxicol. Appl. Pharmacol. 207 (3), 221–231.

Murr, L.E., Garza, K.M., Soto, K.F., Carrasco, A., Powell, T.G., Ramirez, D.A., et al., 2005. Cytotoxicity assessment of some carbon nanotubes and related carbon nanoparticle aggregates and the implications for anthropogenic carbon nanotube aggregates in the environment. Int. J. Environ. Res. Public Health 2 (1), 31–42.

Murphy, F., Chernova, T., Schinwald, A., Sun, X.M., et al., 2014. Long-fibre carbon nanotube- and asbestos-induced pleural lesions exhibit a common molecular signature. Toxicol. Lett. 229, S196.

Oberdorster, G., 2000. Toxicology of ultrafine particles: *in vivo* studies. Philos. Trans. R. Soc. A: Math. Phys. Eng. Sci. 358, 2719–2740.

Oberdorster, G., Oberdorster, E., Oberdorster, J., 2005. Nanotoxicology: an emerging discipline evolving from studies of ultrafine particles. Environ. Health Perspect. 113, 823–839.

Oberdorster, G., Stone, V., Donaldson, K., 2007. Toxicology of nanoparticles: a historical perspective. Nanotoxicology 1, 2–25.

Park, E.-J., Zaharia, N.E.M., Kang, M.-S., et al., 2014. Toxic response of HIPCO single-walled carbon nanotubes in mice and RAW264.7 macrophage cells. Toxicol. Lett. 229 (1), 167–177.

Poland, C.A., Duffin, R., Kinloch, I., Maynard, A., Wallace, W.A., Seaton, A., et al., 2008. Carbon nanotubes introduced into the abdominal cavity of mice show asbestos like pathogenicity in a pilot study. Nat. Nanotechnol. 3, 423–428.

Salvador-Morales, C., Townsend, P., Flahaut, E., Vénien-Bryan, C., Vlandas, A., Green, M.L.H., et al., 2007. Binding of pulmonary surfactant proteins to carbon nanotubes; potential for damage to lung immune defense mechanisms. Carbon 45, 607–617.

Shvedova, A.A., Castranova, V., Kisin, E.R., Murray, A.R., Schwegler-Berry, D., Gandelsman, V.Z., et al., 2003. Exposure to nanotube material: assessment of nanotube cytotoxicity using human keratinocyte cells. J. Toxicol. Environ. Health 66, 1901–1918.

Song, Y., Li, X., Du, X., 2009. Exposure to nanoparticles is related to pleural effusion, pulmonary fibrosis and granuloma. Eur. Respir. J. 34, 559–567.

Stone, V., Donaldson, K., 2006. Signs of stress. Nature 1, 23–24.

Viallat, J.R., Raybuad, F., Passarel, M., Boutin, C., 1986. Pleural migration of chrysotile fibers after intratracheal injection in rats. Arch. Environ. Health 41, 282–286.

Wan, B., Wang, Z.-X., Lv, Q.-Y., Dong, P.-X., et al., 2013. Single-walled carbon nanotubes and graphene oxides induce autophagosome accumulation and lysosome impairment in primarily cultured murine peritoneal macrophages. Toxicol. Lett. 221 (2), 118–127.

Warheit, D.B., Laurence, B.R., Reed, K.L., Roach, D.H., Reynolds, G.A.M., Webb, T.R., 2004. Comparative pulmonary toxicity assessment of single-wall carbon nanotubes in rats. Toxicol. Sci. 77, 117–125.

Yamashita, Y., Yoshioka, Y., Higashisaka, K., et al., 2010. Carbon nanotubes elicit DNA damage and inflammatory response relative to their size and shape. Inflammation 33 (4), 276–280.

CHAPTER 12

Trends in Recent Publications on Mechanics of Carbon Nanotubes

Chapter Outline

12.1 A Review of Recent Publications on Mechanics of Carbon Nanotubes

Recent publications on the mechanics of carbon nanotubes describe molecular dynamics studies (Ajori et al., 2018; Ansari et al., 2015; Eberhardt and Wallmersperger, 2015; Kinoshita et al., 2013; Chowdhury et al., 2012), the buckling behavior of carbon nanotubes (She et al., 2017; Benguediab et al., 2014; Zhao et al., 2011), properties of the boron nitride nanotubes (Li et al., 2017; Zhang and Peng, 2017; Mercan and Civalek, 2016), the relationship between local buckling and atomic elastic stiffness (Nishimura et al., 2017), size dependence (Akgoz and Civalek, 2017), the role of defects (Badjian and Setoodeh, 2017), and the enhanced $\pi-\pi$ interactions (Gorantla et al., 2010) with carbon nanotubes. Publications on the axial buckling of carbon nanotubes have been presented in the appropriate section below, after the section on material properties of carbon nanotubes. Talukdar et al. (2011) have examined the dependence of mechanical characteristics and the fracture and buckling behavior of single-walled carbon nanotubes on their geometry. It has

Mechanics of Carbon Nanotubes.
DOI: https://doi.org/10.1016/B978-0-12-811071-3.00012-3

© 2018 Elsevier Inc. All rights reserved.

been shown that geometry of carbon nanotubes can affect mechanisms of phagocytosis and their toxicity (Harik, 2017) and their intracellular trafficking (Mazzaglia et al., 2018).

Two sections devoted to the publications on various shell models and molecular dynamics simulations have been included below. A section on the latest applications of carbon nanotubes is presented with the publications on nanotechnology applications, sensors, and the electromechanical coupling in nanocomposites. A large section on the carbon nanotube-reinforced nanocomposites (Section 12.2.1) has been presented with a long list of publications on the latest research in this area of nanoscale mechanics. Special attention is given to the interfacial interactions in nanocomposites (Section 12.2.1.1) and the studies on the carbon nanotube pullout (Section 12.2.1.2) from the polymer matrix of nanocomposites.

12.1.1 Publications on Material Properties of Carbon Nanotubes

Material properties of carbon nanotubes have been investigated by Svatos et al. (2018), Thamaraikannan et al. (2017), Shahini et al. (2017), Bagwell et al. (2015), Dilrukshi et al. (2015), Fang et al. (2015), Milowska and Majewski (2013), and others as indicated in the list of references on this topic. An elastica model is shown to predict radial corrugations in a double-walled carbon nanotube (Bagwell et al., 2015). Dilrukshi et al. (2015) have examined the size dependency and the influence of potential field on deriving mechanical properties of carbon nanotubes by using molecular dynamics. Cheng et al. (2009) and Chandraseker and Mukherjee (2007) have presented the atomistic-continuum and ab initio evaluations of the elastic moduli and behavior of single-walled carbon nanotubes. Li et al. (2017) have compared the mechanical and thermal properties of single-walled carbon nanotubes and boron nitride nanotubes. Mechanical characteristics and the fracture and buckling behavior of single-walled carbon nanotubes can depend on their geometry, as demonstrated by Talukdar et al. (2011), which has been also shown in earlier chapters of this book.

12.1.2 Publications of Axial Buckling of Carbon Nanotubes

Axial buckling of carbon nanotubes has been examined by atomistic finite element method and molecular dynamics simulations (Ansari et al., 2010, 2014; Safaei et al., 2016), as well as by other methods as indicated in the section with references for this topic. Atomistic and continuum models for carbon nanotubes have been presented by Singh and Patel (2018), Chowdhury et al. (2014), Ansari et al. (2014), Hollerer (2014), Rouhi et al. (2014), Wang et al. (2014), Cheng et al. (2009), and Chandraseker and Mukherjee (2007). Development of an accurate molecular mechanics model for the buckling behavior of multiwalled carbon nanotubes under axial compression has been presented by Safaei et al. (2016) and Wang et al (2014). The effect of polyethylene molecules wrapped around carbon nanotubes on their buckling has been examined by Wang et al. (2011).

12.1.3 Publications on Shell Models for Carbon Nanotubes

A continuum shell model for the buckling of armchair carbon nanotubes under compression or torsion has been presented by Chowdhury et al. (2014). Eftekhari et al. (2013) have examined the effect of defects on the local shell buckling and post-buckling behavior of single- and multiwalled carbon nanotubes. A continuum shell model has been also used by Korayem et al. (2012) in order to investigate the buckling behavior of short multiwalled carbon nanotubes under axial compression. Coordinated buckling of the thick multiwalled carbon nanotubes under uniaxial compression has been considered by Huang et al. (2010). Kulathunga et al. (2009) have also analyzed the accurate modeling of buckling of single- and double-walled carbon nanotubes based on shell theories.

Peng et al. (2008) have tried to answer the question: "Can a single-wall carbon nanotube be modeled as a thin shell?" An original investigation of this question has been presented in a NASA report (and a related AIAA paper) on the limitations and the ranges of applicability of the continuum shell theories to the mechanics of carbon nanotubes by Harik (2017). Zhang et al. (2009) have also presented an assessment of the continuum mechanics models for the prediction of the buckling strains of single-walled carbon nanotubes. Motevalli et al. (2013) have carried out a comparison of the continuum-based and the atomistic-based modeling of axial buckling of carbon nanotubes subject to hydrostatic pressure.

12.1.4 Molecular Dynamics Simulations of Carbon Nanotubes

Ajori et al. (2018) have presented a molecular dynamics study of the buckling behavior of the crosslinked functionalized carbon nanotubes under physical adsorption of polymer chains. Elastic properties and buckling behavior of the single-walled carbon nanotubes functionalized with diethyltoluenediamines and under chemical adsorption of atomic oxygen and hydroxyl have been also investigated by Ansari et al. (2015) by using molecular dynamics simulations, as well as Ansari et al. (2014). Ansari et al. (2014) have also developed a hybrid continuum and molecular mechanics model for the axial buckling of chiral single-walled carbon nanotubes. Kinoshita et al. (2013) and Cao and Chen (2006) have examined the buckling of single-walled carbon nanotubes with molecular dynamics and finite element simulations.

Molecular dynamics simulations of pristine and defective carbon nanotubes under monotonic and combined loading have been conducted by Chowdhury et al. (2012), and Hao et al. (2008) for the defective single-walled and double-walled carbon nanotubes under axial compression. The buckling of defective carbon nanotubes has been also analyzed by molecular dynamics method by Kulathunga et al. (2010). Eberhardt and Wallmersperger (2015) have presented the energy-consistent modified molecular structural mechanics model for the determination of the elastic properties of single wall carbon nanotubes. Heo and Sinnott (2007) have carried out an investigation of the influence of thermostat configurations on the mechanical properties of carbon nanotubes in molecular dynamics simulations. Huang et al. (2011) have investigated the temperature-dependent thermal conductivity of the bent carbon nanotubes by molecular dynamics simulations.

12.1.5 Publications on Applications of Carbon Nanotubes

Applications of carbon nanotubes are reviewed in the section on publications for this topic. For example, Li et al. (2018) have introduced a supercapacitor based on the porous carbon nanotube/graphene composites. Ahmed et al. (2017), Goh and Chen (2017), and Abbaspour et al. (2016) have considered transport of a fluid and the Lennard-Jones fluid in carbon nanotubes, respectively. Xu et al. (2018) have presented a review of the functionalized carbon nanotubes and graphene being used for heavy metal adsorption from water. Boztepe et al. (2018) have developed a novel interlaminar sensor based on carbon nanotubes for the carbon fiber-reinforced composites. Grabowski et al. (2017) have investigated the multiscale electromechanical coupling and conductivity (Han et al., 2016) in the carbon nanotube-reinforced nanocomposites. Electronic transport through carbon nanotubes has been examined by Pantano (2017) and Ribeiro et al. (2017); He et al. (2017) have proposed the biocompatible carbon nanotube fibers for implantable supercapacitors. Tachizaki et al. (2018) have presented the nanometer optical length measurements by using the near-field scanning optical microscopy with the sharpened single carbon nanotube probe.

12.2 Publications on the Carbon Nanotube-Reinforced Nanocomposites

The carbon nanotube-reinforced nanocomposites have been studied by many researchers (e.g., Shen et al., 2018; Sharma et al., 2015) as demonstrated by the long list of publications presented below. Buckling (Thang et al., 2017) and vibration of the constrained annular sector plates made of the functionally graded carbon nanotube-reinforced composites under thermal loading have been analyzed by Ansari et al. (2017), and under compression and shear by Garcia-Macias et al. (2017). Static analysis of the functionally graded carbon nanotube-reinforced plate and shell structures has been presented by Zghal et al. (2017). Buckling of hybrid nanocomposites with embedded graphene and carbon nanotubes has been also investigated by Chandra et al. (2016). Ashrafi et al. (2017) have considered the multifunctional fiber-reinforced polymer composites, which include carbon and boron nitride nanotubes. Dynamic properties of hybrid composite structures based on multiwalled carbon nanotubes have been investigated by Benyahia et al. (2017). Slobodian et al. (2018)

have shown that carbon nanotubes can be used in the glass fiber/epoxy composites for the integrated layer of carbon nanotubes for deformation detection.

Strengthening behavior of carbon nanotube–graphene hybrids in copper matrix composites has been examined by Chen et al. (2018). Durability of the multiwalled carbon nanotube-reinforced concrete has been investigated by Carrico et al. (2018). Enhancing mode-I and mode-II fracture toughness of epoxy and carbon fiber-reinforced epoxy composites can be done by using multiwalled carbon nanotubes (Quan et al., 2018). Microscopic mechanisms of reinforcement in the single-wall carbon nanotube/polypropylene nanocomposite have been studied by Chang et al. (2005); Hu et al. (2017) have presented a study of the ultrastrong and excellent dynamic mechanical properties of carbon nanotube composites. An investigation of elastic, electrical, and electromechanical properties of polyurethane/grafted carbon nanotubes nanocomposites has been conducted by Jomaa et al. (2015). An evaluation of elastic properties of the multiwalled carbon nanotube-reinforced composites has been presented by Joshi and Upadhyay (2014); Kumar and Srinivas (2017) have examined the vibration, buckling, and bending behavior of the functionally graded multiwalled carbon nanotube-reinforced polymer composite plates using the layer-wise formulation.

Carbon nanotubes and silicon carbide particles may have an effect on the ablative properties of carbon fiber phenolic matrix composites (Saghar et al., 2018). Diameter-dependent elastic properties of carbon nanotube-polymer composites can display the emergence of size effects, as demonstrated by atomistic-scale simulations (Malagu et al., 2017). Another aspect of size effects has been investigated by a coarse-grained model for the elastic properties of the crosslinked short carbon nanotube/polymer composites (Mousavi et al., 2016). Peng and Meguid (2017) have presented molecular dynamics simulations of the buckling behavior of the defective carbon nanotubes embedded in epoxy nanocomposites. Tsai and Jeng (2013) have conducted experimental and numerical investigations into the effect of carbon nanotube buckling on the reinforcement of the CNT/Cu composites.

Rahmat and Hubert (2011) have investigated the carbon nanotube–polymer interactions in nanocomposites, while Seidel and Lagoudas (2006) have conducted micromechanical analysis of the effective elastic properties of carbon nanotube-reinforced composites. Morphology of multiwall carbon nanotube may have an effect on the electrical and mechanical properties of polyurethane nanocomposites (Sethi et al., 2017). Carbon nanotubes also influence the electrical conduction in the aluminum nitride-single-walled carbon nanotube nanocomposites (Singh et al., 2018), which can be used for the detection of damage. Tarfaoui et al. (2017) have presented progressive damage modeling of the carbon fibers/carbon nanotubes-reinforced polymer composites. A review of the analytical, numerical, and experimental approaches in the analysis of carbon nanotubes as reinforcement in composites has been presented by Yengejeh et al. (2017).

12.2.1 Publications on Interfacial Interactions in Nanocomposites

Interfacial properties of carbon nanotube-reinforced nanocomposites and in the multiwall carbon nanotubes (Li et al., 2010) have been often examined by molecular dynamics simulations (Goclon et al., 2018; Duan et al., 2017; Gao and Muller-Plathe, 2017; Zhang et al., 2017; Zhang and Peng, 2017; Lu et al., 2016; Minoia et al., 2012) and other methods (Kulathunga and Ang, 2014). Interfacial interactions in the carbon fiber-reinforced PEEK composites can be enhanced by regulating the graphene oxide complex sizing and Ni-coating at the interface (Chen et al., 2018; Duan et al., 2017). Quantitative determination of the size and properties of interphase in carbon nanotube-based multiscale composites has been presented by Bedi et al. (2018). Interfaces of carbon nanotubes can be modified by amphiphilic glycosylated proteins and other additives (Fang et al., 2018). Morphology, crystallization, and rigid amorphous fraction of polymer adsorbed into carbon nanotubes and graphite has been investigated by Klonos et al. (2018).

Polymer matrix and the fiber/matrix interface can be optimized to achieve combination of strength, ductility and toughness in carbon nanotube-reinforced carbon/carbon composites (Feng et al., 2017). Interface tailoring can be also used in order to enhance mechanical properties of carbon nanotube-reinforced composites (Nai et al., 2014). The covalently bonded carbon nanotubes can yield the highly deformable thermal interface materials (Wang et al., 2016). Joining interfaces between the carbon/carbon composites can be improved by the radially aligned carbon nanotube-reinforced pyrocarbon bonding interlayer (Feng et al., 2017). Functional groups may also have an effect on the properties of the interfaces in the multiwalled carbon nanotubes/polyvinylidenefluoride composite membranes (Fontananova et al., 2017). Si et al. (2018) have introduced the mussel-inspired polymerized molecules for facile functionalization of carbon nanotubes in order to improve their interfacial interactions. Functionalization of carbon nanotubes by combining the mussel-inspired chemistry and RAFT polymerization can also provide nanocomposites with high dielectric parameters and improved thermal management capability (Xie et al., 2018). Interfacial adhesion in nanocomposites and the carbon fiber-reinforced composites can be improved by the carbon nanotubes grafted on carbon fibers (Lavagna et al., 2017). Yao et al. (2018) have conducted a comparison of carbon nanotubes and graphene oxide coated carbon fibers for improving the interfacial properties of the carbon fiber/epoxy composites. Guignier et al. (2016) have investigated the tribological behavior and adhesion of carbon nanotubes grafted on carbon fibers. Characterization of the microscale damage may be required for the carbon nanotube-grafted carbon fiber polymer composites (Zhang et al., 2017). Zhao et al. (2017) have evaluated the mechanical, thermal, and interfacial performances of the carbon fiber-reinforced composites modified by carbon nanotubes along the fiber–matrix interface. Carbon nanotubes can be also used to modify an epoxy adhesive for the carbon fiber reinforced polymer (CFRP)-to-steel interface (Korayem et al., 2015), as well as other interfaces. Multiwalled carbon nanotubes have been used for reinforcement of the mortar aggregate in order to improve its interfacial properties (Song et al., 2017).

Interfacial bonding characteristics of nanotube/polymer composites (Lau, 2003) can be improved by the atomic-level mechanics of the adhesive forces between aromatic molecules and carbon nanotubes (Lechner and Sax, 2017; Liu et al., 2016). Multiscale modeling can be used to improve the crosslinked interfaces in the carbon nanotube-reinforced nanocomposites (Ozden-Yenigun et al., 2017). Rouhi et al. (2014) have demonstrated how to characterize the interface between the polyethylene and the single-walled carbon nanotubes using molecular dynamics simulations. A molecular dynamics approach can be utilized to evaluate the role of carbon nanotube diameter on thermal interfacial resistance through vibrational mismatch analysis (Sarode et al., 2017) and interfacial characteristics of the carbon nanotube-polyethylene nanocomposites (Zhang et al., 2011; Zheng et al., 2009). Atomistic simulations of surface functionalization can be employed to investigate the interfacial properties of the graphene–polymer nanocomposites (Wang et al., 2014). Determination of interfacial properties of the carbon nanotube-reinforced polymer composites can be also done by using the atomistic-based continuum model (Wernik et al., 2012). Adhesion between two carbon nanotubes can be also investigated by molecular dynamics simulations and continuum mechanics (Yuan et al., 2018).

12.2.2 Publications on the Carbon Nanotube Pullout from Nanocomposites

Quantitative determination of size and material properties of the interphase in the carbon nanotube-based nanocomposites is important for improving their load-carrying capabilities (Bedi et al., 2018). Molecular dynamics simulations of the carbon nanotube pullout from polyethylene matrix (Chawla and Sharma, 2017) and other polymers (Yang et al., 2014; Chowdhury and Okabe, 2007) can characterize the interfacial properties in nanocomposites (Li et al., 2011). Theoretical analysis of the pullout behavior of a carbon nanotube in cryogenic environment can characterize the interfacial properties in nanocomposites with the consideration of thermal residual stress (Ma et al., 2017). Pullout simulations can also characterize interfacial properties of the amine functionalized multiwalled carbon nanotube epoxy composites (Sharma et al., 2015), as well as toughness of nanocomposites (Wagner et al., 2013). Helical fiber pullout can be also used for biological materials (Wang et al., 2016). Prediction of the pullout force for the internal walls in the multiwalled carbon nanotubes in sword-in-sheath mode can be also obtained by utilizing the pullout methodology, as well as the direct measurements of the mechanical strength of the carbon nanotube–aluminum interfaces (Yi et al., 2017).

12.3 Conclusions

In this chapter, a short review of recent publications on mechanics of carbon nanotubes is presented, along with a few selections of papers on different topics in mechanics of carbon nanotubes. Publications on axial buckling of carbon nanotubes, shell models, and the carbon nanotube-reinforced nanocomposites are included, along with those on molecular dynamics simulations, interfacial interactions, and the carbon nanotube pullout from polymer matrices.

References

Recent publications on mechanics of carbon nanotubes

Akgoz, B., Civalek, O., 2017. A size-dependent beam model for stability of axially loaded carbon nanotubes surrounded by Pasternak elastic foundation. Compos. Struct. 176 (15), 1028–1038.

Badjian, H., Setoodeh, A.R., 2017. Improved tensile and buckling behavior of defected carbon nanotubes utilizing boron nitride coating—a molecular dynamic study. Physica B: Condens. Matter 507, 156–163.

Gorantla, S., Avdoshenko, S., Börrnert, F., Bachmatiuk, A., Dimitrakopoulou, M., Schäffel, F., et al., 2010. Enhanced $\pi-\pi$ interactions between a C60 fullerene and a buckle bend on a double-walled carbon nanotube. Nano Res. 3, 92–97.

Harik, V.M., 2017. Geometry of carbon nanotubes and mechanisms of phagocytosis and toxic effects. Toxicol. Lett. 273, 69–85.

Mazzaglia, A., Scala, A., Sortino, G., Zagami, R., Piperno, A., 2018. Intracellular trafficking and therapeutic outcome of multiwalled carbon nanotubes modified with cyclodextrins and polyethylenimine. Colloids Surf. B: Biointerfaces 163, 55–63.

Mercan, K., Civalek, O., 2016. DSC method for buckling analysis of boron nitride nanotube (BNNT) surrounded by an elastic matrix. Compos. Struct. 143, 300–309.

Singh, S., Patel, B.P., 2018. Large deformation static and dynamic response of carbon nanotubes by mixed atomistic and continuum models. Int. J. Mech. Sci. 135, 565–581.

She, G.-L., Yuan, F.-G., Ren, Y.-R., Wan-Shen Xiao, W.-S., 2017. On buckling and postbuckling behavior of nanotubes. Int. J. Eng. Sci. 121, 130–142.

Zhao, J., He, M.-R., Dai, S., Huang, J.-Q., Zhu, J., 2011. TEM observations of buckling and fracture modes for compressed thick multiwall carbon nanotubes. Carbon 49 (1), 206–213.

Publications on material properties of carbon nanotubes

Bagwell, R., Leta, J., Golovaty, D., Wilber, J.P., 2015. An elastica model that predicts radial corrugations in a double-walled carbon nanotube. Eur. J. Mech.—A/Solids 54, 74–83.

Benguediab, S., Tounsi, A., Zidour, M., Abdelwahed Semmah, A., 2014. Chirality and scale effects on mechanical buckling properties of zigzag double-walled carbon nanotubes. Compos. B: Eng. 57, 21–24.

Chandraseker, K., Mukherjee, S., 2007. Atomistic-continuum and ab initio estimation of the elastic moduli of single-walled carbon nanotubes. Comput. Mater. Sci. 40, 147–158.

Cheng, H.C., Liu, Y.L., Hsu, Y.C., Chen, W.H., 2009. Atomistic-continuum modeling for mechanical properties of single-walled carbon nanotubes. Int. J. Solids Struct. 46, 1695–1704.

Dilrukshi, K.G.S., Dewapriya, M.A.N., Puswewala, U.G.A., 2015. Size dependency and potential field influence on deriving mechanical properties of carbon nanotubes using molecular dynamics. Theor. Appl. Mech. Lett. 5 (4), 167–172.

Fang, T.-H., Chang, W.-J., Yu-Lun Feng, Y.-L., 2015. Mechanical characteristics of graphene nanoribbons encapsulated in single-walled carbon nanotubes using molecular dynamics simulations. Appl. Surf. Sci. 356, 221–225.

Li, T., Tang, Z., Huang, Z., Yu, J., 2017. A comparison between the mechanical and thermal properties of single-walled carbon nanotubes and boron nitride nanotubes. Physica E: Low-dimension. Syst. Nanostruct. 85, 137–142.

Milowska, K.Z., Majewski, J.A., 2013. Elastic properties of functionalized carbon nanotubes. Phys. Chem. Chem. Phys. 15, 14303–14309.

Shahini, E., Karimi Taheri, K., Karimi Taheri, A., 2017. An investigation on tensile properties of coiled carbon nanotubes using molecular dynamics simulation. Diamond Relat. Mater. 74, 154–163.

Svatos, V., Sun, W., Kalousek, R., Gablech, I., Neuzil, P., 2018. Single measurement determination of mechanical, electrical, and surface properties of a single carbon nanotube via force microscopy. Sens. Actuators A: Phys. 271, 217–222.

Talukdar, K., Agrawala, R., Mitra, A.K., 2011. Dependence of mechanical characteristics and the fracture and buckling behavior of single-walled carbon nanotubes on their geometry. New Carbon Mater. 26 (6), 408–416.

Thamaraikannan, S., Pradhan, S.C., Sunny, M.R., 2017. Parametric study on topology of carbon nanotubes effects on mechanical properties. Mater. Today: Proc. 4 (8), 9117–9125.

Publications of axial buckling of carbon nanotubes

Ansari, R., Rouhi, S., 2010. Atomistic finite element model for axial buckling of single-walled carbon nanotubes. Phys. E. 43, 58–69.

Hollerer, S., 2014. Numerical validation of a concurrent atomistic-continuum multiscale method and its application to the buckling analysis of carbon nanotubes. Comp. Meth. Appl. Mech. Eng. 270, 220–246.

Nishimura, M., Takahashi, N., Takagi, Y., 2017. Relationship between local buckling and atomic elastic stiffness in multi-walled carbon nanotubes under compression and bending deformations. Comp. Mater. Sci. 130, 214–221.

Safaei, B., Naseradinmousavi, P., Rahmani, A., 2016. Development of an accurate molecular mechanics model for buckling behavior of multi-walled carbon nanotubes under axial compression. J. Mol. Graphics Modell. 65, 43–60.

Publications on shell models for carbon nanotubes

Chowdhury, A.N.R., Wang, C.M., Koh, S.J.A., 2014. Continuum shell model for buckling of armchair carbon nanotubes under compression or torsion. Int. J. Appl. Mech. 6 (1), 1450006.

Eftekhari, M., Mohammadi, S., Khoei, A.R., 2013. Effect of defects on the local shell buckling and post-buckling behavior of single and multi-walled carbon nanotubes. Comput. Mater. Sci. 79, 736–744.

Huang, X., Yuan, H.Y., Hsia, K.J., Zhang, S.L., 2010. Coordinated buckling of thick multi-walled carbon nanotubes under uniaxial compression. Nano Res. 3, 32–42.

Korayem, A.H., Duan, W.H., Zhao, X.L., Wang, C.M., 2012. Buckling behaviour of short multi-walled carbon nanotubes under axial compression loads. Int. J. Struct. Stab. Dyn. 12, 1250045.

Kulathunga, D.D.T.K., Ang, K.K., Reddy, J.N., 2009. Accurate modeling of buckling of single- and double-walled carbon nanotubes based on shell theories. J. Phys. Condens. Matter 21, 435301.

Motevalli, B., Montazeri, A., Liu, J.Z., Rafii-Tabar, H., 2013. Comparison of continuum-based and atomistic-based modeling of axial buckling of carbon nanotubes subject to hydrostatic pressure. Comp. Mater. Sci. 79, 619–626.

Peng, J., Wu, J., Hwang, K.C., Song, J., Huang, Y., 2008. Can a single-wall carbon nanotube be modeled as a thin shell?. J. Mech. Phys. Solids 56, 2213–2224.

Wang, C.M., Tay, Z.Y., Chowdhuary, A.N.R., Duan, W.H., Zhang, Y.Y., Silvestre, N., 2011. Examination of cylindrical shell theories for buckling of carbon nanotubes. Int. J. Struct. Stab. Dyn. 11, 1035–1058.

Zhang, Y.Y., Wang, C.M., Duan, W.H., Xiang, Y., Zong, Z., 2009. Assessment of continuum mechanics models in predicting buckling strains of single-walled carbon nanotubes. Nanotechnology 20, 395707.

Molecular dynamics simulations of carbon nanotubes

Ajori, S., Ansari, R., Haghighi, S., 2018. A molecular dynamics study on the buckling behavior of cross-linked functionalized carbon nanotubes under physical adsorption of polymer chains. Appl. Surf. Sci. 427 (Part B), 704–714.

Ansari, R., Ajori, S., Rouhi, S., 2015. Elastic properties and buckling behavior of single-walled carbon nanotubes functionalized with diethyltoluenediamines using molecular dynamics simulations. Superlattices Microstruct. 77, 54–63.

Ansari, R., Rouhi, H., Mirnezhad, M., 2014. A hybrid continuum and molecular mechanics model for the axial buckling of chiral single-walled carbon nanotubes. Curr. Appl. Phys. 14 (10), 1360–1368.

Ansari, R., Ajori, S., Ameri, A., 2014. Elastic and structural properties and buckling behavior of single-walled carbon nanotubes under chemical adsorption of atomic oxygen and hydroxyl. Chem. Phys. Lett. 616–617, 120–125.

Cao, G., Chen, X., 2006. Buckling of single-walled carbon nanotubes upon bending: molecular dynamics and finite element simulations. Phys. Rev. B 73, 155435.

Chowdhury, S.C., (Gama) Haque, B.Z., Gillespie Jr., J.W., Hartman, R., 2012. Molecular simulations of pristine and defective carbon nanotubes under monotonic and combined loading. Comput. Mater. Sci. 65, 133–143.

Eberhardt, O., Wallmersperger, T., 2015. Energy consistent modified molecular structural mechanics model for the determination of the elastic properties of single wall carbon nanotubes. Carbon 95, 166–180.

Hao, X., Qiang, H., Xiaohu, Y., 2008. Buckling of defective single-walled and double-walled carbon nanotubes under axial compression by molecular dynamics simulation. Compos. Sci. Technol. 68 (7–8), 1809–1814.

Heo, S.J., Sinnott, S.B., 2007. Investigation of influence of thermostat configurations on the mechanical properties of carbon nanotubes in molecular dynamics simulations. J. Nanosci. Nanotechnol. 7, 1518–1524.

Huang, Z., Tang, Z., Yu, J., Bai, S., 2011. Temperature-dependent thermal conductivity of bent carbon nanotubes by molecular dynamics simulation. J. Appl. Phys. 109, 104316.

Kinoshita, Y., Kawachi, M., Matsuura, T., Ohno, N., 2013. Axial buckling behavior of wavy carbon nanotubes: a molecular mechanics study. Physica E: Low-dimension. Syst. Nanostruct. 54, 308–312.

Kulathunga, D.D.T.K., Ang, K.K., Reddy, J.N., 2010. Molecular dynamics analysis on buckling of defective carbon nanotubes. J. Phys. Condensed Matter. 22, 345301.

Publications on applications of carbon nanotubes

Abbaspour, M., Akbarzadeh, H., Salemi, S., Abroodi, M., 2016. Equation of state and some structural and dynamical properties of the confined Lennard-Jones fluid into carbon nanotube: a molecular dynamics study. Physica A: Stat. Mech. Appl. 462, 1075–1090.

Ahmed, S.B., Zhao, Y., Fang, C., Su, J., 2017. Transport of a simple liquid through carbon nanotubes: role of nanotube size. Phys. Lett. A 381 (40), 3487–3492.

Boztepe, S., Liu, H., Heider, D., Thostenson, E.T., 2018. Novel carbon nanotube interlaminar film sensors for carbon fiber composites under uniaxial fatigue loading. Compos. Struct. 189, 340–348.

Goh, K., Chen, Y., 2017. Controlling water transport in carbon nanotubes. Nano Today 14, 13–15.

Grabowski, K., Zbyrad, P., Uhl, T., Staszewski, W.J., Packo, P., 2017. Multiscale electro-mechanical modeling of carbon nanotube composites. Comp. Mater. Sci. 135, 169–180.

Han, Y., Li, S., Chen, F., Zhao, T., 2016. Multi-scale alignment construction for strong and conductive carbon nanotube/carbon composites. Mater. Today Comm. 6, 56–68.

He, S., Hu, Y., Wan, J., Gao, Q., Peng, H., 2017. Biocompatible carbon nanotube fibers for implantable supercapacitors. Carbon 122, 162–167.

Li, J., Tang, J., Yuan, J., Zhang, K., Qin, L.-C., 2018. Porous carbon nanotube/graphene composites for high-performance supercapacitors. Chem. Phys. Lett. 693, 60–65.

Pantano, A., 2017. Effects of mechanical deformation on electronic transport through multiwall carbon nanotubes. Int. J. Solids Struct. 122–123, 33–41.

Ribeiro, M.S., Pascoini, A.L., Knupp, W.G., Camps, I., 2017. Effects of surface functionalization on the electronic and structural properties of carbon nanotubes: a computational approach. Appl. Surf. Sci. 426, 781–787.

Tachizaki, T., Nakata, T., Zhang, K., Yamakawa, I., Taniguchi, S.-I., 2018. Nanometer-precise optical length measurement using near-field scanning optical microscopy with sharpened single carbon nanotube probe. Ultramicroscopy 186, 18–22.

Publications on the carbon nanotube-reinforced nanocomposites

Ansari, R., Torabi, J., Shojaei, M.F., 2017. Buckling and vibration analysis of embedded functionally graded carbon nanotube-reinforced composite annular sector plates under thermal loading. Compos. B: Eng. 109, 197–213.

Ashrafi, B., Jakubinek, M.B., Martinez-Rubi, Y., Rahmat, M., Yousefpour, A., 2017. Multifunctional fiber reinforced polymer composites using carbon and boron nitride nanotubes. Acta Astronaut. 141, 57–63.

Benyahia, H., Tarfaoui, M., Datsyuk, V., El Moumen, A., Reich, S., 2017. Dynamic properties of hybrid composite structures based multi walled carbon nanotubes. Compos. Sci. Technol. 148, 70–79.

Carrico, A., Bogas, J.A., Hawreen, A., Guedes, M., 2018. Durability of multi-walled carbon nanotube reinforced concrete. Constr. Building Mater. 164, 121–133.

Chandra, Y., Saavedra Flores, E.I., Scarpa, F., Adhikari, S., 2016. Buckling of hybrid nanocomposites with embedded graphene and carbon nanotubes. Physica E: Low-dimension. Syst. Nanostruct. 83, 434–441.

Chang, T.E., Jensen, L.R., Kisliuk, A., Pipes, R.B., Pyrz, R., Sokolov, A.P., 2005. Microscopic mechanism of reinforcement in single-wall carbon nanotube/polypropylene nanocomposite. Polymer 46, 439–444.

Garcia-Macias, E., Rodriguez-Tembleque, L., Castro-Triguero, R., Saez, A., 2017. Buckling analysis of functionally graded carbon nanotube-reinforced curved panels under axial compression and shear. Compos. B: Eng. 108, 243–256.

Hu, D., Xing, Y., Chen, M., Gu, B., Li, Q., 2017. Ultrastrong and excellent dynamic mechanical properties of carbon nanotube composites. Compos. Sci. Technol. 141, 137–144.

Jomaa, M.H., Masenelli-Varlot, K., Seveyrat, L., Lebrun, L., Cavaille, J.Y., 2015. Investigation of elastic, electrical and electromechanical properties of polyurethane/grafted carbon nanotubes nanocomposites. Compos. Sci. Technol. 121, 1–8.

Joshi, P., Upadhyay, S.H., 2014. Evaluation of elastic properties of multi walled carbon nanotube reinforced composite. Compos. Mater. Sci. 81, 332–338.

Kumar, P., Srinivas, J., 2017. Vibration, buckling and bending behavior of functionally graded multi-walled carbon nanotube reinforced polymer composite plates using the layer-wise formulation. Compos. Struct. 177, 158–170.

Malagu, M., Goudarzi, M., Lyulin, A., Benvenuti, E., Simone, A., 2017. Diameter-dependent elastic properties of carbon nanotube-polymer composites: Emergence of size effects from atomistic-scale simulations. Compos. B: Eng. 131, 260–281.

Mousavi, A.A., Arash, B., Zhuang, X., Rabczuk, T., 2016. A coarse-grained model for the elastic properties of cross linked short carbon nanotube/polymer composites. Compos. B: Eng. 95, 404–411.

Peng, X., Meguid, S.A., 2017. Molecular dynamics simulations of the buckling behavior of defective carbon nanotubes embedded in epoxy nanocomposites. Eur. Polym. J. 93, 246–258.

Quan, D., Urdaniz, J.L., Ivankovic, A., 2018. Enhancing mode-I and mode-II fracture toughness of epoxy and carbon fibre reinforced epoxy composites using multi-walled carbon nanotubes. Mater. Des. 143, 81–92.

Rahmat, M., Hubert, P., 2011. Carbon nanotube–polymer interactions in nanocomposites: a review. Compos. Sci. Technol. 72, 72–84.

Saghar, A., Khan, M., Sadiq, I., Subhani, T., 2018. Effect of carbon nanotubes and silicon carbide particles on ablative properties of carbon fiber phenolic matrix composites. Vacuum 148, 124–126.

Seidel, G.D., Lagoudas, D.C., 2006. Micromechanical analysis of the effective elastic properties of carbon nanotube reinforced composites. Mech. Mater. 38, 884–907.

Sethi, J., Sarlin, E., Meysami, S.S., Suihkonen, R., Vuorinen, J., 2017. The effect of multi-wall carbon nanotube morphology on electrical and mechanical properties of polyurethane nanocomposites. Compos. A: Appl. Sci. Manuf. 102, 305–313.

Singh, R., Chakravarty, A., Chowdhury, U., Bhattacharya, D., Biswas, S.K., 2018. Electrical conduction in aluminum nitride-single-walled carbon nanotube nanocomposites. Mater. Lett. 215, 144–147.

Sharma, S., Chandra, R., Kumar, P., Kumar, N., 2015. Molecular dynamics simulation of polymer/carbon nanotube composites. Acta Mech. Solida Sin. 28 (4), 409–419.

Shen, Q., Li, H., Li, W., Qiang Song, Q., 2018. Realizing the synergy of carbon nanotubes and matrix microstructure for improved flexural behavior of laminated carbon/carbon composites. J. Alloys Compd. 738, 49–55.

Slobodian, P., Pertegas, S.L., Riha, P., Matyas, J., Kovar, M., 2018. Glass fiber/epoxy composites with integrated layer of carbon nanotubes for deformation detection. Compos. Sci. Technol. 156, 61–69.

Tarfaoui, M., El Moumen, A., Lafdi, K., 2017. Progressive damage modeling in carbon fibers/carbon nanotubes reinforced polymer composites. Compos. B: Eng. 112, 185–195.

Thang, P.T., Nguyen, T.-T., Lee, J., 2017. A new approach for nonlinear buckling analysis of imperfect functionally graded carbon nanotube-reinforced composite plates. Compos. B: Eng. 127, 166–174.

Yengejeh, S.I., Kazemi, S.A., Ochsner, A., 2017. Carbon nanotubes as reinforcement in composites: A review of the analytical, numerical and experimental approaches. Comput. Mater. Sci. 136, 85–101.

Zghal, S., Frikha, A., Dammak, F., 2017. Static analysis of functionally graded carbon nanotube-reinforced plate and shell structures. Compos. Struct. 176, 1107–1123.

Publications on interfacial interactions in nanocomposites

Bedi, H.S., Tiwari, M., Agnihotri, P.K., 2018. Quantitative determination of size and properties of interphase in carbon nanotube based multiscale composites. Carbon 132, 181–190.

Chen, J., Wang, K., Zhao, Y., 2018. Enhanced interfacial interactions of carbon fiber reinforced PEEK composites by regulating PEI and graphene oxide complex sizing at the interface. Compos. Sci. Technol. 154, 175–186.

Chen, X., Tao, J., Yi, J., Liu, Y., Bao, R., 2018. Strengthening behavior of carbon nanotube-graphene hybrids in copper matrix composites. Mater. Sci. Eng.: A 718 (7), 427–436.

Duan, K., Li, L., Hu, Y., Wang, X., 2017. Enhanced interfacial strength of carbon nanotube/copper nanocomposites via Ni-coating: Molecular-dynamics insights. Physica E: Low-dimension. Syst. Nanostruct. 88, 259–264.

Fang, W., Linder, M.B., Laaksonen, P., 2018. Modification of carbon nanotubes by amphiphilic glycosylated proteins. J. Colloid Interface Sci. 512, 318–324.

Feng, L., Li, K., Xue, B., Fu, Q., Leilei Zhang, L., 2017. Optimizing matrix and fiber/matrix interface to achieve combination of strength, ductility and toughness in carbon nanotube-reinforced carbon/carbon composites. Mater. Des. 113, 9–16.

Fontananova, E., Grosso, V., Aljlil, S.A., Bahattab, M.A., Di Profio, G., 2017. Effect of functional groups on the properties of multi walled carbon nanotubes/polyvinylidenefluoride composite membranes. J. Membr. Sci. 541, 198–204.

Gao, Y., Muller-Plathe, F., 2017. Effect of grafted chains on the heat transfer between carbon nanotubes in a polyamide-6.6 matrix: a molecular dynamics study. Polymer 129, 228–234.

Goclon, J., Panczyk, T., Winkler, K., 2018. Investigation of the interfacial properties of polyurethane/carbon nanotube hybrid composites: a molecular dynamics study. Appl. Surf. Sci. 433, 213–221.

Guignier, C., Bueno, M.-A., Camillieri, B., Huu, T.L., Durand, B., 2016. Tribological behaviour and adhesion of carbon nanotubes grafted on carbon fibres. Tribol. Int. 100, 104–115.

Klonos, P., Sulym, I.Y., Sternik, D., Konstantinou, P., Pissis, P., 2018. Morphology, crystallization and rigid amorphous fraction in PDMS adsorbed onto carbon nanotubes and graphite. Polymer 139, 130–144.

Korayem, A.H., Li, C.Y., Zhang, Q.H., Zhao, X.L., Duan, W.H., 2015. Effect of carbon nanotube modified epoxy adhesive on CFRP-to-steel interface. Compos. B: Eng. 79, 95–104.

Kulathunga, D.D.T.K., Ang, K.K., 2014. Modeling and simulation of buckling of embedded carbon nanotubes. Compos. Mater. Sci. 81, 233–238.

Lavagna, L., Massella, D., Pavese, M., 2017. Preparation of hierarchical material by chemical grafting of carbon nanotubes onto carbon fibers. Diamond Relat. Mater. 80, 118–124.

Lau, K.T., 2003. Interfacial bonding characteristics of nanotube/polymer composites. Chem. Phys. Lett. 370 (3–4), 399–405.

Lechner, C., Sax, A.F., 2017. Towards atomic-level mechanics: adhesive forces between aromatic molecules and carbon nanotubes. Appl. Surf. Sci. 420, 606–617.

Li, Y., Hu, N., Yamamoto, G., et al., 2010. Molecular mechanics simulation of the sliding behavior between nested walls in a multi-walled carbon nanotube. Carbon 48, 2934–2940.

Liu, J., Tang, X., Chen, F., Huang, H., Chen, D., 2016. Molecular dynamics of adhesion force of single-walled carbon nanotubes. Diamond Relat. Mater. 69, 74–80.

Lu, Y.-B., Yang, Q.-S., He, X.-Q., Liew, K.-M., 2016. Modeling the interfacial behavior of carbon nanotube fiber/polyethylene composites by molecular dynamics approach. Comput. Mater. Sci. 114, 189–198.

Minoia, A., Chen, L., Beljonne, D., Lazzaroni, R., 2012. Molecular modeling study of the structure and stability of polymer/carbon nanotube interfaces. Polymer 53 (24), 5480–5490.

Nai, M.H., Wei, J., Gupta, M., 2014. Interface tailoring to enhance mechanical properties of carbon nanotube reinforced magnesium composites. Mater. Des. 60, 490–495.

Ozden-Yenigun, E., Atilgan, C., Elliott, J.A., 2017. Multi-scale modelling of carbon nanotube reinforced crosslinked interfaces. Comput. Mater. Sci. 129, 279–289.

Rouhi, S., Alizadeh, Y., Ansari, R., 2014. On the interfacial characteristics of polyethylene/single-walled carbon nanotubes using molecular dynamics simulations. Appl. Surf. Sci. 292, 958–970.

Sarode, A., Ahmed, Z., Basarkar, P., Bhargav, A., Banerjee, D., 2017. A molecular dynamics approach of the role of carbon nanotube diameter on thermal interfacial resistance through vibrational mismatch analysis. Int. J. Therm. Sci. 122, 33–38.

Si, S., Gao, T., Wang, J., Liu, Q., Zhou, G., 2018. Mussel inspired polymerized P(TA-TETA) for facile functionalization of carbon nanotube. Appl. Surf. Sci. 433, 94–100.

Song, X., Shang, S., Chen, D., Gu, X., 2017. Multi-walled carbon nanotube reinforced mortar-aggregate interfacial properties. Constr. Building Mater. 133, 57–64.

Tsai, P.-C., Jeng, Y.-R., 2013. Experimental and numerical investigation into the effect of carbon nanotube buckling on the reinforcement of CNT/Cu composites. Compos. Sci. Technol. 79, 28–34.

Wang, M.C., Lai, Z.B., Galpaya, D., et al., 2014. Atomistic simulation of surface functionalization on the interfacial properties of graphene-polymer nanocomposites. J. Appl. Phys. 115, 123520.

Wernik, J.M., Cornwell-Mott, B.J., Meguid, S.A., 2012. Determination of the interfacial properties of carbon nanotube reinforced polymer composites using atomistic-based continuum model. Int. J. Solids Struct. 49, 1852–1863.

Xie, A., Wang, Y., Jiang, P., Li, S., Huang, X., 2018. Nondestructive functionalization of carbon nanotubes by combining mussel-inspired chemistry and RAFT polymerization: Towards high dielectric nanocomposites with improved thermal management capability. Compos. Sci. Technol. 154, 154–164.

Xu, J., Cao, Z., Zhang, Y., Yuan, Z., Wang, X., 2018. A review of functionalized carbon nanotubes and graphene for heavy metal adsorption from water: preparation, application, and mechanism. Chemosphere 195, 351–364.

Yao, X., Gao, X., Jiang, J., Xu, C., Wang, J., 2018. Comparison of carbon nanotubes and graphene oxide coated carbon fiber for improving the interfacial properties of carbon fiber/epoxy composites. Compos. B: Eng. 132, 170–177.

Yuan, X., Wang, Y., Zhu, B., 2018. Adhesion between two carbon nanotubes: insights from molecular dynamics simulations and continuum mechanics. Int. J. Mech. Sci. 138–139, 323–336.

Zhang, D., Yang, H., Liu, Z., Liu, A., Li, Y., 2017. Interfacial interaction between polypropylene and nanotube: a molecular dynamics simulation. J. Mol. Struct. 1144, 260–264.

Zhang, J., Peng, X., 2017. Superior interfacial mechanical properties of boron nitride-carbon nanotube reinforced nanocomposites: a molecular dynamics study. Mater. Chem. Phys. 198, 250–257.

Zhang, Z.Q., Ward, D.K., Xue, Y., et al., 2011. Interfacial characteristics of carbon nanotube-polyethylene composites using molecular dynamics simulations. ISRN Mater. Sci. 2011, 1–10.

Zheng, Q., Xia, D., Xue, Q., et al., 2009. Computational analysis of effect of modification on the interfacial characteristics of a carbon nanotube-polyethylene composite system. Appl. Surf. Sci. 255, 3534–3543.

Zhao, Z., Teng, K., Li, N., Li, X., Ya Liu, Y., 2017. Mechanical, thermal and interfacial performances of carbon fiber reinforced composites flavored by carbon nanotube in matrix/interface. Compos. Struct. 159, 761–772.

Publications on the carbon nanotube pull-out from nanocomposites

Bedi, H.S., Tiwari, M., Prabhat, K., Agnihotri, P.K., 2018. Quantitative determination of size and properties of interphase in carbon nanotube based multiscale composites. Carbon 132, 181–190.

Chawla, R., Sharma, S., 2017. Molecular dynamics simulation of carbon nanotube pull-out from polyethylene matrix. Compos. Sci. Technol. 144, 169–177.

Chowdhury, S.C., Okabe, T., 2007. Computer simulation of carbon nanotube pull-out from polymer by the molecular dynamics method. Compos. A: Appl. Sci. Manuf. 38 (3), 747–754.

Li, Y., Liu, Y., Peng, X., Yan, C., Hu, N., 2011. Pull-out simulations on interfacial properties of carbon nanotube-reinforced polymer nanocomposites. Comput. Mater. Sci. 50 (6), 1854–1860.

Ma, H.-K., Lau, K.-T., Hui, D., Shi, S.-Q., Poon, C.-K., 2017. Theoretical analysis on the pullout behavior of carbon nanotube at cryogenic environment with the consideration of thermal residual stress. Compos. B: Eng. 128, 67–75.

Sharma, K., Kaushalyayan, K.S., Shukla, M., 2015. Pull-out simulations of interfacial properties of amine functionalized multi-walled carbon nanotube epoxy composites. Comput. Mater. Sci. 99, 232–241.

Wagner, H.D., Ajayan, P.M., Schulte, K., 2013. Nanocomposite toughness from a pull-out mechanism. Compos. Sci. Technol. 83, 27–31.

Wang, L., Cui, Y., Qin, Q., Wang, H., Wang, J., 2016. Helical fiber pull-out in biological materials. Acta Mech. Solida Sin. 29 (3), 245–256.

Yang, L., Tong, L., He, X., Wagner, H.D., Wang, R., 2014. Molecular dynamic simulation of oblique pullout of carbon nanotube from resin. Comput. Mater. Sci. 83, 504–512.

Yi, C., Chen, X., Gou, F., Dmuchowski, C.M., Ke, C., 2017. Direct measurements of the mechanical strength of carbon nanotube—aluminum interfaces. Carbon 125, 93–102.

Additional References

Advani, S.G., 2007. Processing and Properties of Nanocomposites.. World Scientific Publishing Co, London, United Kingdom.

Ajori, S., Ansari, R., Haghighi, S., 2018. A molecular dynamics study on the buckling behavior of cross-linked functionalized carbon nanotubes under physical adsorption of polymer chains. Appl. Surf. Sci. 427 (B), 704–714.

Akbarov, S.D., 2013. Microbuckling of a double walled carbon nanotube embedded in an elastic matrix. Int. J. Solids Struct. 50 (1), 2584–2596.

Ansari, R., Rouhi, H., Mirnezhad, M., 2014. A hybrid continuum and molecular mechanics model for axial buckling of chiral single-walled carbon nanotubes. Curr. Appl. Phys. 14, 1360–1368.

Ansari, R., Rouhi, S., Aryayi, M., Mirnezhad, M., 2012. On the buckling behavior of single-walled silicon carbide nanotubes. Sci. Iranica 19 (6), 1984–1990.

Bakir, M., Iwona Jasiuk, I., 2017. Novel metal-carbon nanomaterials: A review on covetics. Adv. Mater. Lett. 8 (9), 884–890.

Batra, R.C., Sears, A., 2007. Continuum models of multi-walled carbon nanotubes. Int. J. Solids Struct. 44, 7577–7596.

Buehler, M.J., Kong, J., Gao, H.J., 2004. Deformation mechanism of very long single-wall carbon nanotubes subject to compressive loading. J. Eng. Mater. Technol. 126, 245–249.

Byrne, E.M., McCarthy, M.A., Xia, Z., Curtin, W.A., 2009. Multiwall nanotubes can be stronger than single wall nanotubes and implications for nanocomposite design. Phys. Rev. Lett. 103, 045502.

Cao, C., Reiner, A., Chung, C., Chang, S.-H., Korach, C.S., 2011. Buckling initiation and displacement dependence in compression of vertically aligned carbon nanotube arrays. Carbon 49 (10), 3190–3199.

Chan, Y., Thamwattana, N., Hill, J.M., 2011. Axial buckling of multi-walled carbon nanotubes and nanopeapods. Eur. J. Mech.—A/Solids 30 (6), 794–806.

Chang, T.C., Geng, J.Y., Guo, X.M., 2005. Chirality- and size-dependent elastic properties of single-walled carbon nanotubes. Appl. Phys. Lett. 87 (25), 251929.

Chang, T., Li, G., Guo, X., 2005. Elastic axial buckling of carbon nanotubes via a molecular mechanics model. Carbon 43, 287–294.

Chang, T., Gao, H., 2003. Size dependent elastic properties of a single walled carbon nanotube via a molecular mechanics model. J. Mech. Phys. Solids 51, 1059–1074.

Chang, T., Hou, J., 2006. Molecular dynamics simulations on buckling of multiwalled carbon nanotubes under bending. J. Appl. Phys. 100, 114327.

Chang, T., Guo, W., Guo, X., 2005. Buckling of multi-walled carbon nanotubes under axial compression and bending via a molecular mechanics model. Phys. Rev. B 72, 064101.

Chen, X., Tao, J., Yi, J., Liu, Y., Bao, R., 2018. Strengthening behavior of carbon nanotube-graphene hybrids in copper matrix composites. Mater. Sci. Eng.: A 718, 427–436.

Cornwell, C.F., Wille, L.T., 1997. Elastic properties of single-walled carbon nanotubes in compression. Solid State Commun. 101, 555.

Dai, H.L., Wang, L., Abdelkefi, A., Ni, Q., 2015. On nonlinear behavior and buckling of fluid-transporting nanotubes. Int. J. Eng. Sci. 87, 13–22.

De Oliveira, A.B., Chacham, H., Soares, J.S., Manhabosco, T.M., Batista, R.J.C., 2016. Vibrational G peak splitting in laterally functionalized single wall carbon nanotubes: theory and molecular dynamics simulations. Carbon 96, 616–621.

Duan, X.J., Tang, C., Zhang, J., Guo, W.L., Liu, Z.F., 2007. Two distinct buckling modes in carbon nanotube bending. Nano Lett. 7, 143–148.

Elliott, J.A., Sandler, L.K.W., Windle, A.H., Young, R.J., Shaffer, M.S.P., 2004. Collapse of single-wall carbon nanotubes is diameter dependent. Phys. Rev. Lett. 92, 095501.

Fang, S.C., Chang, W.J., Wang, Y.H., 2007. Computation of chirality- and size-dependent surface Young's moduli for single-walled carbon nanotubes. Phys. Lett. A 371, 499–503.

Feliciano, J., Chun Tang, C., Zhang, Y.Y., Chen, C.F., 2011. Aspect ratio dependent buckling mode transition in single-walled carbon nanotubes under compression. J. Appl. Phys. 109, 084323.

Feng, L., Li, K.-Z., Xue, B., Song, Q., Fu, Q.-G., 2017. Joining carbon/carbon composites with radially-aligned carbon nanotube-reinforced pyrocarbon bonding interlayer. Mater. Lett. 187, 158–161.

Gao, X., Isayev, A.I., Yi, C., 2016. Ultrasonic treatment of polycarbonate/carbon nanotubes composites. Polymer 84, 209–222.

Ghasemi, A., Dardel, M., Ghasemi, M.H., Barzegari, M.M., 2013. Analytical analysis of buckling and post-buckling of fluid conveying multi-walled carbon nanotubes. Appl. Math. Modell. 37 (7), 4972–4992.

Gou, J., Minaie, B., Wang, B., et al., 2004. Computational and experimental study of interfacial bonding of single-walled nanotube reinforced composites. Comp. Mater. Sci. 31, 225–236.

Guo, X., Leung, A.Y.T., He, X.Q., Jiang, H., Huang, Y., 2008. Bending buckling of single-walled carbon nanotubes by atomic-scale finite element. Compos. B: Eng. 39 (1), 202–208.

Guo, X., Leung, A.Y.T., Jiang, H., He, X.Q., Huang, Y., 2007. Critical strain of carbon nanotubes: an atomic-scale finite element study. J. Appl. Mech. 74, 347–351.

Hadjiev, V.G., Lagoudas, D.C., Oh, E.-S., Thakre, P., Tour, J.M., 2006. Buckling instabilities of octadecylamine functionalized carbon nanotubes embedded in epoxy. Compos. Sci. Technol. 66 (1), 128–136.

Han, Y., Elliott, J., 2007. Molecular dynamics simulations of the elastic properties of polymer/carbon nanotube composites. Comput. Mater. Sci. 39, 315–323.

Hao, M.J., Guo, X.M., Wang, Q., 2010. Small-scale effect on torsional buckling of multi-walled carbon nanotubes. Eur. J. Mech.—A/Solids 29 (1), 49–55.

Harik, V.M., Luo, L.-S. (Eds.), 2004. Micromechanics and Nanoscale Effects.. Kluwer Academic Publishers. Reprinted by Springer, Dordrecht, The Netherlands.

Harik, V.M., Salas, M.D. (Eds.), 2003. Trends in Nanoscale Mechanics.. Kluwer Academic Publishers. Reprinted by Springer, Dordrecht, The Netherlands.

Harris, P.J.F., 2004. Carbon nanotube composites. Int. Mater. Rev. 49, 31–43.

He, X.Q., Kitipornchai, S., Liew, K.M., 2005. Buckling analysis of multi-walled carbon nanotubes: a continuum model accounting for van der Waals interaction. J. Mech. Phys. Solids 53, 303–326.

Hokkanen, M.J., Lautala, S., Flahaut, E., Ahlskog, M., 2017. Experimental studies on the detachment of multi-walled carbon nanotubes by a mobile liquid interface. Colloids Surf. A: Physicochem. Eng. Aspects 533, 109–115.

Hsiao, K.-T., Alms, J., Advani, S.G., 2003. Use of epoxy/multi walled carbon nanotubes as adhesives to join graphite fibre reinforced polymer composites. Nanotechnology 14, 791.

Hu, N., Nunoya, K., Pan, D., Okabe, T., Fukunaga, H., 2007. Prediction of buckling characteristics of carbon nanotubes. Int. J Solids Struct. 44, 6535–6550.

Huang, Y., Wu, J., Hwang, K., 2006. Thickness of graphene and single-wall carbon nanotubes. Phys. Rev. B 74, 245413.

Jafari Mehrabadi, S., Sobhani Aragh, B., Khoshkhahesh, V., Taherpour, A., 2012. Mechanical buckling of nanocomposite rectangular plate reinforced by aligned and straight single-walled carbon nanotubes. Compos. B: Eng. 43 (4), 2031–2040.

Jeong, B.W., Lim, J.K., Sinnott, S.B., 2007. Turning stiffness of carbon nanotube systems. Appl. Phys. Lett. 91, 093102.

Jiang, H., Zhang, P., Liu, B., Huang, Y., Geubelle, P.H., Gao, H., et al., 2003. The effect of nanotube radius on the constitutive model for carbon nanotubes. Comput. Mater. Sci. 28, 429–442.

Jin, S.-Y., Young, R.J., Eichhorn, S.J., 2014. Hybrid carbon fibre–carbon nanotube composite interfaces. Compos. Sci. Technol. 95, 114–120.

Jones, S.L.T., Greene-Diniz, G., Haverty, M., Shankar, S., Greer, J.C., 2014. Effect of structure on electronic properties of the iron-carbon nanotube interface. Chem. Phys. Lett. 615, 11–15.

Kang, J.W., Lee, K.W., 2014. Molecular dynamics study of carbon-nanotube shuttle-memory on graphene nanoribbon array. Comput. Mater. Sci. 93, 164–168.

Korayem, A.H., Duan, W.H., Zhao, X.L., 2011. Investigation on buckling behavior of short MWCNT. Proc. Eng. 14, 250–255.

Kuang, Y.D., He, X.Q., Chen, C.Y., Li, G.Q., 2009. Buckling of functionalized single-walled nanotubes under axial compression. Carbon 47, 279–285.

Kudin, K.N., Scuseria, G.E., Yakobson, B.I., 2001. C2F, BN, and C nanoshell elasticity from *ab initio* computations. Phys. Rev. B 64, 235406.

Lee, C., Wei, X.D., Kysar, J.W., Hone, J., 2008. Measurement of the elastic properties and intrinsic strength of monolayer graphene. Science 321, 385–388.

Lee, H.-L., Chang, W.-J., 2009. A closed-form solution for critical buckling temperature of a single-walled carbon nanotube. Physica E: Low-dimension. Syst. Nanostruct. 41 (8), 1492–1494.

Leung, A.Y.T., Guo, X., He, X.Q., Jiang, H., Huang, Y., 2006. Postbuckling of carbon nanotubes by atomic-scale finite element. J. Appl. Phys. 99 (12), 124308.

Li, C.Y., Chou, T.-W., 2003. Single-walled carbon nanotubes as ultrahigh frequency nanomechanical resonators. Phys. Rev. B 68, 073405/1–073405/4.

Li, C.Y., Chou, T.-W., 2004. Vibrational behaviors of multi-walled carbon nanotube-based nanomechancial resonators. Appl. Phys. Lett. 84, 121–123.

Li, C., Chou, T.-W., 2004. Modeling of elastic buckling of carbon nanotubes by molecular structural mechanics approach. Mech. Mater. 36, 1047–1055.

Li, R., Kardomateas, G.A., 2006. Thermal buckling of multi-walled carbon nanotubes by nonlocal elasticity. J. Appl. Mech. 74 (3), 399–405.

Li, Y., Qiu, X.M., Yang, F., Wang, X.-S., Qinshan Fan, Q., 2008. Chirality independence in critical buckling forces of super carbon nanotubes. Solid State Commun. 148 (1–2), 63–68.

Li, Y., Wang, S., Wang, Q., Xing, M., 2018. A comparison study on mechanical properties of polymer composites reinforced by carbon nanotubes and graphene sheet. Compos. B: Eng. 133, 35–41.

Li, Y., Wang, S., Wang, Q., Xing, M., 2018. Enhancement of fracture properties of polymer composites reinforced by carbon nanotubes: a molecular dynamics study. Carbon 129, 504–509.

Liao, K., Li, S., 2001. Interfacial characteristics of a carbon nanotube polystyrene composite system. Appl. Phys. Lett. 79, 4225.

Lim, M.-J., Lee, H.K., Nam, I.-W., Kim, H.-K., 2017. Carbon nanotube/cement composites for crack monitoring of concrete structures. Compos. Struct. 180, 741–750.

Lordi, V., Yao, N., 2000. Molecular mechanics of binding in carbon nanotube-polymer composites. J. Mater. Res. 15, 2770–2779.

Lu, J., Luo, M., Yakobson, B.I., 2018. Glass composites reinforced with silicon-doped carbon nanotubes. Carbon 128, 231–236.

Lu, J.-M., Hwang, C.-C., Kuo, Q.-Y., Wang, Y.-C., 2008. Mechanical buckling of multi-walled carbon nanotubes: the effects of slenderness ratio. Physica E: Low-dimensional. Syst. Nanostruct. 40 (5), 1305–1308.

Lv, Q., Wang, Z., Chen, S., Li, C., Hu, S., 2017. Effects of single ad-atom and Stone–Wales defects on the elastic properties of carbon nanotube/polypropylene composites: a molecular simulation study. Int. J. Mech. Sci. 131–132, 527–534.

Ma, J., Wang, J.N., Wang, X.X., 2009. Large-diameter and water-dispersible single-walled carbon nanotubes: synthesis, characterization and applications. J. Mater. Chem. 19, 3033–3041.

Malekie, S., Ziaie, F., 2015. Study on a novel dosimeter based on polyethylene–carbon nanotube composite. Nuclear Instr. Meth. Phys. Res., Sec. A: Accel. Spectrom. Detectors Assoc. Equip. 791, 1–5.

Malekzadeh, P., Shojaee, M., 2013. Buckling analysis of quadrilateral laminated plates with carbon nanotubes reinforced composite layers. Thin-Walled Struct. 71, 108–118.

Misra, A., Tyagi, P.K., Rai, P., Mahopatra, D.R., Ghatak, J., Satyam, P.V., et al., 2007. Axial buckling and compressive behavior of nickel-encapsulated multiwalled carbon nanotubes. Phys. Rev. B. 76, 014108.

Monteiro, A.O., Costa, P.M.F.J., Cachim, P.B., Holec, D., 2014. Buckling of ZnS-filled single-walled carbon nanotubes—the influence of aspect ratio. Carbon 79, 529–537.

Ni, B., Sinnott, S.B., Mikulski, P.T., Harrison, J.A., 2002. Compression of carbon nanotubes filled with C60, CH4, or Ne: predictions from molecular dynamics simulations. Phys. Rev. Lett. 88, 205505.

Ounaies, Z., Park, C., Harrison, J., Lillehei, P., 2008. Evidence of piezoelectricity in SWNT-polyimide and SWNT-PZT-polyimide composites. J. Thermoplastic Compos. Mater. 21 (5), 393–409.

Pantano, A., Boyce, M.C., Parks, D.M., 2004. Mechanics of axial compression of single-and multi-wall carbon nanotubes. J. Eng. Mater. Technol. 126, 279–284.

Pantano, A., Parks, D.M., Boyce, M.C., 2004. Mechanics of deformation of single- and multi-wall carbon nanotubes. J. Mech. Phys. Solids 52, 789–821.

Potts, J.R., Dreyer, D.R., Bielawski, C.W., Ruoff, R.S., 2011. Graphene-based polymer nanocomposites. Polymer 52, 5–25.

Qian, D., Dickey, E.C., Andrews, R., et al., 2000. Load transfer and deformation mechanisms in carbon nanotube-polystyrene composites. Appl. Phys. Lett. 76, 2868–2870.

Ramazani, A., Reihani, A., Soleimani, A., Larson, R., Sundararaghavan, V., 2017. Molecular dynamics study of phonon transport in graphene nanotubes. Carbon 123, 635–644.

Rizzuto, C., Pugliese, G., Bahattab, M.A., Aljlil, S.A., Tocci, E., 2018. Multiwalled carbon nanotube membranes for water purification. Sep. Purif. Technol. 193, 378–385.

Ru, C.Q., 2000. Column buckling of multi walled carbon nanotubes with interlayer radial displacements. Phys. Rev. B 62, 16962–16967.

Ru, C.Q., 2001. Axially compressed buckling of a double walled carbon nanotube embedded in an elastic medium. J. Mech. Phys. Solids 49 (6), 1265–1279.

Saavedra Flores, E.I., Adhikari, S., Friswell, M.I., Scarpa, F., 2011. Hyperelastic axial buckling of single wall carbon nanotubes. Physica E: Low-dimension. Syst. Nanostruct. 44 (2), 525–529. 2011.

Schadler, L.S., Giannaris, S.C., Ajayan, P.M., 1998. Load transfer in carbon nanotube epoxy composites. Appl. Phys. Lett. 73, 3842–3844.

Sears, A., Batra, R.C., 2006. Buckling of multi walled carbon nanotubes under axial compression. Phys. Rev. B 73, 085410.

Seifoori, S., 2015. Molecular dynamics analysis on impact behavior of carbon nanotubes. Appl. Surf. Sci. 326, 12–18.

Shen, H.S., 2004. Postbuckling prediction of double-walled carbon nanotubes under hydrostatic pressure. Int. J. Solids Struct. 41, 2643–2657.

Shen, L., Li, J., 2004. Transversely isotropic elastic properties of single-walled carbon nanotubes. Phys. Rev. B 69, 045414.

Silvestre, N., 2008. Length dependence of critical measures in single-walled carbon nanotubes. Int. J. Solids Struct. 45, 4902–4920.

Silvestre, N., Wang, C.M., Zhang, Y.Y., Xiang, Y., 2011. Sanders shell model for buckling of single-walled carbon nanotubes with small aspect ratio. Compos. Struct. 93, 1683–1691.

Slobodian, P., Riha, P., Olejnik, R., Matyas, J., Kovar, M., 2018. Poisson effect enhances compression force sensing with oxidized carbon nanotube network/polyurethane sensor. Sens. Actuators A: Phys. 271, 76–82.

Sun, Y.Z., Liew, K.M., 2014. Effect of higher-order deformation gradients on buckling of single-walled carbon nanotubes. Compos. Struct. 109, 279–285.

Tahir, M.A., Hameed, S., Munawar, A., Amin, I., Sadia Zafar Bajwa, S.Z., 2017. Investigating the potential of multi walled carbon nanotubes based zinc nanocomposite as a recognition interface towards plant pathogen detection. J. Virol. Methods 249, 130–136.

Tarfaoui, M., Lafdi, K., El Moumen, A., 2016. Mechanical properties of carbon nanotubes based polymer composites. Compos. B: Eng. 103, 113–121.

Thostenson, E.T., Chou, T.-W., 2002. Aligned multi-walled carbon nanotube-reinforced composites: Processing and mechanical characterization. J. Phys. D: Appl. Phys. 35 (16), L77–L80.

Thostenson, E.T., Chou, T.-W., 2003. On the elastic properties of carbon nanotube-based composites: modeling and characterization. J. Phys. D: Appl. Phys. 36, 573–582.

Thostenson, E.T., Ren, Z.F., Chou, T.-W., 2001. Advances in the science and technology of carbon nanotubes and their composites: a review. Comp. Sci. Technol. 61, 1899–1912.

Toth, P.S., Rabiu, A.K., Dryfe, R.A.W., 2015. Controlled preparation of carbon nanotube-conducting polymer composites at the polarisable organic/water interface. Electrochem. Commun. 60, 153–157.

Tsai, P.-C., Yeau-Ren Jeng, Y.-R., 2016. In-situ nanomechanical study on bending characteristics of individual multi-walled carbon nanotubes. Mater. Sci. Eng.: A 658, 389–392.

Vakili-Nezhaad, G., Al-Wadhahi, M., Gujrathi, A.M., Al-Maamari, R., Mohammadi, M., 2017. Effect of temperature and diameter of narrow single-walled carbon nanotubes on the viscosity of nanofluid: a molecular dynamics study. Fluid Phase Equilib. 434, 193–199.

Vijayaraghavan, V., Wong, C.H., 2014. Transport characteristics of water molecules in carbon nanotubes investigated by using molecular dynamics simulation. Comput. Mater. Sci. 89, 36–44.

Vijayaraghavan, V., Jacob, F., Dethan, N., Garg, A., 2018. Nanomechanics and modelling of hydrogen stored carbon nanotubes under compression for PEM fuel cell applications. Comput. Mater. Sci. 146, 176–183.

Wackerfuß, J., 2009. Molecular mechanics in the context of the finite element method. Int. J. Numer. Meth. Eng. 77, 969–997.

Wagner, H.D., Lourie, O., Feldman, Y., Tenne, R., 1998. Stress-induced fragmentation of multiwall carbon nanotubes in a polymer matrix. Appl. Phys. Lett. 72, 188–190.

Wang, C.M., Chowdhury, A.N.R., Koh, S.J.A., Zhang, Y.Y., 2014. Molecular dynamics simulation and continuum shell model for buckling analysis of carbon nanotubes. In: Tserpes, K.I., Silvestre, N. (Eds.), Modeling of Carbon Nanotubes, Graphene and their Composites, 188. Springer Series in Materials Science, p. 239.

Wang, C.M., Tan, V.B.C., Zhang, Y.Y., 2006. Timoshenko beam model for vibration analysis of multi-walled carbon nanotubes. J. Sound Vib. 294, 1060–1072.

Wang, C.Y., Ru, C.Q., Mioduchowski, A., 2003. Axially compressed buckling of pressured multiwall carbon nanotubes. Int. J. Solids Struct. 40 (15), 3893–3911.

Wang, C.Y., Ru, C.Q., Mioduchowski, A., 2003. Elastic buckling of multiwall carbon nanotubes under high pressure. J. Nanosci. Nanotechnol. 3, 199–208.

Wang, C.Y., Zhang, Y.Y., Wang, C.M., Tan, V.B.C., 2007. Buckling of carbon nanotubes: A literature survey. J. Nanosci. Nanotechnol. 7, 4221–4247.

Wang, D., Song, P., Liu, C., Wu, W., Fan, S., 2008. Highly oriented carbon nanotube papers made of aligned carbon nanotubes. Nanotechnology 19, 075609.

Wang, H., Tazebay, A.S., Yang, G., Lin, H.T., Yu, C., 2016. Highly deformable thermal interface materials enabled by covalently-bonded carbon nanotubes. Carbon 106, 152–157.

Wang, L., Prasad, A.K., Advani, S.G., 2012. Composite membrane based on SiO_2-MWCNTs and Nafion for PEMFCs. J. Electrochem. Soc. 159 (8), F490–F493.

Wang, Q., 2011. Buckling of carbon nanotubes wrapped by polyethylene molecules. Phys. Lett. A 375 (3), 624–627.

Wang, Q., Liew, K.M., He, X.Q., Xiang, Y., 2007. Local buckling of carbon nanotubes under bending. Appl. Phys. Lett. 73, 093128.

Wang, S., Zhang, S., Wang, Y., Sun, X., Sun, K., 2017. Reduced graphene oxide/carbon nanotubes reinforced calcium phosphate cement. Ceramics Int 43 (16), 13083–13088.

Wang, X., Lu, G.X., Lu, Y.J., 2007. Buckling of embedded multi-walled carbon nanotubes under combined torsion and axial loading. Int. J. Solids Struct. 44, 336–351.

Wang, X., Sun, B., Yang, H.K., 2006. Stability of multi-walled carbon nanotubes under combined bending and axial compression loading. Nanotechnology 17, 815–823.

Wang, Y., Wang, X.X., Ni, X.G., Wu, H.A., 2005. Simulation of the elastic response and the buckling modes of single-walled carbon nanotubes. Comput. Mater. Sci. 32, 141–146.

Wang, C.M., Zhang, Y.Y., Xiang, Y., Reddy, J.N., 2010. Recent studies on buckling of carbon nanotubes. Appl. Mech. Rev. 63, 030804.

Wang, Q., Liew, K.M., Duan, W.H., 2008. Modeling of the mechanical instability of carbon nanotubes. Carbon 46 (2), 285–290.

Waters, J.F., Riester, L., Jouzi, M., Guduru, P.R., Xu, J.M., 2004. Buckling instabilities in multiwalled carbon nanotubes under uniaxial compression. Appl. Phys. Lett. 85, 1787–1789.

Waters, J.F., Guduru, P.R., Jouzi, M., Xu, J.M., Hanlon, T., Suresh, S., 2005. Shell buckling of individual multi-walled carbon nanotubes using nanoindentation. Appl. Phys. Lett. 87, 103109.

Waters, J.F., Gudurua, P.R., Xu, J.M., 2006. Nanotube mechanics—recent progress in shell buckling mechanics and quantum electromechanical coupling. Compos. Sci. Technol. 66, 1141–1150.

Weick, G., von Oppen, F., Pistolesi, F., 2011. Euler buckling instability and enhanced current blockade in suspended single-electron transistors. Phys. Rev. B 83, 035420.

Wu, H., Kitipornchai, S., Yang, J., 2017. Imperfection sensitivity of thermal post-buckling behaviour of functionally graded carbon nanotube-reinforced composite beams. Appl. Math. Modell. 42, 735–752.

Wu, J., Hwang, K.C., Huang, Y., 2009. A shell theory for carbon nanotubes based on the interatomic potential and atomic structure. *Advances in Applied Mechanics*. Elsevier, pp. 1–68, Chapter 1

Xiao, J.R., Gama, B.A., Gillespie, J.W., 2005. An analytical molecular structural mechanics model for the mechanical properties of carbon nanotubes. Int. J. Solids Struct. 42, 3075–3092.

Xin, H., Han, Q., Yao, X.H., 2007. Buckling and axially compressive properties of perfect and defective single-walled carbon nanotubes. Carbon 45, 2486–2495.

Yamakov, V., Park, C., Kang, J.H., Chen, X., Fay, C., 2017. Piezoelectric and elastic properties of multiwall boron-nitride nanotubes and their fibers: a molecular dynamics study. Comput. Mater. Sci. 135, 29–42.

Yamamoto, G., Liu, S., Hu, N., Hashida, T., Wu, L., 2012. Prediction of pull-out force of multi-walled carbon nanotube (MWCNT) in sword-in-sheath mode. Comput. Mater. Sci. 60, 7–12.

Yao, X., Han, Q., Xin, H., 2008. Bending buckling behaviors of single- and multi-walled carbon nanotubes. Comput. Mater. Sci. 43, 579–590.

Yap, H.W., Lakes, R.S., Carpick, R.W., 2007. Mechanical instabilities of individual multi walled carbon nanotubes under cyclic axial compression. Nano Lett. 7, 1149–1154.

Zamfirova, G., Gaydarov, V., Faraguna, F., Vidovic, E., Jukic, A., 2016. Influence of carbon nanotubes functionalization on the mechanical properties of polymethacrylate nanocomposites. Colloids Surf. A: Physicochem. Eng. Aspects 510, 169–175.

Zhang, C.L., Shen, H.S., 2007. Buckling and postbuckling of single-walled carbon nanotubes under combined axial compression and torsion in thermal environments. Phys. Rev. B 75, 045408.

Zhang, J., Peng, X., 2017. Superior interfacial mechanical properties of boron nitride-carbon nanotube reinforced nanocomposites: a molecular dynamics study. Mater. Chem. Phys. 198, 250–257.

Zhang, L., De Greef, N., Kalinka, G., Van Bilzen, B., Seo, J.W., 2017. Carbon nanotube-grafted carbon fiber polymer composites: damage characterization on the micro-scale. Compos. B: Eng. 126, 202–210.

Zhang, P., Huang, Y., Geubelle, P.H., Klein, P.A., Hwang, K.C., 2002. The elastic modulus of single-wall carbon nanotubes: a continuum analysis incorporating interatomic potentials. Int. J. Solids Structr. 39, 3893–3906.

Zhang, S., Khare, R., Belytschko, T., Hsia, K.J., Mielke, S.L., Schatz, G.C., 2006. Transition states and minimum energy pathways for the collapse of carbon nanotubes. Phys. Rev. B 73, 075423.

Zhang, Y.Y., Wang, M., Tan, V.B.C., 2008. Examining the effects of wall numbers on buckling behavior and mechanical properties of multi walled carbon nanotubes via molecular dynamics simulations. J. Appl. Phys. 103, 053505.

Zhang, Y.Y., Tan, V.B.C., Wang, C.M., 2006. Effect of chirality on buckling behavior of single-walled carbon nanotubes. J. Appl. Phys. 100 (7), 074304.

Zhang, Y.Y., Xiang, Y., Wang, C.M., 2009. Buckling of defective carbon nanotubes. J. Appl. Phys. 106, 113503.

Zhao, J., He, M.R., Dai, S., Huang, J.Q., Wei, F., Zhu, J., 2010. TEM observations of buckling and fracture modes for compressed thick multiwall carbon nanotubes. Carbon 49, 206–213.

Zheng, Q., Jiang, Q., 2002. Carbon nanotubes as oscillators. Phys. Rev. Lett. 88, 045503/1–045503/3.

Zhu, J., Pan, Z.Y., Wang, Y.X., Zhou, L., Jiang, Q., 2007. The effects of encapsulating C60 fullerenes on the bending flexibility of carbon nanotubes. Nanotechnology 18, 275702.

Exercises for Work With References

1. Find a few examples of journal papers on nanostructures related to your topic of research or interests. Describe the contents of their summaries or abstracts and provide a main figure from each paper.
2. Find a few examples of journal papers about graphene sheets or nanoribbons related to your topic of research or interests. Describe the contents of their summaries or abstracts and provide a main figure from each paper.
3. Find a few examples of journal papers about carbon nanotubes related to your topic of research or interests. Describe the contents of their summaries or abstracts and provide a main figure from each paper.
4. Find a few examples of journal papers about the carbon nanotube-reinforced nanocomposites related to your topic of research or interests. Describe the contents of their summaries or abstracts and provide a main figure from each paper.
5. Find a few examples of journal papers about the polymer–nanotube interfaces related to your topic of research or interests. Describe the contents of their summaries or abstracts and provide a main figure from each paper.

Index

Note: Page numbers followed by "*f*" and "*t*" refer to figures and tables, respectively.

O

P

Q

R

S

T

U

V

W

X

Y

Z

Printed in the United States
By Bookmasters